# Springer Tracts in Modern Physics
# Volume 160

Managing Editor: G. Höhler, Karlsruhe

Editors: J. Kühn, Karlsruhe
Th. Müller, Karlsruhe
R. D. Peccei, Los Angeles
F. Steiner, Ulm
J. Trümper, Garching
P. Wölfle, Karlsruhe

Honorary Editor: E. A. Niekisch, Jülich

# Springer
Berlin
Heidelberg
New York
Barcelona
Hong Kong
London
Milan
Paris
Singapore
Tokyo

# Springer Tracts in Modern Physics

Springer Tracts in Modern Physics provides comprehensive and critical reviews of topics of current interest in physics. The following fields are emphasized: elementary particle physics, solid-state physics, complex systems, and fundamental astrophysics.
Suitable reviews of other fields can also be accepted. The editors encourage prospective authors to correspond with them in advance of submitting an article. For reviews of topics belonging to the above mentioned fields, they should address the responsible editor, otherwise the managing editor.
See also http://www.springer.de/phys/books/stmp.html

## Managing Editor
Gerhard Höhler

Institut für Theoretische Teilchenphysik
Universität Karlsruhe
Postfach 69 80
76128 Karlsruhe, Germany
Phone: +49 (7 21) 6 08 33 75
Fax: +49 (7 21) 37 07 26
Email: gerhard.hoehler@physik.uni-karlsruhe.de
http://www-ttp.physik.uni-karlsruhe.de/

## Elementary Particle Physics, Editors
Johann H. Kühn

Institut für Theoretische Teilchenphysik
Universität Karlsruhe
Postfach 69 80
76128 Karlsruhe, Germany
Phone: +49 (7 21) 6 08 33 72
Fax: +49 (7 21) 37 07 26
Email: johann.kuehn@physik.uni-karlsruhe.de
http://www-ttp.physik.uni-karlsruhe.de/~jk

Thomas Müller

Institut für Experimentelle Kernphysik
Fakultät für Physik
Universität Karlsruhe
Postfach 69 80
76128 Karlsruhe, Germany
Phone: +49 (7 21) 6 08 35 24
Fax: +49 (7 21) 6 07 26 21
Email: thomas.muller@physik.uni-karlsruhe.de
http://www-ekp.physik.uni-karlsruhe.de

Roberto Peccei

Department of Physics
University of California, Los Angeles
405 Hilgard Avenue
Los Angeles, CA 90024-1547, USA
Phone: +1 310 825 1042
Fax: +1 310 825 9368
Email: peccei@physics.ucla.edu
http://www.physics.ucla.edu/faculty/ladder/peccei.html

## Solid-State Physics, Editor
Peter Wölfle

Institut für Theorie der Kondensierten Materie
Universität Karlsruhe
Postfach 69 80
76128 Karlsruhe, Germany
Phone: +49 (7 21) 6 08 35 90
Fax: +49 (7 21) 69 81 50
Email: woelfle@tkm.physik.uni-karlsruhe.de
http://www-tkm.physik.uni-karlsruhe.de

## Complex Systems, Editor
Frank Steiner

Abteilung Theoretische Physik
Universität Ulm
Albert-Einstein-Allee 11
89069 Ulm, Germany
Phone: +49 (7 31) 5 02 29 10
Fax: +49 (7 31) 5 02 29 24
Email: steiner@physik.uni-ulm.de
http://www.physik.uni-ulm.de/theo/theophys.html

## Fundamental Astrophysics, Editor
Joachim Trümper

Max-Planck-Institut für Extraterrestrische Physik
Postfach 16 03
85740 Garching, Germany
Phone: +49 (89) 32 99 35 59
Fax: +49 (89) 32 99 35 69
Email: jtrumper@mpe-garching.mpg.de
http://www.mpe-garching.mpg.de/index.html

Achim Stahl

# Physics with Tau Leptons

With 236 Figures

 Springer

Dr. Achim Stahl
Universität Bonn
Physikalisches Institut
Nussallee 12
53115 Bonn
Germany
E-mail: stahl@physik.uni-bonn.de

Library of Congress Cataloging-in-Publication Data

Stahl, Achim, 1962-
   Physics with tau-leptons / Achim Stahl.
     p. cm. -- (Springer tracts in modern physics, ISSN 0081-3869 ; v. 160)
   Includes bibliographical references and index.
   ISBN
     1. Leptons (Nuclear physics) 2. Leptons (Nuclear physics)--Decay. I. Title. II.
   Springer tracts in modern physics ; 160.

QC1 .S797 vol. 160
[QC793.5.L42]
539 s--dc21
[539.7'211]
                                                              99-047498

Physics and Astronomy Classification Scheme (PACS): 14.60.Fg, 13.35.Dx, 13.10.+q, 12.38.Qk, 12.60.-i, 11.30.Er, 14.60.Lm

ISSN 0081-3869
ISBN 3-540-66267-7 Springer-Verlag Berlin Heidelberg New York

This work is subject to copyright. All rights are reserved, whether the whole or part of the material is concerned, specifically the rights of translation, reprinting, reuse of illustrations, recitation, broadcasting, reproduction on microfilm or in any other way, and storage in data banks. Duplication of this publication or parts thereof is permitted only under the provisions of the German Copyright Law of September 9, 1965, in its current version, and permission for use must always be obtained from Springer-Verlag. Violations are liable for prosecution under the German Copyright Law.

© Springer-Verlag Berlin Heidelberg 2000
Printed in Germany

The use of general descriptive names, registered names, trademarks, etc. in this publication does not imply, even in the absence of a specific statement, that such names are exempt from the relevant protective laws and regulations and therefore free for general use.

Typesetting: Camera-ready copy by the author using a Springer LATEX macro package
Cover design: *design & production* GmbH, Heidelberg

Printed on acid-free paper   SPIN: 10730869   56/3144/tr   5 4 3 2 1 0

# Preface

The $\tau$ lepton, being the heaviest lepton known to date, offers some unique features which make it an excellent tool for challenging our current understanding of particle physics. This book reviews the many aspects of experimental investigations performed with $\tau$ leptons.

Although the electron and the muon – the charged leptons of the first two generations – have been studied more extensively and with higher precision than the $\tau$, there has always been some prejudice that deviations from the Standard Model of particle physics are more likely to become visible in the third generation. Such a prejudice can, of course, be justified only in the framework of a new model and is indeed true for many of them. This makes the $\tau$ lepton a good candidate in the search for physics beyond the Standard Model. Another unique feature of the $\tau$ lepton is the possibility to access its spin through the dynamics of its parity-violating, weak decay. Although techniques of spin measurements for electrons or muons are well established, the $\tau$ lepton is the only elementary fermion for which this is experimentally feasible with a modern high-energy-physics collider detector. With this information at hand, testing the predictions of a given model becomes more stringent.

The $\tau$ is also the only lepton heavy enough to decay into hadrons. These decays offer an ideal laboratory for the study of strong interactions, including the transition from the perturbative to the nonperturbative regime of QCD in the simplest possible reaction. This might explain the tremendous efforts ongoing in $\tau$ physics (for other reviews see [1–8]).

This review starts with a short look back to the discovery of the $\tau$ lepton and its identification as the first member of a new generation (Chap. 1), followed by some experimental aspects of $\tau$ physics (Chap. 2) and a thorough description of the measurements of its static properties (i.e. mass, lifetime, branching ratios, etc.) in Chap. 3. The next chapter deals with the production of $\tau$ leptons in high-energy $e^+e^-$ collisions and the contribution of $\tau$ physics to the precision tests of the electroweak theory at the $Z^0$ pole and above. Chapter 5 describes the impact of strong interactions on hadronic decays of the $\tau$. Chapter 6 is devoted to results achieved at hadron colliders. Intimately linked to the $\tau$ lepton is its neutrino, the topic of Chap. 7. Finally, there are three chapters describing searches for indications of physics beyond the scope of the Standard Model. The book ends with a consideration of the outlook for the future of $\tau$ physics.

$\tau$ physics is a very lively field with a lot of discussions between the various groups. The highlights are the biennial $\tau$ workshops, where many of the new results are presented. There have been five of them:

- 1990 in Orsay, France, organized by M. Davier and B. Jean-Marie [9]
- 1992 in Columbus, Ohio, organized by K.K. Ghan [10]
- 1994 in Montreux, Switzerland, organized by L. Rolandi [11]
- 1996 in Estes Park, Colorado, organized by J.G. Smith and W. Toki [12]
- 1998 in Santander, Spain, organized by A. Pich and A. Ruiz [13]

and we are looking forward to the next workshop in the year 2000 in Victoria, Canada. The proceedings are a good source of information to start from, too [9–13].

The preparation of such a book is impossible without the help of many. I want to especially thank our $\tau$ working group at the University of Bonn. They have produced some of the results described here and carefully checked the manuscript: N. Wermes, M. Kobel, V. Cremers, A. Höcker, M. Thiergen, J. Colberg, M. Schumacher, B. Kunst, H. Voss, R. Sieberg, A. Posthaus, N. Tesch, U. Müller, F. Scharf, S. Menke, R. Bartoldus, K. Linowsky, A. Hauke, R. Kemp, W. Mader, and A. David.

Bonn, August 1999 *Achim Stahl*

# Contents

1. **The $\tau$ Lepton and the Third Family** .................... 1
   1.1 The Discovery of the $\tau$ .................................. 1
   1.2 A Member of a New Family? ............................ 3
   1.3 Direct Observation ........................................ 8

2. **Experimental Aspects** ........................................ 11
   2.1 Overview of Experiments ................................ 11
   2.2 Kinematics ................................................. 12
   2.3 Event Displays ............................................ 20
   2.4 Selection of $\tau$ Pairs .................................... 21
   2.5 Identification of the Decays ............................ 28
   2.6 Monte Carlo Simulation ................................. 38

3. **The Static Properties of the $\tau$** ........................... 45
   3.1 The Mass ................................................... 45
   3.2 The Lifetime ............................................... 53
   3.3 Form Factors of the Electromagnetic and Weak Currents .... 64
   3.4 Branching Ratios ......................................... 72

4. **Electroweak Physics at the $Z^0$ Pole** ...................... 95
   4.1 Precision Tests of the Standard Model ................ 95
   4.2 $\tau$ Production at the $Z^0$ Pole ........................... 97
   4.3 Cross Sections and Asymmetries ....................... 102
   4.4 Electroweak Physics at the SLC ........................ 105
   4.5 Analyzing the Spin of a $\tau$ Lepton ..................... 108
   4.6 $\tau$ Polarization ............................................ 120
   4.7 Results ..................................................... 132

5. **Strong Interactions in $\tau$ Decays** ........................... 137
   5.1 Selection Rules ........................................... 137
   5.2 Theoretical Description of Hadronic $\tau$ Decays ....... 143
   5.3 Experimental Studies .................................... 163
   5.4 Inclusive Decays ......................................... 173

## VIII  Contents

**6. $\tau$ Physics at Hadron Colliders** .......................... 195
   6.1 Identification of $\tau$ Leptons ............................ 195
   6.2 The $\tau$ and the Top Quark ............................ 197
   6.3 Searches ............................................. 199
   6.4 W Decays ............................................ 201

**7. The $\tau$ Neutrino** ........................................ 203
   7.1 The Mass ............................................ 203
   7.2 The Helicity ......................................... 212
   7.3 Electromagnetic Moments ............................. 217

**8. The Lorentz Structure of the Charged Current** ......... 223
   8.1 Generalization of the Weak Current ..................... 223
   8.2 Hadronic Decays ..................................... 235
   8.3 Spin-Dependent Terms ................................ 237
   8.4 The Current Experimental Situation .................... 246

**9. Searching for $\mathcal{CP}$ Violation** ........................... 253
   9.1 $\mathcal{CP}$ Violation in $\tau$ Production ........................... 253
   9.2 $\mathcal{CP}$ Violation in $\tau$ Decays ............................... 264

**10. Rare and Forbidden Decays** ............................ 269
   10.1 Second-Class Currents ............................... 269
   10.2 Forbidden $\tau$ Decays ................................. 272
   10.3 Flavor-Changing Neutral Currents ................... 277
   10.4 Excited Leptons ..................................... 279
   10.5 New Heavy Leptons ................................. 283

**11. Summary and Outlook** ................................. 285

**References** ............................................... 287

**Index** .................................................... 311

# 1. The $\tau$ Lepton and the Third Family

## 1.1 The Discovery of the $\tau$

The story of the $\tau$ started in the early 1960s. At that time physicists were still puzzled by the unexpected appearance of the muon. There was no apparent reason for a second lepton besides the electron. This electron–muon problem was one of the driving questions in high-energy physics at that time.

Three basic strategies were being followed to gather insight into the e–$\mu$ problem:

- Comparing $\mu$–p inelastic scattering with e–p data, looking for any difference between electron and muon [14, 15]. No differences were found and the comparison was finally limited by systematic uncertainties in the cross sections.
- Searching for new charged leptons in photoproduction. None were found and limits between 0.5 and 1 GeV,[1] depending on the lifetime of the new lepton and other assumptions, were set [16].
- Searches for new charged leptons in e$^+$e$^-$ colliding-beam experiments. The best limit of 1.15 GeV came from the ADONE collider at Frascati [17].

It was the latter of the three strategies which eventually led to the discovery of the $\tau$ by Martin Perl and the Stanford Linear Accelerator Center–Lawrence Berkeley Laboratories (SLAC–LBL) collaboration, later called MARK I [18].

The basic idea is simple: once the energy of an e$^+$e$^-$ collider is larger than twice the mass of the new lepton, the new lepton will be pair produced (e$^+$e$^-$ $\to$ $\tau^+\tau^-$) just like ordinary muons. These leptons will be unstable and decay independently of each other. Among other decay modes, there will be a decay to the lighter leptons $\tau \to$ e $\nu_e\nu_\tau$ and $\tau \to \mu\,\nu_\mu\nu_\tau$. In some events one of the $\tau$ leptons will decay via the first reaction and the other via the second. In such events the only visible particles will be an electron and a muon, acollinear and of opposite charge. This apparent violation of lepton number conservation will be accompanied by missing energy and momentum. If the energy of the collider is set below the production threshold these events will disappear.

---
[1] Natural units will be used throughout the book.

# 1. The τ Lepton and the Third Family

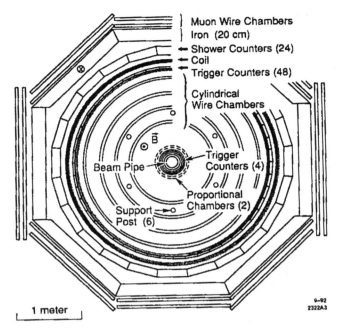

**Fig. 1.1.** Cross section of the SLAC–LBL detector in its initial setup [19]

It was in 1964 that an electron–positron colliding-beam storage ring was proposed at SLAC at Stanford, California [20]. But, owing to funding problems, it took until the end of the decade before the construction of SPEAR could be started. During that time the method to search for a new lepton was developed and the cross section and decay rates were calculated [21–23]. In 1971 a detector for SPEAR was proposed [24] by physicists from SLAC and LBL. It was one of the first large-solid-angle, multipurpose detectors (see Fig. 1.1). The search for new heavy leptons was one out of four topics in the proposal. SPEAR and the SLAC–LBL detector finally went into operation in 1973.

At that time the maximum energy of SPEAR with useful luminosity was 4.8 GeV. The first e–$\mu$ events showed up in 1974 (see Fig. 1.2). It took Martin Perl and his colleagues a year to convince themselves that the events they were seeing could not be explained by conventional sources. They published a paper with the title 'Evidence for anomalous lepton production in $e^+$–$e^-$ annihilation' [18], based on 24 events. It took another year to establish that the source of these events was indeed a new heavy lepton and that is the content of their second paper, where they conclude 'The simplest hypothesis compatible with all the data is that these events come from the production of a pair of heavy leptons' [25]. The discovery was finally confirmed by two

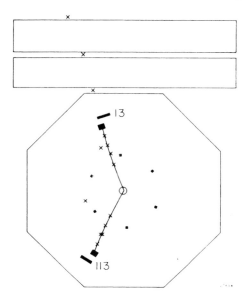

**Fig. 1.2.** The first signals of the τ: one of the early e–μ events as seen by the SLAC–LBL detector [29]. The detector is now equipped with a muon tower on top to improve muon identification. The numbers indicate pulse heights in the shower counters

other experiments running at the DORIS storage ring at DESY in Hamburg, Germany: PLUTO [26, 27] and DASP [28].

The biggest experimental challenge was the separation of the e–μ events from all kinds of background sources, the most serious one being charm events. Remember, the J/Ψ had just been discovered [30, 31] and not much was known about the properties of charm events, especially open charm. The detectors were equipped with electron and muon identification, but the probabilities for misidentification of hadrons were high compared to today's standards. They covered a large solid angle, but they were not as hermetic as multipurpose detectors are nowadays. Particle losses were not negligible. All these problems turned the proper identification of the e–μ events into an experimental challenge (the reader is referred to the original publications for more details [18, 25–28]).

Martin Perl and his collaborators named the new heavy lepton in their 1977 publication [32]. They choose τ as the first letter of the Greek word τριτον, which means 'the third', in allusion to the three charged leptons e, μ, and τ. The story of the discovery of the τ eventually culminated in 1995 with Martin Perl being awarded the Nobel Prize for physics for his contributions.

## 1.2 A Member of a New Family?

In 1974, just before the discovery of the τ, the world of particle physics consisted of four leptons (e, $\nu_e$, μ, $\nu_\mu$) and three quarks (u,d,s). A fourth quark

4   1. The τ Lepton and the Third Family

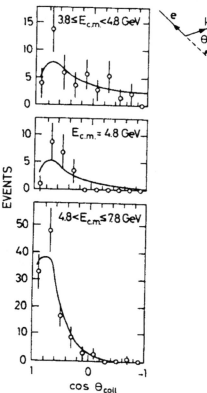

Fig. 1.3. Acollinearity distribution of e–μ events from the SLAC–LBL detector at different center-of-mass energies [25]. *Open circles*: data; *solid lines*: expectation from τ decays

was predicted. The first speculations on the charm quark were of a 'why not?' type [33]: why not have equal numbers of quarks and leptons? Then more profound theoretical arguments came up: an equal number of leptons and quarks would make the divergent triangle anomalies in the $SU(2) \times U(1)$ gauge theory vanish [34, 35] and the absence of flavor-changing neutral currents could be explained by a fourth quark [36], too. In November 1974 the J/Ψ was discovered [30, 31] and almost immediately explained as a state of hidden charm built from a new, so-called charm quark and its antiquark. Open charm production was expected to be observed as a step in the total hadronic cross section at slightly higher center-of-mass energies. Nobody was really astonished when this step was indeed observed in early 1975 [37]. But ...

The picture appeared to be complete: two generations of leptons and quarks. There was absolutely no need for another lepton. And this might explain why it was so hard to convince people that only part of the observed step in the cross section was due to open charm and the other part was due to a new lepton which just happened to have roughly the same mass as the

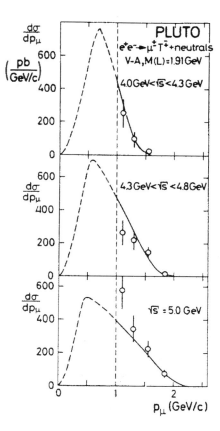

**Fig. 1.4.** The spectrum of muons from $\tau$ decays at different center-of-mass energies measured by PLUTO [26]. *Open circles*: data; *solid* and *dashed lines*: expectation from $\tau$ decays. The minimum energy for muon identification was 1 GeV (*vertical lines*)

charm mesons. The first doubts about the threshold being caused by open charm only came with a closer look at the properties of these events. Charm mesons were expected to decay predominantly to strange particles, but no increase in the K/$\pi$ ratio was observed. An increase in the multiplicity of the events was expected, but there was no visible change across the threshold. There were too many e$\nu_e$ and $\mu\nu_\mu$ decays and no mass peaks in the K$\pi$ and K$\pi\pi$ spectra were observed. The expectations were correct, but they were compensated by the admixture of $\tau$ events with few strange particles, low multiplicities, and many leptonic decays.

Finally, detailed comparisons between the observed events and the theoretical expectations for $\tau$ events convinced people that there was indeed a new lepton present in the data. They used the e–$\mu$ events and also events with one $\tau$ decaying leptonically and the other to any singly charged particle. Some examples are shown in Figs. 1.3–1.6.

Once the existence of a third charged lepton was established, the question arose as to whether it was directly linked to one of the leptons of the first two generations or whether it was the first observed member of a third generation,

1. The $\tau$ Lepton and the Third Family

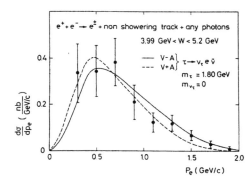

**Fig. 1.5.** The spectrum of electrons from $\tau$ decays measured by DASP [28]. (Data points above 1 GeV also include muons)

which would imply another neutrino and two more quarks. The question was answered in a chain of three arguments. The first deals with the lepton spectra; two examples are shown in Figs. 1.4 and 1.5. The spectra exclude two-body decays of the $\tau$ but they are consistent with three-body decays.[2] The second argument is experimental: from the knowledge of particle detection efficiencies it can be excluded that the two unobserved particles in the decays are either charged particles, photons, or neutral kaons, leaving neutrinos as the only possibility. The appearance of two neutrinos implies that the $\tau$ is carrying some kind of a lepton number.

So the remaining question was what kind of lepton number the $\tau$ is carrying. Several possibilities have been suggested [39–41]: in the case of an ortho- or a para-lepton the $\tau$ will carry the lepton number of one of the first two generations and will not be accompanied by its own neutrino. A sequential lepton, however, forms a new generation with its own lepton number and neutrino.

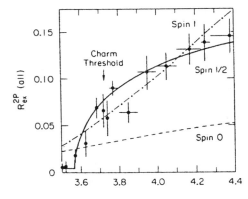

**Fig. 1.6.** The threshold behavior of the cross section for $\tau$ production measured from inclusive electron events (events with one electron and one other charged track) by the DELCO collaboration. The data is normalized to $\mu$ pair production and compared to expectations for pair production of point-like particles of spin 0, 1/2, and 1 [38]

---

[2] For $\tau$ leptons almost at rest near threshold, a 2-body decay would create a peak in the spectrum.

## 1.2.1 Para-lepton

If the $\tau$ is a para-lepton, it carries the lepton number of an 'old' lepton of opposite charge. Let us assume it is of electron type; then $L_e(e^-) = +1$ and $L_e(\tau^-) = -1$. The leptonic decays of the $\tau^-$ read

$$\tau^- \to e^- \bar{\nu}_e \bar{\nu}_e,$$

$$\tau^- \to \mu^- \bar{\nu}_\mu \bar{\nu}_e.$$

Because of the two identical neutrinos in the first decay, it is favored over the second by a statistical factor of two.[3] This is in clear disagreement with the measurements, which give almost equal branching ratios for the two decays (see Sect. 3.4.2), so that a para-lepton can be excluded.

## 1.2.2 Ortho-lepton

If the $\tau$ is an ortho-lepton, it carries the lepton number of an 'old' lepton of the same charge, i.e. $L_e(e^-) = L_e(\tau^-) = +1$ for an electron-type ortho-lepton. It is kind of an excited electron, carrying the same quantum numbers. In that case decays of the type

$$\tau^- \to e^- e^- e^+,$$

$$\tau^- \to e^- \pi^0,$$

and many more should be possible. None of these have been observed (see Sect. 10.2), so that an ortho-lepton can be excluded.

## 1.2.3 Sequential Lepton

This is the only possibility left, the $\tau$ lepton being the first observed member of a third generation, accompanied by a new neutrino $\nu_\tau$ distinct from $\nu_e$ and $\nu_\mu$. And then, by the same arguments that lead to the prediction of the charm quark, there are two more quarks missing...

---

[3] A para-lepton of muon type would give a factor of two in the other direction.

## 1.3 Direct Observation

### 1.3.1 The $\tau$ Lepton

For most experiments studying $\tau$ physics the lifetime of the $\tau$ lepton is so short that it decays long before it reaches the detector. It needs an emulsion target to actually see the track of a $\tau$. The CHORUS experiment at the CERN muon neutrino beam has such a target. The collaboration is searching for neutrino oscillations of the type $\nu_\mu \to \nu_\tau$, a process impossible with massless neutrinos. Their detector is specifically designed to identify $\tau$ leptons produced from the interaction of $\tau$ neutrinos in the target. They have not found any $\tau$ leptons from neutrino oscillations and have derived limits on the parameters describing the oscillation [43].

Neutrino oscillations are not the only processes that produce $\tau$ leptons in the CHORUS detector. Diffractive scattering of muon neutrinos off the nuclei in the target material produces, amongst many other particles, $D_s$ mesons. These $D_s$ mesons decay with a 7 % branching ratio [44] into a $\tau$ lepton and its neutrino.

The CHORUS collaboration has identified such an event [42]. It is a background to the search for oscillations, but it is also the first direct observation of the track of a $\tau$ lepton. Figure 1.7 shows the reconstructed track in the emulsion in the vicinity of the primary vertex. The reaction observed is

$$\nu_\mu n \to \mu^- D_s^{*+} n$$
$$D_s^{*+} \to D_s^+ \gamma$$
$$D_s^+ \to \tau^+ \nu_\tau$$
$$\tau^+ \to \mu^+ \nu_\mu \bar{\nu}_\tau.$$

### 1.3.2 The $\tau$ Neutrino

It is even more difficult to observe the interaction of a $\tau$ neutrino. A $\tau$ neutrino beam is needed and also a detector able to identify the $\tau$ lepton produced by a charged-current interaction by one of the neutrinos.

A dedicated experiment set out at FermiLab to make the first direct observation of a $\nu_\tau$ interaction. It is called DONUT, for 'direct observation of nu tau'. 800 GeV protons are dumped into a large lead block. Amongst many other particles $D_s$ mesons will be produced, which decay into $\tau$ neutrinos through the decay chain mentioned before. The detector is located 36 m downstream of the dump behind massive shielding and sweeping magnets. Essentially only neutrinos reach the detector. Neutrino interactions are recorded in four nuclear-emulsion targets surrounded by external tracking and veto walls. Charged-current interactions of $\tau$ neutrinos produce a $\tau$ lepton, which is identified in the emulsion by a kink in the track where the $\tau$ decayed.

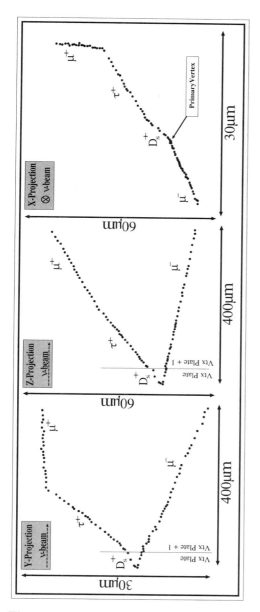

**Fig. 1.7.** The track of a $\tau$ lepton in the CHORUS emulsion target [42]. Views from the top and side and along the neutrino beam

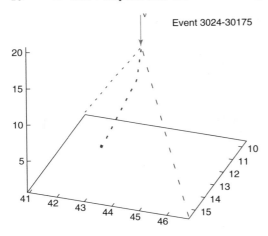

**Fig. 1.8.** A candidate for a $\tau$ neutrino interaction in a nuclear emulsion observed by the DONUT collaboration [45]. Each *dash* is a section of a track measured in the emulsion. The *central track* shows a 100 mrad kink 4.5 mm from the interaction vertex characteristic of a $\tau$ lepton. All units are mm

The DONUT collaboration completed its first run in 1997 and a second one is planned in 1999 [46]. There are no results yet, but an initial analysis of some of the data has produced the first candidate. This is shown in Fig. 1.8 [45].

# 2. Experimental Aspects

In this section a few experimental aspects will be briefly summarized which are common to most analyses: the kinematics of the events, the identification of $\tau$ pairs, and the classification of the decay channels. But first the experiments will be listed that have contributed to $\tau$ physics.

## 2.1 Overview of Experiments

Since its discovery the $\tau$ lepton has been studied with ever-increasing precision at every new $e^+e^-$ collider that has gone into operation. The appearance of the events changed with increasing energy of the machines and improvements in detectors. The samples became more numerous and more and more decay modes became available.

It started with a handful of events at SPEAR (MARK I), confirmed by the first events from DORIS (PLUTO). The energy of DORIS was increased in steps to the $\Upsilon(4s)$ resonance to produce B mesons, but it also was the first machine to provide a large number of $\tau$ pairs, recorded by ARGUS. Meanwhile the next generation of accelerators, PEP and PETRA, were built with center-of-mass energies in the continuum around 30 to 40 GeV and with a number of experiments. Also CESR started running at the $\Upsilon(4s)$ with the CLEO experiment, which today has the largest $\tau$ sample. With TRISTAN the Japanese joined, with a machine again in the continuum at 50 to 60 GeV. Despite good luminosity the $\tau$ production rate is low, as the cross section falls like $1/s$. In 1989 SLC, and shortly afterwards, LEP began running at the $Z^0$ peak, producing $\tau$ pairs polarized by the parity violation of the $Z^0$ boson. BEPC was constructed in Beijing to go back to the $\tau$ production threshold and to precisely remeasure, amongst other things, the $\tau$ mass. Today CESR and CLEO are being upgraded, and the asymmetric b factories PEP-II and KEK-B are starting to run. Table 2.1 lists the accelerators that have produced $\tau$ pairs and Table 2.2 summarizes the experiments that have analyzed them.

**Table 2.1.** Accelerators for $\tau$ physics

| Accelerator<br>Laboratory | Energy<br>in GeV | Years of<br>operation |
|---|---|---|
| SPEAR<br>SLAC, USA | 3 – 8 | 73 – 88 |
| DORIS<br>DESY, Germany | 8 – 11 | 77 – 92 |
| PETRA<br>DESY, Germany | 10 – 47 | 78 – 86 |
| PEP<br>SLAC, USA | 29 | 80 – 90 |
| CESR<br>Cornell, USA | 9 – | 79 – |
| TRISTAN<br>KEK, Japan | 50 – 62 | 86 – 95 |
| SLC<br>SLAC, USA | 91 | 89 – |
| LEP<br>CERN, Europe | 88 – 200 | 89 – 00 |
| BEPC<br>Beijing, China | 3 – 4 | 91 – |
| PEP-II<br>SLAC, USA | 11 | 99 – |
| KEK-B<br>KEK, Japan | 11 | 99 – |

## 2.2 Kinematics

### 2.2.1 Lorentz Boost

The $\tau$ leptons are produced in pairs in $e^+e^-$ collisions. Above the production threshold they are not produced at rest. They are moving away from each other with an energy equal to the beam energy. Table 2.3 gives the Lorentz factors $\beta$ and $\gamma$ for some energies. With a lifetime of $\tau_\tau = (290.0 \pm 1.2)$ fs [44], each $\tau$ will travel on average a distance of $\bar{\ell} = \beta\gamma\, c\tau_\tau$ until it decays. This distance is plotted in Fig. 2.1.

### 2.2.2 Two-Body Decays

For all hadronic channels, the decay of a $\tau$ proceeds through a two-body reaction into the $\tau$ neutrino and a hadronic resonance, which further decays to pions, kaons, and other 'stable' mesons. The decay can be described as $\tau \to \text{had}\,\nu_\tau$, where the 4-momentum of 'had' is the sum of that of the 'stable'

**Table 2.2.** Experiments in $\tau$ physics. The number of $\tau$ pairs produced is given in units of 1000. 'Int. $\mathcal{L}$' is the integrated luminosity recorded by the experiment. A typical efficiency for the identification of a $\tau$ pair is quoted as $\epsilon$. Only the running periods relevant to the $\tau$ results are listed. These are, for the LEP experiments, the data taken at LEP-I around the $Z^0$ peak. At LEP-II the cross section is too small to produce a reasonable $\tau$ sample. Crystal Ball (CB) traveled from SPEAR at SLAC to DORIS in Hamburg; the $\tau$ results are from DORIS. PLUTO produced $\tau$ results first at DORIS and then at PETRA. DELCO ran at SPEAR and at PEP. MARK II ran a few years at SPEAR studying $\tau$ leptons, then it took data at the SLC until the SLD detector became operational

| Name of experiment | Accelerator facility | Years of operation | Typical $E_{cm}$ in GeV | $N_{\tau^+\tau^-}$ produced | Int. $\mathcal{L}$ in $pb^{-1}$ | Typical $\epsilon$ in % | Remarks | Publications |
|---|---|---|---|---|---|---|---|---|
| ALEPH | LEP | 89 – 95 | 91 | 200 | 170 | 90 | $Z^0$ | [47–73] |
| AMY | TRISTAN | 86 – 94 | 50 – 62 | 4 | 150 | 40 | Cont. | [74, 75] |
| ARGUS | DORIS | 82 – 92 | 10.58 | 400 | 500 | 10 | $\Upsilon(4s)$ | [76–97] |
| BaBar | PEP-II | 99 – | 10.58 | 50000 | $10^5$ | ? | | |
| BELLE | KEK-B | 99 – | 10.58 | 50000 | $10^5$ | ? | | |
| BES | BEPC | 91 – | 3.4 – 3.6 | 1.5 | 5 | 5 | $\tau$ thres. | [98, 99] |
| CB | DORIS | 82 – 86 | 10.58 | 250 | 300 | 5 | $\Upsilon(4s)$ | [100–102] |
| CELLO | PETRA | 80 – 86 | 14 – 47 | 10 | 140 | 35 | Cont. | [103–109] |
| CLEO | CESR | 79 – | 10.58 | 4300 | 4700 | 10 | $\Upsilon(4s)$ | [110–144] |
| DASP | DORIS | 78 – 78 | 3 – 5 | 20 | 7 | 1 | $\tau$ thres. | [145] |
| DELCO | PEP | 81 – 84 | 29 | 15 | 150 | 20 | Cont. | [146–149] |
| DELPHI | LEP | 89 – 95 | 91 | 200 | 170 | 90 | $Z^0$ | [150–159] |
| HRS | PEP | 78 – 86 | 29 | 30 | 300 | 20 | Cont. | [160–164] |
| JADE | PETRA | 78 – 86 | 12 – 47 | 6 | 100 | 50 | Cont. | [165–169] |
| L3 | LEP | 89 – 95 | 91 | 200 | 170 | 90 | $Z^0$ | [170–180] |
| MAC | PEP | 80 – 86 | 29 | 20 | 200 | 30 | Cont. | [181–183] |
| MARK I | SPEAR | 73 – 77 | 3 – 8 | | | | | [18, 25, 32] |
| MARK II | PEP | 79 – 84 | 29 | 20 | 200 | 20 | Cont. | [184–201] |
| MARK III | SPEAR | 82 – 88 | 3.77 | 25 | 10 | 5 | $\tau$ thres. | [202, 203] |
| MARK J | PETRA | 78 – 82 | 12 – 47 | 15 | 200 | 15 | Cont. | [204–208] |
| OPAL | LEP | 89 – 95 | 91 | 200 | 170 | 90 | $Z^0$ | [209–239] |
| PLUTO | DORIS | 77 – 78 | 3 – 9 | | | | | [26, 27] |
|  | PETRA | 78 – 79 | 35 | 3 | 40 | 2 | Cont. | [240–244] |
| SLD | SLC | 89 – 99 | 91 | 20 | 10 | 90 | $Z^0$ | [245–247] |
| TASSO | PETRA | 79 – 86 | 14 – 47 | 10 | 200 | 8 | Cont. | [248–253] |
| TOPAZ | TRISTAN | 90 – 95 | 52 – 62 | 7 | 280 | 20 | Cont. | [254, 255] |
| TPC/2$\gamma$ | PEP | 82 – 90 | 29 | 14 | 140 | 5 | Cont. | [256–260] |
| VENUS | TRISTAN | 86 – 95 | 50 – 62 | 7 | 270 | 35 | Cont. | [261, 262] |

14   2. Experimental Aspects

**Table 2.3.** Lorentz factors for some center-of-mass energies (units GeV)

| $E_{cm}$ | $\beta$ | $\gamma$ | $\beta\gamma$ |
|---|---|---|---|
| 3.67 | 0.2493 | 1.033 | 0.257 |
| 10.58 | 0.9419 | 2.977 | 2.804 |
| 29 | 0.9925 | 8.160 | 8.098 |
| 35 | 0.9948 | 9.848 | 9.797 |
| 58 | 0.9981 | 16.319 | 16.289 |
| 91.2 | 0.9992 | 25.661 | 25.641 |

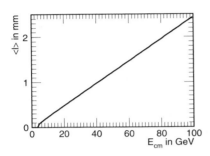

**Fig. 2.1.** Average decay length of a $\tau$ lepton at various center-of-mass energies

mesons participating in the decay. It is irrelevant here, from the point of view of kinematics, whether there really is a formation of an intermediate resonance or not.

In the rest frame of the decaying $\tau$ the energy of the hadron is completely determined by energy and momentum conservation. Neglecting a possible neutrino mass, we have

$$E^*_{had} = \frac{m_\tau^2 + m_{had}^2}{2\,m_\tau},$$
$$p^*_{had} = \frac{m_\tau^2 - m_{had}^2}{2\,m_\tau}. \tag{2.1}$$

(All quantities with an asterisk refer to the $\tau$ rest frame.) The functions are displayed graphically in Fig. 2.2. The direction of emission of the hadron, however, is not fixed by the kinematics. There are two angles $\theta^*$ and $\phi^*$ necessary to specify this direction. They are defined in Fig. 2.3 with respect to the $z$ axis, which is the direction of the Lorentz boost into the laboratory frame. The angle $\theta^*$ is called the Gottfried–Jackson angle. The neutrino is emitted back to back with the hadron.

After the boost into the laboratory frame (along $+z$) the angle $\phi^*$ becomes unobservable. $\theta^*$, however, is reflected in the laboratory energy of the hadron. A hadron emitted forward ($\theta^* = 0$) will always have the full beam energy. Backward emission creates lower energies, depending on the mass of the hadron. There is a linear relation between $\cos\theta^*$ and $E_{had}$:

$$E_{had} = \gamma\,E^*_{had} + \beta\,\gamma\,p^*_{had}\,\cos\theta^*. \tag{2.2}$$

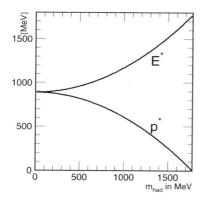

**Fig. 2.2.** Energy and momentum of the hadron in $\tau \to \text{had}\,\nu_\tau$ in the rest frame of the $\tau$ lepton

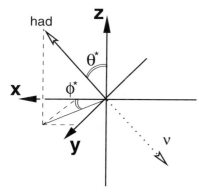

**Fig. 2.3.** Definition of the decay angles $\theta^*$ and $\phi^*$ in the rest frame of the $\tau$. The $z$ axis points in the direction of the boost into the laboratory frame

As $p^*_\text{had}$ is smaller than $E^*_\text{had}$ for massive hadrons, the latter receive a minimum energy in the laboratory. It is convenient to express these relations in the quantity $x_\text{had}$ defined by

$$x_\text{had} = \frac{E_\text{had}}{\max(E_\text{had})} \tag{2.3}$$

with $\max(E_\text{had}) = \gamma\, E^*_\text{had} + \beta\,\gamma\, p^*_\text{had}$. Then for $\beta \approx 1$ the minimum $x_\text{had}$ becomes independent of the center-of-mass energy. This can be seen in Fig. 2.4 at $\cos\theta^* = -1$.

Another effect of the boost is the collimation of the decay products in the direction of the $\tau$. The higher the boost, the more the hadron lines up with the $\tau$. This effect also depends on the mass of the hadron as can be seen from Fig. 2.5, which shows the angle $\psi$ between the directions of the hadron and the $\tau$ in the laboratory. For example, a 1230 MeV $a_1$ meson is never more than $8°$ away from the $\tau$ at the $\Upsilon(4s)$ and never more than $0.8°$ at LEP. This becomes an experimental problem at higher energies, when the decay products of the $\tau$ leptons move so close together that their signals begin to overlap in the detector.

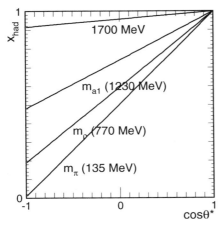

**Fig. 2.4.** The laboratory energy of hadrons as a function of the Gottfried–Jackson angle $\theta^*$. The relation is plotted for four different hadron masses in the limit $\beta \to 1$

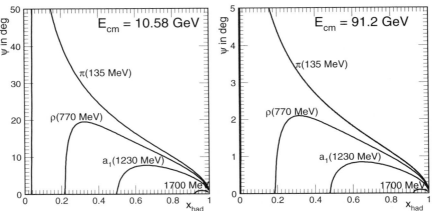

**Fig. 2.5.** The opening angle $\psi$ between the directions of the hadron and the $\tau$ in the laboratory at the $\Upsilon(4s)$ (*left*) and the $Z^0$ pole (*right*)

It is interesting to realize that the angle $\psi$ is completely determined by the mass and energy of the hadron. Both of them are measurable and therefore $\psi$ can be calculated for each hadronic decay. Hence, even if one cannot determine the direction of the $\tau$ from vertex information, just from the measured hadron one knows that the $\tau$ is somewhere on a cone around the hadron direction. The opening angle of the cone is given by

$$\sin \psi = \frac{p^*_{\text{had}}}{p_{\text{had}}} \sin \theta^* \qquad (2.4)$$

with $\theta^*$ calculated from (2.2).

## 2.2.3 Acollinearity

In the absence of radiation, the two $\tau$ leptons in $e^+e^- \to \tau^+\tau^-$ are produced back to back, i.e. $\boldsymbol{p}_{\tau^+} = -\boldsymbol{p}_{\tau^-}$. The visible decay products, however, are not back to back, as there are neutrinos that escape detection. The orientation of the visible decay products with respect to each other is described by two angles, the acollinearity $\eta$ and the aplanarity $\Phi$. The acollinearity is the angle between the decay products of the two $\tau$ leptons in space, and the aplanarity describes the two decay planes (see Fig. 2.6). Another angle, $\Delta\phi$, is commonly used. It is the difference in azimuth between the decay products of the two $\tau$ leptons (see Sect. 3.2.3). If there is more than one visible decay product from a $\tau$, the momentum sum of all of them is used to calculate the angles.

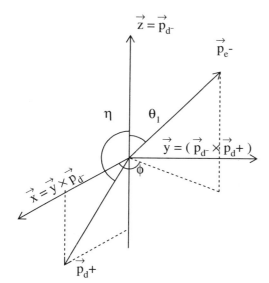

Fig. 2.6. Definition of the aplanarity angle $\Phi$. $\boldsymbol{p}_{d^-}$ and $\boldsymbol{p}_{d^+}$ are the momenta of the visible decay products of the $\tau^-$ and $\tau^+$, respectively. $\boldsymbol{p}_{e^-}$ is the direction of the initial electron beam [52]

## 2.2.4 Reconstruction of the $\tau$ Direction

For many analyses it would be very helpful to know the direction of the $\tau$ in the event, but the neutrinos that escape detection prevent a simple reconstruction. The thrust axis can be used to approximate the $\tau$ direction,[1] a good approximation at high energies (see Fig. 2.5). In events where both $\tau$ leptons decay to multiprongs the line connecting the secondary vertices gives the $\tau$ direction. This method works in principle, but with the short distance of flight of the $\tau$ leptons the resolution of the secondary vertices is in most cases not sufficient to be of practical use.

---

[1] The thrust axis is defined as the axis which minimizes $\sum p_T$ of all particles, where $p_T$ is taken with respect to this axis.

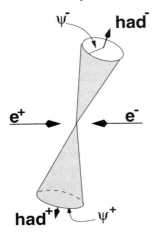

**Fig. 2.7.** Kinematics of an event with both $\tau$ leptons decaying as $\tau \to$ had $\nu_\tau$. The hadron vectors are the momentum sums of all particles observed in each hemisphere. The $\tau$ momentum must fall on the surface of the respective cone. The opening angles of the cones are given by (2.4)

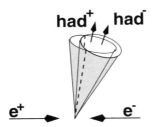

**Fig. 2.8.** Kinematics of an event with both $\tau$ leptons decaying as $\tau \to$ had $\nu_\tau$. The cone of the $\tau^+$ has been reflected at the origin. There are two intersections between the cones (*dashed lines*). One of them has to be the true $\tau$ direction

Another method that works when both $\tau$ leptons decay semihadronically has been described in [263]. It has been successfully used (for example [218, 264]). In the previous sections it has been described how to determine the opening angle $\psi$ between the hadron and the $\tau$ in the laboratory, i.e. the true $\tau$ direction must be somewhere on a cone around the observed direction of the hadron and the opening angle of this cone is known from the measurement of the hadron (see Fig. 2.7). In addition, ignoring photon radiation, we know that the two $\tau$ leptons were produced back to back. So if the cone of the positive hadron is mirrored at the origin, the true $\tau$ direction should be common to both cones. Figure 2.8 shows the situation. In general there are two solutions, the two intersections of the cones.

The situation with two solutions for the $\tau$ direction is most easily visualized with respect to the plane spanned by the two hadrons. Imagining this plane in Fig. 2.8, one realizes that the two solutions are symmetric with respect to this plane. The projections into the plane are the same for both solutions, but one is above the plane and the other below by the same amount. This plane with two solutions is shown in ordinary space in Fig. 2.9. A good approximation of the $\tau$ direction can already be achieved by ignoring the component of the $\tau$ direction perpendicular to the plane, i.e. averaging the two solutions. But the ambiguity can be resolved if a point on the track of the hadron in space can be determined. In $\tau \to \pi \nu_\tau$ this can be the point

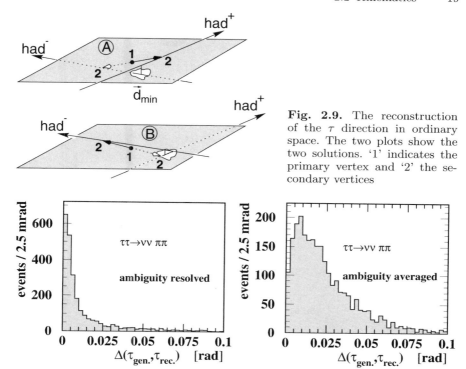

**Fig. 2.9.** The reconstruction of the $\tau$ direction in ordinary space. The two plots show the two solutions. '1' indicates the primary vertex and '2' the secondary vertices

**Fig. 2.10.** Reconstruction of the $\tau$ direction in the channel $\tau^+\tau^- \to \pi^+\pi^-\bar{\nu}_\tau\nu_\tau$ (OPAL collaboration [218]). The plots show the difference in angle between the true and the reconstructed $\tau$ direction

of closest approach of the pion to the primary vertex, which is given by the impact parameter of the track. For $\tau \to 3\pi\nu_\tau$ the secondary vertex serves as the point on the virtual track of the $3\pi$ system. An orientation can be assigned to the plane by defining its normal as

$$\boldsymbol{n} = \frac{\boldsymbol{p}(\mathrm{had}^-) \times \boldsymbol{p}(\mathrm{had}^+)}{|\boldsymbol{p}(\mathrm{had}^-)|\,|\boldsymbol{p}(\mathrm{had}^+)|}. \tag{2.5}$$

Then, in the case of solution A, the positive hadron is below the plane and the negative above.[2] Calculating the closest approach of the tracks of the two hadrons $\boldsymbol{d}_{\min}$ and assigning to it a direction pointing from the negative hadron to the positive, its orientation with respect to $\boldsymbol{n}$ distinguishes the two solutions.

Figure 2.10 shows the resolution in the $\tau$ direction achieved by the OPAL collaboration [218] with and without resolving the ambiguity in $\tau^+\tau^- \to \pi^+\pi^-\bar{\nu}_\tau\nu_\tau$ events.

---

[2] The normal is pointing downwards in this picture.

20    2. Experimental Aspects

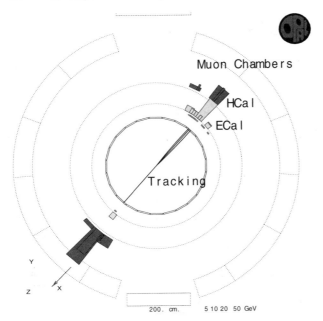

**Fig. 2.11.** Display of a $\tau\tau$ event in the OPAL detector

## 2.3 Event Displays

The characteristics of a $\tau^+\tau^-$ event at $Z^0$ energies are two highly collimated, back-to-back jets with low multiplicity, accompanied by missing energy and momentum due to the neutrinos that escape detection. At lower center-of-mass energies the jets become wider and the acollinearity of the event increases, but they otherwise look similar.

Figure 2.11 shows a $\tau$ pair viewed along the beam axis in the OPAL detector. A single track is recoiling against a jet of three. There is substantial activity in the hadron calorimeter in both jets, identifying them as hadronic $\tau$ decays. The lower jet shows an energy deposition consistent with a minimum-ionizing particle in the electromagnetic calorimeter. The upper jet has much more electromagnetic energy, due to either a hadron starting its shower early or photons from $\pi^0 \to \gamma\gamma$ decays. Only a more detailed analysis could tell. There are no additional particles outside the two narrow jets. The total energy of the event is 72.1 GeV, i.e. roughly 20 GeV of energy is missing.

Figure 2.12 shows a closer look at a similar event in the SLD detector. It is a view in the same projection, zoomed in on the primary vertex. The primary vertex is indicated by the cross in the middle. The size of the beam spot in this view is only 2.5 μm. The main picture shows the five silicon detectors with hits from the tracks (small dots). The inset shows an even stronger zoom on the primary vertex. On this scale the displacement of the common vertex of the 3-prongs from the primary vertex is clearly visible.

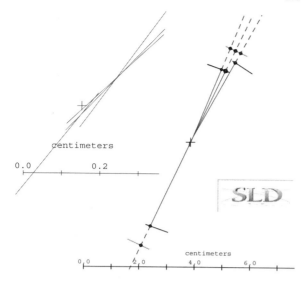

**Fig. 2.12.** Display of the vertex region of a $\tau\tau$ event in the SLD detector. The *inset* in the *upper left corner* is a zoom into the region around the primary vertex. It shows the displaced vertices

Often the most difficult task in the reconstruction of $\tau$ decays is that of the photons. Figure 2.13 shows a $\tau \to \pi^\pm \pi^0 \pi^0 \nu_\tau$ decay in the ALEPH detector. The photons are identified by their energy depositions in the electromagnetic calorimeter (ECal), a sampling calorimeter with longitudinal segmentation (three sections). In this event the four photons are clearly separated from each other and from the charged pion, which does not interact in the electromagnetic calorimeter in this event.

## 2.4 Selection of $\tau$ Pairs

### 2.4.1 Background Sources

The first major step in every analysis is the identification of $\tau$ pairs from the whole data sample. The Feynman diagrams for the major background processes in $e^+e^-$ collisions are shown in Figs. 2.14 and 2.15. These are:

- $e^+e^- \to \mu^+\mu^-$
  The cross section and production mechanism are identical for $\mu$ pairs and $\tau$ pairs, except for very close to the $\tau$ threshold, where the phase space suppression due to the $\tau$ mass is relevant (Fig. 2.14). As in $\tau$ events, a pair of back-to-back leptons each with the full beam energy is produced. The muons, however, are stable at these energies,[3] so that a pair of highly energetic back-to-back tracks appears in the detector with the characteristic signature of muons.

---
[3] The mean flight path is already 11 km at the $\tau$ threshold.

22  2. Experimental Aspects

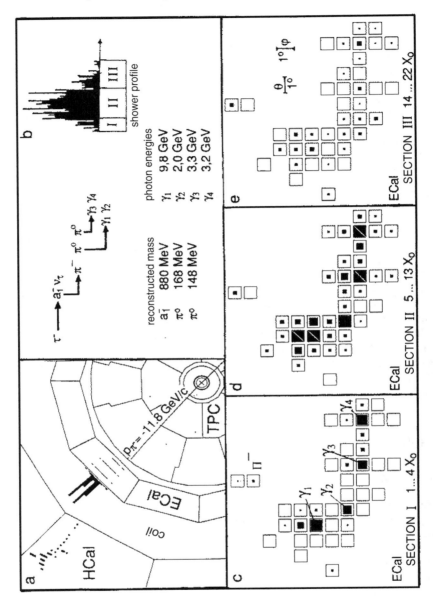

**Fig. 2.13.** Display of a $\tau \to \pi^{\pm} \pi^0 \pi^0 \nu_\tau$ decay in the ALEPH detector. (**a**) Overview of the event shows the charged track coming from the primary vertex, passing through ECal and interacting in the hadron calorimeter. (**b**) The energy profile in depth in the electromagnetic calorimeter from the readout of the wire planes. (**c**) The energy deposition from the pads in the first section of the ECal. The points of entry of the photons and the track are indicated. (**d**), (**e**) Same for the second and third section in depth [265]

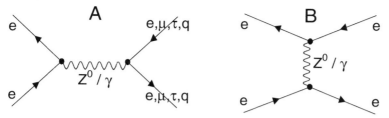

**Fig. 2.14.** Feynman diagrams for the production of lepton and quark pairs in $e^+e^-$ collisions. **A**: s channel; **B**: t channel

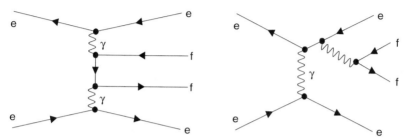

**Fig. 2.15.** Two Feynman diagrams of two-photon processes. The *left* diagram gives the largest contribution to the background in $e^+e^- \to \tau^+\tau^-$. The *right* diagram is one out of four, where the photon that makes the f$\bar{\text{f}}$ pair is coupled to different electron legs

- $e^+e^- \to e^+e^-$

  For the production of electron pairs there is the s-channel production which is the same for all three lepton species, and there is in addition the t-channel reaction (Fig. 2.14). Both produce a pair of back-to-back electrons with the full beam energy. The cross section for the s-channel process is the same as that for $\tau$ production; that for the t channel steeply rises in the forward direction. It dominates away from the $Z^0$ resonance (see Fig. 2.16). The situation changes on the $Z^0$ peak, as only the s channel resonates. Here the s channel dominates and the t channel contributes only for scattering angles below $\cos\theta = 0.8$.

  Like the muons, an $e^+e^- \to e^+e^-$ reaction leaves a pair of highly energetic back-to-back tracks in the detector. And both tracks are associated with high energy depositions in the electromagnetic calorimeter.

- $q\bar{q}$ production

  Hadronic events are a serious background at lower energies. Hadrons are produced as the quarks fragment. Typically their number is large, which distinguishes these events from $\tau$ pairs. But there are tails to low multiplicities, which create background. The average multiplicity increases with the center-of-mass energy, increasing the separation at higher energies (see Figs. 2.17 and 2.18).

24    2. Experimental Aspects

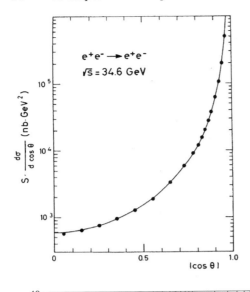

**Fig. 2.16.** Angular distribution of electron pairs produced in $e^+e^-$ collisions at $\sqrt{s} = 35$ GeV (MARK J collaboration [206])

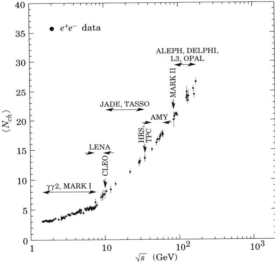

**Fig. 2.17.** Average charged multiplicity in $e^+e^- \to q\bar{q}$ events [44]

The cross section for hadron production is larger than for $\tau$ production. It changes with energy as more quark flavors become available and it changes on the $Z^0$ resonance (see Fig. 2.19).

- **Two-photon events**
  Figure 2.15 shows Feynman diagrams of two-photon processes. The initial electron and positron are deflected only little and are often lost in the beam pipe. Only the $f\bar{f}$ pair is visible in the detector. It is typically produced with low mass and low $p_T$ with respect to the beam. The low mass largely

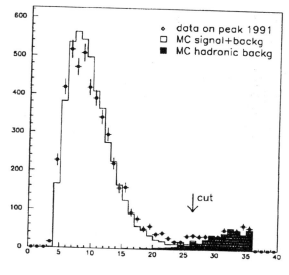

**Fig. 2.18.** Multiplicity (tracks plus clusters) of $\tau$ candidates as measured by the L3 collaboration [266]. The cut to reject $q\bar{q}$ events is indicated by the *arrow*

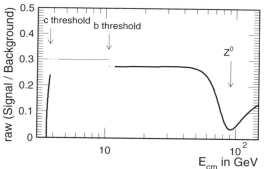

**Fig. 2.19.** The ratio of cross sections for the production of $\tau$ pairs to hadron production (signal/background). The area of the charm and beauty resonances corresponds to the *dotted sections* of the line. There the ratio changes very rapidly and is in most places worse than in the neighboring non-resonant areas

suppresses the two-photon production of $\tau$ pairs and heavy quarks. Low $p_T$ and high acollinearity are the characteristics of these events, on which basis they can be rejected.

### 2.4.2 Event Selection

A typical selection of $\tau$ pairs starts from a loose preselection. It ensures that the event was created by a true $e^+e^-$ interaction, not for example by an off-momentum particle hitting the detector or by a cosmic ray. This preselection is applied for technical reasons to reduce the data sample that has to be handled. An initial rejection of obvious $q\bar{q}$ background can also be applied at this stage. Then one assigns the visible particles to two $\tau$ candidates. One can use a plane perpendicular to the thrust axis to cast the event into two hemispheres or simple jet-finding algorithms (cone jets) to sort the daughters

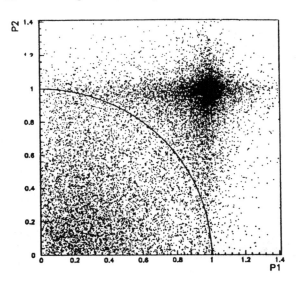

**Fig. 2.20.** The momenta of the leading tracks normalized to the beam energy in the two hemispheres of $\tau$ candidate events (DELPHI collaboration [266]). The peak at (1,1) comes from the electron and muon pair background. The events inside the *circle* were used for further analysis

of the $\tau$ leptons. Now the $\tau$ candidates are required to point into the fiducial volume of the detector and to be roughly consistent with a $\tau$ pair.

The largest initial background from q$\bar{\text{q}}$ production can be rejected on the basis of the multiplicity of the events. One can use either the charged or the neutral multiplicity or a combination of both. As mentioned above, the multiplicity of the q$\bar{\text{q}}$ events increases with the available $Q^2$, i.e. with the center-of-mass energy (see Fig. 2.17). The multiplicity of the $\tau$ pairs, however, is fixed, given by their branching ratios (see Sect. 3.4.1). Figure 2.18 shows the multiplicity measured.

The background from electron and muon pairs is rejected by cuts on the momenta of the decay products of the two $\tau$ candidates. Figure 2.20 shows the distribution of the momentum of the positive versus the negative candidate. The background stands out as a peak at ($E_{\text{beam}}, E_{\text{beam}}$). Its width is determined by the detector resolution. The tails to lower momenta are caused by photon radiation. The signal populates the whole plane, owing to the momentum carried away by the neutrinos. It is separated from the background by cuts in this plane. The cuts can be made more efficient by combining them with particle identification. More stringent cuts are applied if one or two electrons or muons are identified in the event. The rejection of electron pairs can be supported by cuts on the energy deposition in the electromagnetic calorimeter. The calorimeters provide an independent measurement of the energy of electrons which includes final-state photon radiation.

Acollinearity and acoplanarity are other interesting quantities for rejecting electron and muon pairs. In the absence of radiation, these are expected to be exactly back to back (i.e. $\eta = 0$ and $\Delta\phi = 0$), whereas the $\tau$ pairs have nonvanishing angles due to the missing neutrinos. Figure 2.21 shows

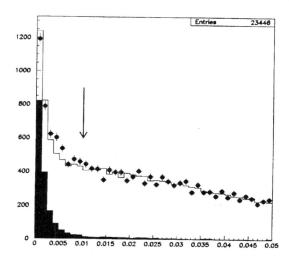

**Fig. 2.21.** Acoplanarity $\Delta\phi = |\phi_1 - \phi_2|$ between the two most energetic tracks in $\tau$ events, in radians (L3 collaboration [267]). The *arrow* indicates the cut in this quantity. Other cuts were applied. The *points* are data, the *open histogram* is the $\tau\tau$ Monte Carlo result, and the *dark histogram* is the background. The peak in the background at zero comes from electron and muon pairs

the acoplanarity measured by the L3 collaboration. The L3 collaboration placed a cut at 10 mrad to reject the background in this analysis [267].

Finally, two-photon background is eliminated by requiring some minimum $p_T$ and/or $p$ in the event or by restricting the acollinearity to values not too large.

### 2.4.3 Tagging

At lower energies (at the $\Upsilon(4s)$ and below) the background from $q\bar{q}$ events becomes so serious that the above method of selecting $\tau$ events is no longer appropriate. An additional problem arises at these low energies: the opening angles between the $\tau$ lepton and its daughters become larger, so that the decay products of the two $\tau$ leptons start to overlap. It is no longer obvious which particle originates from which $\tau$.

A different method of selecting $\tau$ events is applied which solves both problems at the same time: the $\tau$ events are tagged. Stringent cuts are applied to select single $\tau$ leptons in a specific decay channel. For example, a clearly identified muon, not accompanied by any photons close by, is a clean signal of a $\tau \to \mu\nu_\mu\nu_\tau$ decay, if the momentum of the muon is not too low, to exclude two-photon production, and if it is not too high either, to ensure it did not originate from a muon pair. Such a muon is called a tag of the $\tau$ and everything else in the event belongs to the recoiling $\tau$, the signal side. It can then be studied for the question under consideration. Various tags have been used, mainly the two leptonic channels, the hadronic decay $\tau \to \pi\pi\nu_\tau$, and the decay to three charged pions $\tau \to 3\pi\nu_\tau$. Which tag is appropriate depends on the analysis. A three-pion tag can well be used for the measurement of the leptonic branching ratio on the signal side, but it is not good for the determination of the neutrino mass in $\tau \to 5\pi^\pm\nu_\tau$, because of $q\bar{q}$ background.

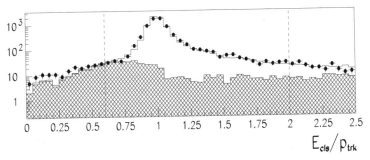

**Fig. 2.22.** The ratio of the energy deposition in the electromagnetic calorimeter associated with a charged track to its momentum in one-prong $\tau$ decays (OPAL collaboration [268]). The *points* are data, the *open histogram* is the Monte Carlo simulation of $\tau \to e\,\nu_e\nu_\tau$, and the *shaded histogram* the background from other $\tau$ decays. Electrons deposit all their energy in an electromagnetic shower in the calorimeter, resulting in a peak at 1. The tail extending to higher values is generated by photon radiation which overlaps with the cluster of the electron. In this case $E$ measures the sum of the electron and photon, but $p$ only the electron. The distribution is shown on a log scale, to make the background visible

## 2.5 Identification of the Decays

### 2.5.1 Classification of the Decays

Now that $\tau$ pairs have been selected from the data, the next step is to identify their decays, i.e. to select those $\tau$ leptons which decayed into the channel under investigation or, in a more global analysis, to categorize the decays into classes such as $\tau \to e\,\nu_e\nu_\tau$, $\tau \to \mu\,\nu_\mu\nu_\tau$, $\tau \to \pi\,\nu_\tau$, or $\tau \to \pi\,n\,\pi^0\,\nu_\tau$. The information provided by the detector has to be exploited in great detail to

- determine the proper number of charged particles in the decay
- identify them as electrons, muons, pions, or kaons
- reconstruct the photons and combine them to neutral pions and maybe $\eta$ mesons
- search for the presence of neutral kaons, decaying as $K^0_S \to \pi^+\pi^-$ or interacting as $K^0_L$ in the hadron calorimeter.

The basic methods for classifying decay channels will be discussed in this section. The distinction of $\tau \to e\,\nu_e\nu_\tau$ and $\tau \to \mu\,\nu_\mu\nu_\tau$ from hadronic decays will serve as the example. The following subsections are devoted to the special problems of reconstructing $\pi^0$s, separating charged kaons from pions, and finding neutral kaons.

Typical variables used to identify electrons and muons are shown in Figs. 2.22–2.24. They are based on the characteristic properties of these particles, e.g. the high penetration of muons, leading to signals even in the outermost

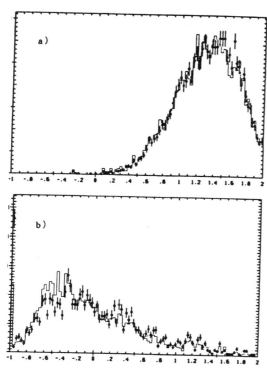

Fig. 2.23. The shape variable measuring the longitudinal shower development in the CELLO electromagnetic calorimeter [269]: (a) electrons, (b) hadrons. The *points* are data, the *histogram* is the Monte Carlo result

layers of the hadron calorimeters and the muon chambers. Electrons are classified by their compact showers in the electromagnetic calorimeter, where they deposit all their energy.

There are various methods of combining the information from different variables into the final identification of a decay, as follows.

**Application of Cuts.** The simplest way to select the signal is to apply cuts on each variable individually. This method is appropriate if the signal is clearly distinguishable from the background by a few variables. It is no longer sufficient if there are many variables with low discrimination power. The disadvantage of applying cuts in such a situation can easily be seen from two examples. A decay which is perfectly consistent with the signal in all but one variable will be rejected if this one variable just fails the cut, although it is most likely a good event. On the other hand, an event which is close but within the cuts on all variables is accepted, although it is doubtful whether it is a good event. This disadvantage can be avoided with the next two methods.

**Likelihood Method.** The likelihood method is based on a statistical interpretation of distributions like those shown in Figs. 2.22–2.24. After a preselection each variable is histogrammed for each class of decay channels and the histograms are normalized. They are interpreted as the probability density

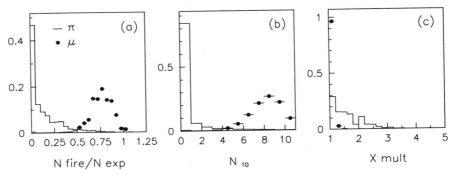

**Fig. 2.24.** Some variables based on the signals of the hadron calorimeter used for muon identification (ALEPH collaboration [270]). (a) The ratio of planes fired (out of 23) to the number expected to fire for a fully penetrating particle. (b) The number of planes fired out of the last 10 expected for a fully penetrating particle. (c) The number of hits in a single plane, averaged over all planes fired. The *points* refer to muons, the *histogram* to pions. These are particles from hadronic events, but the distributions look very similar for $\tau$ decays

function of the variable $x_i$ for class $j$. For example, if the shape variable of Fig. 2.23 is measured in a data event to be 0.9, then the value of the histogram in the upper plot (after normalization) gives the probability that this event originated from a $\tau \to e\nu_e\nu_\tau$ decay, whereas the value of the lower histogram at 0.9 gives the probability that it came from a hadronic decay. Thus we have

$$p_i^j(x_i) = \frac{f_i^j(x_i)}{\sum_{j=1}^{N_{\text{cl}}} f_i^j(x_i)}. \tag{2.6}$$

The superscript $j$ specifies a certain class and $i$ a variable, so that $p_i^j$ is the probability that the event belongs to class $j$ on the basis of the information from variable $i$. It is normalized, so that the probability of the event belonging to any of the $N_{\text{cl}}$ classes is 1. Then, if a total of $N_{\text{var}}$ variables are available for the identification, they are combined by multiplying the probabilities and renormalizing them:

$$P^j = \frac{\prod_{i=1}^{N_{\text{var}}} p_i^j(x_i)}{\sum_{j=1}^{N_{\text{cl}}} \left( \prod_{i=1}^{N_{\text{var}}} p_i^j(x_i) \right)}. \tag{2.7}$$

$P^j$ is the total probability, or the likelihood, for the event to belong to class $j$. Depending on the application, one can either put a cut on $P^j$ and select the events above the cut as belonging to class $j$, or classify the event into the class with the highest probability.[4] The first method gives better purity, but some events will be left unidentified. The second method identifies all events, even the doubtful ones, where the probabilities for several classes are similar.

Such likelihood methods have been used, for example, in [72, 103, 228]. In the case that all variables are uncorrelated, $P^j$ is a probability in the statistical sense and this is the perfect method to use.

**Artificial Neural Nets.** If a likelihood method is invoked with a set of variables of which a few are correlated and the correlation is known, this can be removed by a transformation of variables [271]. With more complicated correlations this procedure becomes unpractical and it is better to use a method that handles correlations properly. A neural net represents such a method. The net of [272] has been used, for example, in [67]. For more information see [272–280].

### 2.5.2 Reconstruction of $\pi^0$s

In $\tau$ decays a large number of neutral pions is produced. They decay predominantly through $\pi^0 \to \gamma\gamma$ with a lifetime invisible to the experiments. The $\pi^0$ mesons have to be reconstructed from the energy depositions of the photons in the electromagnetic calorimeters.

At lower center-of-mass energies the identification of the photons is relatively simple. Each photon makes a neutral cluster, so that each reconstructed neutral cluster will be called a photon candidate.[5] A minimum energy and a minimum separation from the nearest charged track have to be required to reject noise clusters and fake photons. Fake photons at these energies are produced by split-offs from the showers of the hadrons in the event. The inefficiency introduced by these cuts makes a low energy threshold for the photons desirable.

With the photons identified, there remains the problem of how to combine them into $\pi^0$s, if more than two photons are present in the event. Figure 2.25 shows an example from an analysis of the $\tau \to 3\,\pi^\pm\,2\,\pi^0\,\nu_\tau$ decay by the CLEO collaboration [135]. Events are selected with four photons identified from one $\tau$. There are three possibilities for combining the four photons into two pairs. For each of the three combinations the invariant masses of both pairs are calculated and plotted against each other. The peak in the center of the plot gives the proper combinations.

At higher center-of-mass energies the identification of the photons is a much more serious problem, as the photons and the showers of the charged tracks overlap in the calorimeter. A high granularity of the calorimeter will be an advantage at these energies. It is no longer sufficient to call each neutral cluster a photon candidate. The clusters have to be investigated for the

---

[4] The value of the cut should be larger than 0.5. Otherwise it might happen that events are selected in several classes at the same time.
[5] A neutral cluster is a cluster with no track pointing to it.

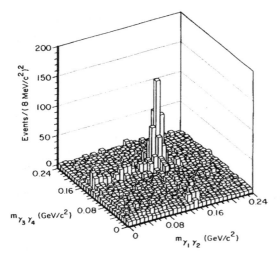

**Fig. 2.25.** Invariant mass of photon pairs in events with four identified photons (from a study of $\tau \to 3\pi^{\pm} 2\pi^0 \nu_\tau$ by the CLEO collaboration [135]). There are three entries per event, corresponding to the three possible combinations of four photons into two pairs. The combinatorial background can be extrapolated from the side bands

presence of several photons. Given enough granularity, each photon causes a local maximum in the energy deposition in the cells within a cluster. The ALEPH calorimeter might serve as an example (for details see [270]; see also Fig. 2.13). It is a sampling calorimeter (lead with proportional chambers) with pad and wire readout. The pads are used for the photon identification. They are organized in projective towers of approximately $30 \times 30$ mm$^2$ at a radius of about 2 m. After the local maxima are identified the energy in the remaining cells is associated with the largest neighbors and the energy of the photons is calculated from the four central cells. Fake photons are the largest problem. Local fluctuations in the energy depositions of photonic and hadronic showers are the origin of these fakes. Figure 2.26 shows some of the variables used by the ALEPH collaboration to separate true from fake photons [56].

Then the $\pi^0$s are reconstructed from the photons. But at high energies it is not sufficient to simply pair the identified photons. The opening angle between the two photons decreases with the energy of the $\pi^0$ so that the two photons merge at higher energies. The shower resulting from the two photons is eccentric in the transverse plane and it is still possible to estimate the invariant mass of the underlying photons. Figure 2.27 shows the distribution for three different $\pi^0$ energies.

Finally there are $\pi^0$s where one of the photons has been lost. Some of these $\pi^0$s can still be recovered if the reconstructed photon can be distinguished clearly enough from fake photons, initial- or final-state radiation, bremsstrahlung of the tracks in the detector material, and genuine photons from $\omega \to \pi^0 \gamma$ or $\eta \to \gamma \gamma$.

In a final step the energy resolution for the $\pi^0$s can be improved, in the case of two reconstructed photons, by constraining their invariant mass in a

2.5 Identification of the Decays    33

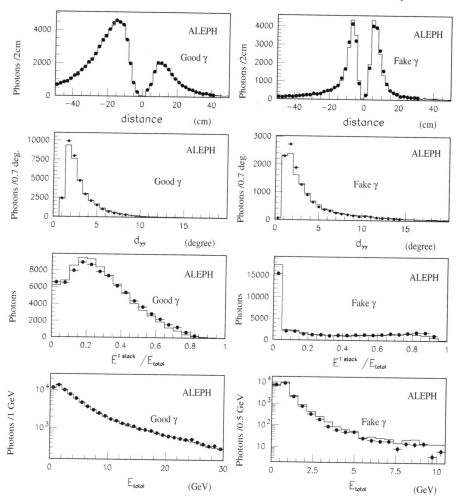

**Fig. 2.26.** Data and Monte Carlo distributions of some of the discriminating variables used for photon identification (ALEPH collaboration [56]). *1st row*: distance of the photon to the closest charged track. The sign reflects the position of the photon with respect to the bending of the track. *2nd row*: distance to the closest photon. *3rd row*: fraction of energy deposited in the first four radiation lengths. *4th row*: energy of the photon. *Left* column shows the good photons, *right* column the fakes. *Points* refer to data, histograms to Monte Carlo results

kinematic fit to the $\pi^0$ mass. Similar techniques can be used to identify the decay $\eta \to \gamma\gamma$, but the background is much larger.

34     2. Experimental Aspects

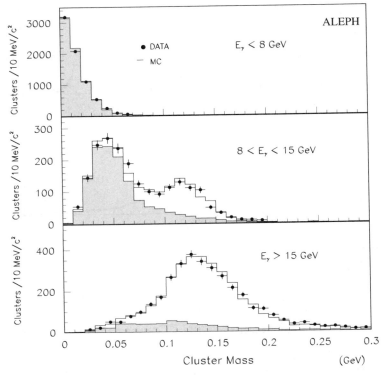

**Fig. 2.27.** Invariant-mass distribution for the unresolved $\pi^0$s in three different cluster energy ranges (ALEPH collaboration [56]). The *points* are data, the histogram is the Monte Carlo prediction. The *white areas* refer to clusters emerging from true $\pi^0$s, the *shaded area* to single photons

### 2.5.3 $\pi/K$ Separation

Many analyses require a separation between charged kaons and pions. This can be achieved either by a measurement of the specific energy loss of the particle in the tracking volume or by a Čerenkov detector. For both techniques the discrimination is better at low momenta. A time-of-flight measurement is another possibility, but it only works at center-of-mass energies close to the $\tau$ production threshold.[6] The TPC/$2\gamma$ experiment was one of the first with a reasonably accurate measurement of the specific energy loss [281]. Figure 2.28 shows a scatter plot of the measured energy loss versus the particles' momentum in one-prong $\tau$ decays. Pions and kaons are nicely separated for energies above 1.5 GeV. The measurement also supports the identification of electrons.

---

[6] Right at the threshold $\tau \to \pi \nu_\tau$ and $\tau \to K \nu_\tau$ are kinematically separated. As the $\tau$ leptons are at rest even in the laboratory frame, the pion and kaon have a different momentum (see (2.1)).

2.5 Identification of the Decays   35

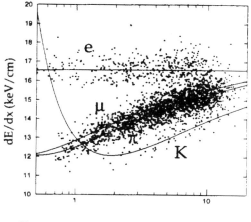

**Fig. 2.28.** Measurement of the specific energy loss (d$E$/d$x$) in one-prong $\tau$ decays by the TPC/2$\gamma$ collaboration [281]. The four graphs represent the expected energy loss for electrons, muons, pions, and kaons (*top* to *bottom*)

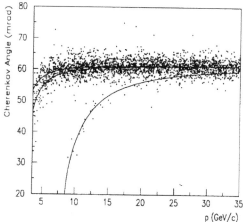

**Fig. 2.29.** Measured Čerenkov angle versus the momentum in one-prong $\tau$ decays (DELPHI collaboration [282]). The three graphs correspond to the expected angles for muons, pions, and kaons (*top* to *bottom*)

Only a few experiments in $\tau$ physics are equipped with Čerenkov detectors (SLD, DELPHI, TASSO[7]). Figure 2.29 shows the measured Čerenkov angle versus the momentum in one-prong $\tau$ decays (DELPHI experiment [282]). There is a separation of more than four standard deviations between pions and kaons up to about 15 GeV.

### 2.5.4 Neutral Kaons

Special identification procedures are also needed for neutral kaons in $\tau$ decays. There are three different cases

---

[7] SLD and DELPHI have ring-imaging Čerenkov detectors and TASSO has threshold counters.

36   2. Experimental Aspects

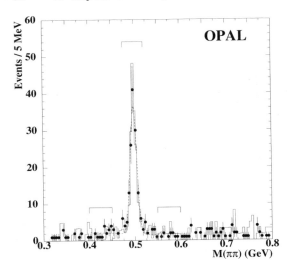

Fig. 2.30. Invariant mass of $K_S^0 \to \pi^+ \pi^-$ candidates (OPAL collaboration [233]). The events underneath the *central bracket* are accepted. The *outer brackets* indicate the side bands for the subtraction of the combinatorial background

- $K_S^0 \to \pi^+ \pi^-$
- $K_S^0 \to \pi^0 \pi^0$
- $K_L^0$.

Most of the results to date have been achieved with $K_S^0 \to \pi^+ \pi^-$ decays. The $K_S^0 \to \pi^0 \pi^0$ is very hard to identify on top of the much larger combinatorial background from $\tau \to \pi^\pm \pi^0 \pi^0 \nu_\tau$. Only a few analyses identifying the $K_L^0$ as a neutral cluster in the hadronic calorimeter have been published [63, 176].

The $K_S^0 \to \pi^+ \pi^-$ is identified as a pair of oppositely charged tracks which make a secondary vertex. The mean flight length of the $K_S^0$ from the primary vertex to its decay depends on its energy. Typical values are 5 cm at the $\Upsilon(4s)$ and 50 cm at the $Z^0$. The invariant mass between the two tracks can be measured with good resolution. The narrow $K_S^0$ peak stands out over a small combinatorial background. Figure 2.30 shows the mass distribution as measured by the OPAL collaboration [233]. A secondary vertex between those tracks has been identified before.

The $K_L^0$ has a lifetime so long that it does not decay within the detectors. It interacts in the hadronic calorimeter, creating an additional hadronic shower, additional to those of the charged pions/kaons in the event. Hadronic showers are larger in size and fluctuate more than electromagnetic showers and therefore the neutral cluster of the $K_L^0$ in the hadronic calorimeter is harder to identify than those of the $\pi^0$s in the electromagnetic calorimeter.

The ALEPH collaboration [63] identifies $K_L^0$ in hadronic one-prong decays on the basis of two distinct quantities:

- The excess of energy over the energy deposition expected from the charged hadron in units of the expected energy resolution,

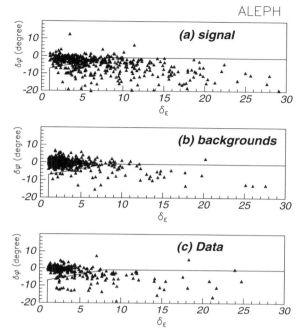

**Fig. 2.31.** Distribution of the calorimetric variables $\Delta_E$ and $\Delta\varphi$ ((2.8) and (2.9)) for the identification of $K_L^0$ in hadronic one-prong $\tau$ decays (ALEPH collaboration [63]). The plots show the class without additional $\pi^0$. **(a)** Monte Carlo simulation of the signal $\tau \to \pi^\pm K^0 \nu_\tau$ and $\tau \to K^\pm K^0 \nu_\tau$ and **(b)** background; **(c)** data. Events below the *line* are accepted

$$\Delta_E = \frac{E_{\text{HCal}} - P_h}{\sigma_h}. \qquad (2.8)$$

- The angular offset between the position of the hadronic shower and that expected from a single charged hadron with no $K_L^0$. The bending of the charged track due to the magnetic field dominates over the expected opening angle between h and $K_L^0$ from the decay. Therefore only the offset in the plane perpendicular to the beam is used, with a sign $\xi$ according to the direction of the bending:

$$\Delta\phi = \xi |\phi_{\text{barycenter}} - \phi_{\text{track impact}}|. \qquad (2.9)$$

Figure 2.31 shows the distribution of the two variables in the data and the Monte Carlo simulation.

## 2.6 Monte Carlo Simulation

### 2.6.1 Simulation of $\tau$ production

An essential tool in any analysis is the investigation of computer-simulated events, where the result can be checked against the input to the simulation. Monte Carlo simulation is often necessary to determine efficiencies, backgrounds, and resolutions. To avoid systematic biases of the results, the simulation should be as accurate and as close as possible to the real events.

The commonly accepted Monte Carlo generator for the production of $\tau$ pairs is KORALB/KORALZ [283–285]. The two versions differ in the treatment of spin correlations and electroweak corrections. The KORALB version refers to the $\Upsilon(4s)$ and KORALZ to the $Z^0$ pole. The KORALB version is optimized for center-of-mass energies where the influence of the $Z^0$ is not yet visible ($2m_\tau \leq \sqrt{s} \leq 30$ GeV) and KORALZ can be used from energies far above the $\tau$ production threshold to below the $W^\pm$ pair threshold, including the $Z^0$ peak (20 GeV $\leq \sqrt{s} \leq 150$ GeV).

The programs include interfaces to libraries for the simulation of QED radiation (YFS3 [286–289], PHOTOS [290, 291], MUSTRAAL[292]), electroweak corrections (DIZET [293], Z0POLE [294]), and $\tau$ decays (TAUOLA [295–297]). The TAUOLA library will be discussed in the next section; for the others the reader is referred to the literature.

Figure 2.32 shows the basic flow chart of the KORALB and KORALZ programs. The three shaded blocks are the core of the program, where events are generated with three different kinematics: without any photons (left), with photon radiation from the initial electrons (center), and from the $\tau$ leptons (right).

Each block has the same structure: any cross section has to be isotropic with respect to a rotation around the beam axis. Therefore, the azimuthal angle specifying this rotation can be chosen uniformly between 0 and $2\pi$ up front. Then the remaining kinematic quantities (the polar angle in the case of no radiation, and three angles with radiation) are generated according to the differential cross section with all radiative corrections by a 'hit and miss' technique. These are the loops from the acceptance decision at the end of the box back to the beginning, if the event is not accepted.

To make the program more efficient, the kinematic quantities are not generated uniformly, but according to a simplified cross section. This simplified cross section is chosen to be close to but always larger than the full cross section. It is refined to the full cross section by the 'hit and miss' technique.

The generation of an event starts with the generation of a photon energy according to the spectra calculated.[8] A photon cutoff energy $k_0$ is introduced. Below that energy the photon is not taken into account in the kinematics of

---

[8] The technique of a simplified cross section refined by a 'hit and miss' decision is also used here.

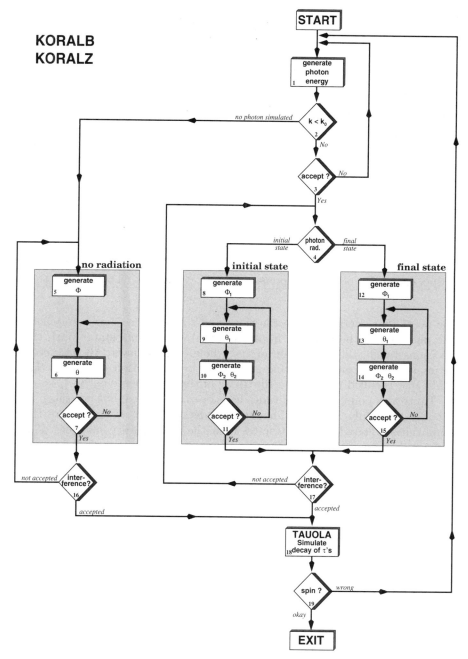

**Fig. 2.32.** Flow chart of the Monte Carlo generators KORALB and KORALZ [283–285]. *Rectangular boxes* denote the generation of variables and *diamond-shaped boxes* mark decision points

**Table 2.4.** Properties of KORALB and KORALZ

|  | KORALB | KORALZ |
|---|---|---|
| Inclusion of the finite $\tau$ mass | Yes | Yes |
| Inclusion of the finite electron mass | No | No |
| QED radiative corrections | $\mathcal{O}(\alpha^3)$ | $\mathcal{O}(\alpha^3)$ |
| Hard bremsstrahlung from initial/final state | Single | Multiple |
| Initial/final-state interference | Yes | Not for multiple $\gamma$ |
| Inclusion of $Z^0$ exchange | Yes | Yes |
| Electroweak corrections | None | $\mathcal{O}(\alpha)$ |
| Polarization of initial beams | Arbitrary | Longitudinal only |
| Spin effects in $\tau$ decays | Yes | Yes<br>Except for $\tau \to n\,\pi\,\nu_\tau,\, n \geq 5$ |
| Spin correlation between $\tau$ leptons | All | Longitudinal only |

the event. For energies above the threshold the next decision is to whether the photon shall be radiated from the initial or the final state. With these decisions taken, the generation of the whole event follows. All interference terms are added only after this generation by a rejection technique. Then TAUOLA is called to simulate the decay of the two $\tau$ leptons.

Up to this point spin amplitudes have been calculated, but not taken into account yet. Only after the final step of decaying the $\tau$ leptons are these amplitudes squared and contracted with the spin density matrices of the external fermions. A final 'hit and miss' decision imprints the spin information on the generated events. This might not be the most efficient way in terms of computer time, but it simplifies the calculations significantly.

Table 2.4 lists some of the basic features of the program. Until the introduction of KORALB the generators for $e^+e^- \to \mu^+\mu^-$ had to be misused to make $\tau$ pairs. These generators ignore the mass of the final fermions, which limits their applicability for $\tau$ production. The accuracy of the muon generators is limited to a few percent, whereas KORALB and KORALZ achieve 1%. Figures 2.33 and 2.34 demonstrate the effect of the finite $\tau$ mass.

### 2.6.2 Simulation of $\tau$ Decays

The decays of the $\tau$ leptons in KORALB and KORALZ are generated by the TAUOLA library [295–297, 299]. At present (version 2.4), 22 different decay channels are implemented. They are summarized in Table 2.5.

2.6 Monte Carlo Simulation    41

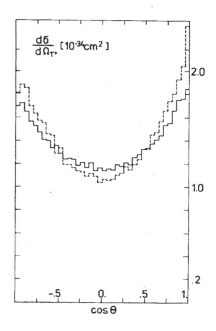

**Fig. 2.33.** Differential cross section for $\tau$ production as a function of the polar angle $\theta$ of the positive $\tau$ [298]. *Solid line*: full KORALB. *Dashed line*: $\tau$ mass set to zero

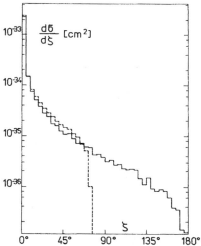

**Fig. 2.34.** Acollinearity distribution [298]. *Solid line*: full KORALB. *Dashed line*: $\tau$ mass set to zero

**Table 2.5.** Decay channels of the $\tau$ decay library TAUOLA [297]. The number in the left column is the internal ID of the decay

| | |
|---|---|
| 1  | $\tau \to e\,\nu_e\nu_\tau$ |
| 2  | $\tau \to \mu\,\nu_\mu\nu_\tau$ |
| 3  | $\tau \to \pi\,\nu_\tau$ |
| 4  | $\tau \to \pi\,\pi\,\nu_\tau$ |
| 5  | $\tau \to 3\,\pi\,\nu_\tau$ |
| 6  | $\tau \to K\,\nu_\tau$ |
| 7  | $\tau \to K\,\pi\,\nu_\tau$ |
| 8  | $\tau \to 3\,\pi^\pm\,\pi^0\,\nu_\tau$ |
| 9  | $\tau \to \pi^\pm\,3\,\pi^0\,\nu_\tau$ |
| 10 | $\tau \to 3\,\pi^\pm\,2\,\pi^0\,\nu_\tau$ |
| 11 | $\tau \to 5\,\pi^\pm\,\nu_\tau$ |
| 12 | $\tau \to 5\,\pi^\pm\,\pi^0\,\nu_\tau$ |
| 13 | $\tau \to 3\,\pi^\pm\,3\,\pi^0\,\nu_\tau$ |
| 14 | $\tau \to K^+\,K^-\,\pi^\pm\,\nu_\tau$ |
| 15 | $\tau \to K^0\,\overline{K}^0\,\pi^\pm\,\nu_\tau$ |
| 16 | $\tau \to K^\pm\,K^0\,\pi^0\,\nu_\tau$ |
| 17 | $\tau \to K^\pm\,\pi^0\,\pi^0\,\nu_\tau$ |
| 18 | $\tau \to K^\pm\,\pi^+\,\pi^-\,\nu_\tau$ |
| 19 | $\tau \to K^0\,\pi^\pm\,\pi^0\,\nu_\tau$ |
| 20 | $\tau \to \eta\,\pi^\pm\,\pi^0\,\nu_\tau$ |
| 21 | $\tau \to \pi^\pm\,\pi^0\,\gamma\,\nu_\tau$ |
| 22 | $\tau \to K^\pm\,K^0\,\nu_\tau$ |

All channels are implemented to the best available knowledge, combining theoretical expectation and experimental information. The charged-current interaction is written as

$$g_v\gamma_\mu - g_a\gamma_\mu\gamma_5 \qquad (2.10)$$

with the parameters $g_v$ and $g_a$ changeable by the user (default: $g_v$, $g_a = 1$). Also, the mass of the $\tau$ neutrino and all branching ratios can be set by the user. A few details about the channels are given below:

- $\tau \to e\,\nu_e\nu_\tau$ and $\tau \to \mu\,\nu_\mu\nu_\tau$
  These are simulated according to the Standard Model matrix element with full $\mathcal{O}(\alpha)$ QED radiative corrections [296].
- $\tau \to \pi\,\nu_\tau$ and $\tau \to K\,\nu_\tau$
  These decays can be calculated from theory up to the normalization, which has to be specified by the user anyhow. Radiative corrections are included in these and the other hadronic channels in the leading logarithmic approximation.
- $\tau \to \pi\,\pi\,\nu_\tau$
  This decay is simulated through the $\rho$ resonance with a 14.5% admixture of $\rho'$ (see [300] for details).

**Table 2.6.** Form factors of the three-meson final states in TAUOLA [297]. $F_1$ and $F_2$ axial-vector, $F_4$ pseudoscalar, and $F_5$ Wess–Zumino anomaly. The number in the left column is the internal ID of the channel

| | Mode | $F_1$ | $F_2$ | $F_4$ | $F_5$ |
|---|---|---|---|---|---|
| 5 | $\tau \to 3\pi\nu_\tau$ | $a_1 \to \rho\pi$ | $a_1 \to \pi\rho$ | $\pi'(1300)$ | – |
| | | $\rho \to \pi\pi$ | $\rho \to \pi\pi$ | optional | |
| 14 | $\tau \to K^+ K^- \pi^\pm \nu_\tau$ | $a_1 \to K^* K$ | $a_1 \to \pi\rho$ | | $\rho \to K^* K$ |
| | | $K^* \to K\pi$ | $\rho \to KK$ | | $K^* \to K\pi$ |
| 15 | $\tau \to K^0 \overline{K}^0 \pi^\pm \nu_\tau$ | $a_1 \to K^* K$ | $a_1 \to \pi\rho$ | | $\rho \to K^* K$ |
| | | $K^* \to K\pi$ | $\rho \to KK$ | | $K^* \to K\pi$ |
| 16 | $\tau \to K^\pm K^0 \pi^0 \nu_\tau$ | – | $a_1 \to \pi\rho$ | – | – |
| | | | $\rho \to KK$ | | |
| 17 | $\tau \to K^\pm \pi^0 \pi^0 \nu_\tau$ | $K_1 \to K^* \pi$ | $K_1 \to \pi K^*$ | – | – |
| | | $K^* \to K\pi$ | $K^* \to K\pi$ | | |
| 18 | $\tau \to K^\pm \pi^+ \pi^- \nu_\tau$ | $K_1 \to K\rho$ | $K_1 \to \pi K^*$ | | $K^* \to \rho K$ |
| | | $\rho \to \pi\pi$ | $K^* \to K\pi$ | | $\rho \to \pi\pi$ |
| 19 | $\tau \to K^0 \pi^\pm \pi^0 \nu_\tau$ | – | $K_1 \to K\rho$ | | $K^* \to \rho K$ |
| | | | $\rho \to \pi\pi$ | | $\rho \to \pi\pi$ |
| 20 | $\tau \to \eta \pi^\pm \pi^0 \nu_\tau$ | – | – | – | $\rho \to \rho\eta$ |
| | | | | | $\rho \to \pi\pi$ |

- $\tau \to 3\pi\nu_\tau$
  The three-pion channel is simulated through the decay chain $a_1 \to \rho\pi$ and $\rho \to \pi\pi$ following the Kühn/Santamaria model [300]. This channel includes both charge combinations $\tau^- \to \pi^- \pi^- \pi^+ \nu_\tau$ and $\tau^- \to \pi^- \pi^0 \pi^0 \nu_\tau$ with equal weights.

- $\tau \to K\pi\nu_\tau$
  This decay proceeds through the $K^*(892)$ resonance with the final states $K^-\pi^0$, $K^0_S\pi^-$, and $K^0_L\pi^-$.

- **Three-Meson Final States**
  There are three possible contributions to the form factors of the three-meson final states: axial-vector current ($F_1$ and $F_2$ [301, 302]), pseudoscalar ($F_4$ [303]), and the Wess–Zumino anomaly ($F_5$ [304, 305]). Table 2.6 summarizes the form factors used in the various channels.

- **Four-Pion Channels**
  There are two final states with different charge combinations, $\tau^- \to \pi^- \pi^- \pi^+ \pi^0 \nu_\tau$ and $\tau^- \to \pi^- 3\pi^0 \nu_\tau$, corresponding to channels 8 and 9 in Table 2.5. The overall resonance structure ($Q^2$ distribution) is described by the vector form factor. The vector form factor is described by the same parametrization of the $\rho$ resonances as in the two-pion final state. Intermediate $\rho \to \pi\pi$ resonances are implemented wherever appropriate. There is also a contribution from the Wess–Zumino anomaly included through the coupling $\rho \to \omega\pi$. It gives a contribution to $\tau^- \to \pi^-\pi^-\pi^+\pi^0\nu_\tau$ through the decay $\omega \to \pi^+ \pi^- \pi^0$, but not to $\tau^- \to \pi^- 3\pi^0 \nu_\tau$. It is also the source of channel 21 ($\tau \to \pi^\pm \pi^0 \gamma \nu_\tau$) through $\omega \to \pi^0 \gamma$.

- **Five-Pion Channels**
  For the generation of the five-pion final states (10 and 11 in Table 2.5), a four-pion state is generated with a structure taken from $e^+e^- \to 4\pi$ via the conserved vector current hypothesis (CVC). The four pions within this subsystem are distributed according to a flat phase space. Then a fifth pion is added using the soft-pion technique [306]. No polarization effects are taken into account.
- **Six-Pion Channels**
  The six-pion final states (12 and 13) are generated with a structure taken from $e^+e^- \to 6\pi$ via CVC. No polarization effects are taken into account.

# 3. The Static Properties of the $\tau$

## 3.1 The Mass

The simplest and most precise technique to determine the mass of the $\tau$ lepton is a scan of the production threshold. An experiment running at the threshold scans the $\tau$ pair cross section to find the center-of-mass energy at which $\tau$ production sets in. This is $2m_\tau$, with some small corrections. This technique needs a precise calibration of the accelerator and a good understanding of the cross section near threshold. The early experiments in $\tau$ physics determined the $\tau$ mass using this technique [38, 145, 307] and recently it has been remeasured by the BES collaboration [98, 99]. The cross section and the measurement are described in the next two subsections, followed by two methods that can be applied to data taken far above threshold [84, 137].

### 3.1.1 The Cross Section at and near the Threshold

The cross section for $\tau$ production near the threshold is given to lowest order by

$$\sigma_{\tau\tau} = \frac{4\pi\alpha^2}{3s} \beta \left( \frac{3-\beta^2}{2} \right). \tag{3.1}$$

$\beta = p_\tau/E_\tau$ is the velocity of a $\tau$, $s = E_{\text{cm}}^2$, the center-of-mass energy squared, and $\alpha$ is the fine structure constant.

It is not sufficient to use the lowest-order cross section for this measurement. Corrections are applied that modify it into [308–310]

$$\sigma_{\tau\tau} = \frac{1}{\sqrt{2\pi}\Delta} \int_{2m_\tau}^{\infty} dE' \left( \exp\frac{-(E_{\text{cm}} - E')^2}{2\Delta^2} \right.$$

$$\left. \times \int_0^{1-4m_\tau^2/E'^2} dx\, F_{\text{I}}(x, E')\, \tilde{\sigma}(\sqrt{1-x}E') \right) \tag{3.2}$$

with

$$\tilde{\sigma}(E) = \frac{4\pi\alpha^2}{3E^2} \beta \left( \frac{3-\beta^2}{2} \right) \frac{F_{\text{C}}(\beta)\, F_\tau(\beta)}{[1-\Pi(E)]^2}. \tag{3.3}$$

46    3. The Static Properties of the τ

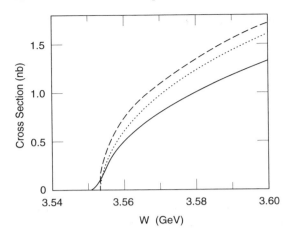

**Fig. 3.1.** The cross section for τ production near threshold [98]. The *dotted curve* is the lowest-order cross section, the *dashed curve* includes the Coulomb correction and final-state radiation ($\tilde{\sigma}$ in (3.2)), and the *solid curve* shows the full cross section with all corrections. The input τ mass is 1776.9 MeV and the beam energy spread about 1.4 MeV

The corrections are described below. Their impact on the cross section can be seen from Fig. 3.1.

- **Final-State Radiation**
  Corrections to the cross section due to photon radiation from the τ leptons are taken into account by the function $F_\tau(\beta)$ [311].
- **Coulomb Correction**
  At the threshold a τ pair is produced at rest. The two τ leptons are not moving away from each other. They have some time to bind into a τ atom before they decay. As a result of the binding energy the cross section at $E_{cm} = 2\,m_\tau$ becomes a finite value. It would approach zero otherwise. The correction is called the Coulomb correction, as it is the Coulomb interaction that binds the τ leptons. It is described by $F_C(\beta)$ [311].
- **Vacuum Polarization**
  The QED corrections to the photon propagator due to the insertion of quark and lepton loops is given by $\Pi(E)$. It has been calculated in [312].
- **Initial-State Radiation**
  The cross section is also modified by photon radiation from the initial electrons. The radiation effectively reduces the center-of-mass energy of the $e^+e^-$ collision, so that the initial cross section has to be replaced by the cross section at the reduced energy $E\sqrt{1-x}$. The probability of such a radiation is described by the function $F_I(x, E)$ and the correction is introduced by integrating the cross section over all possible radiations from the full beam energy down to the threshold [313–315].
- **Beam Energy Spread**
  The last correction is an experimental one. Not all electrons in a collider beam carry exactly the central beam energy. Therefore the center-of-mass energy is smeared out over a certain range $\Delta$ (approximately $\pm 1.4$ MeV in case of the Beijing Electron Positron Collider (BEPC)). This can lead to

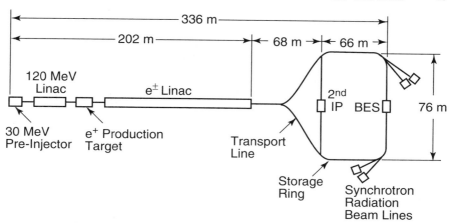

**Fig. 3.2.** The Beijing Electron Positron Collider (BEPC). The 202 m injection linac on the *left* leads to the 240 m circumference storage ring on the *right*. The detector BES is installed at the interaction point, opposite to the injection

production of $\tau$ pairs even if the central beam energy is below the threshold. It is taken into account by folding the cross section over the beam energy spread (the first integral in (3.2)).

### 3.1.2 Measurements at Threshold

At the Beijing Institute of High Energy Physics a new accelerator was constructed in the late 1980s with a more precise measurement of the $\tau$ mass as one of the major goals of the experiment. Figure 3.2 shows a sketch of the accelerator, called the Beijing Electron Positron Collider. It operates at center-of-mass energies in the range of 3 to 5 GeV with a peak luminosity of $5 \times 10^{30}$ cm$^{-2}$s$^{-1}$ in the $\tau$ threshold region.

A total of 5 pb$^{-1}$ of data was collected near threshold during a 2 month running period starting in November 1991. An initial result, based on e$\mu$ events only, was published in [99]. The final result includes all combinations of leptonic decays and $\tau \to$ had $\nu_\tau$, where 'had' is a pion or a kaon [98]. The result is now limited by systematics and no more data will be collected near threshold.

The range of center-of-mass energies in which the $\tau^+\tau^-$ cross section is most sensitive to the $\tau$ mass is of the order of the beam energy spread around the threshold. A running strategy was devised to locate the threshold region and maximize the integrated luminosity there (see Fig. 3.3). The beam energy was set initially to 1784.1 MeV, which was the world average at that time [316]. The data was scanned for e$\mu$ events, which provide a very clean signal for $\tau$ production. Each time, after another 250–400 nb$^{-1}$ of data were collected, a new estimate of the $\tau$ mass was derived. A new prediction of the most sensitive center-of-mass energy was obtained. The beam energy was

48     3. The Static Properties of the $\tau$

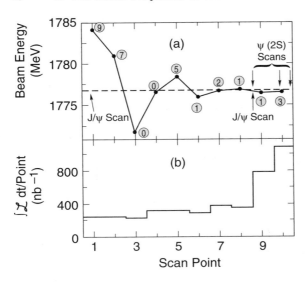

Fig. 3.3. (a) The variation of the beam energy with scan point, showing the convergence to the $\tau^+\tau^-$ threshold. The *arrows* indicate the sequence of the J/$\Psi$ and $\Psi$(2s) scans for calibration. The *numbers in the circles* are the numbers of $\tau$ pairs identified in the final analysis. (b) The integrated luminosity at each scan point (BES collaboration [98])

set to this new value if the difference from the old value was greater than 0.4 MeV. Figure 3.3 shows the history of the scan, with the energies BEPC had been running at and the luminosity collected at each point. There were two more scan points far above threshold, where the cross section is much higher. These were taken to study the selection efficiency.

The calibration and energy spread of the machine were determined by scanning the nearby J/$\Psi$ and $\Psi$(2s) resonances in between the $\tau^+\tau^-$ threshold scan points. Their masses are well known ($m(\mathrm{J}/\Psi) = (3096.88 \pm 0.04)$ MeV and $m(\Psi(2\mathrm{s})) = (3686.00 \pm 0.09)$ MeV), and their natural widths (87 and 277 keV, respectively) are negligible compared to the beam energy spread [44].

The $\tau$ mass was extracted from the data by a binned maximum-likelihood fit to the number of events observed at each scan point with Poisson statistics. The number of expected events at each scan point is given by

$$N_i = [\epsilon\, r_i\, \sigma_{\tau\tau}(m_\tau) + \sigma_\mathrm{B}] \times \mathcal{L}_i. \tag{3.4}$$

Here $\sigma_{\tau\tau}(m_\tau)$ is the cross section from (3.2) and $\mathcal{L}_i$ the luminosity at each scan point. The efficiency is divided into an overall factor $\epsilon$, which incorporates the uncertainties in the detection and trigger efficiencies and the luminosity scale, and a relative factor $r_i$ describing the point-by-point variation; $\sigma_\mathrm{B}$ is the background cross section. In the fit the overall efficiency $\epsilon$ and $\sigma_\mathrm{B}$ are kept floating, in addition to $m_\tau$. The result is

$$m_\tau = 1776.96^{+0.18}_{-0.21}{}^{+0.25}_{-0.17}\ \mathrm{MeV}, \tag{3.5}$$

where the first error is statistical and the second systematic. The contributions to the systematic error are listed in Table 3.1.

## 3.1 The Mass

**Table 3.1.** The contributions to the systematic error on the $\tau$ mass (BES collaboration [98])

| Source | $\Delta_{m_\tau}$ in MeV |
|---|---|
| Overall efficiency $\epsilon$ | $^{+0.09}_{-0.10}$ |
| Background | $^{+0.19}_{-0.00}$ |
| Bias in scanning procedure | $\pm 0.10$ |
| Beam energy calibration | $\pm 0.09$ |
| Beam energy spread $\Delta$ | $\pm 0.02$ |
| Total | $^{+0.25}_{-0.17}$ |

### 3.1.3 The ARGUS Method

The method described above is only applicable to experiments running near the $\tau$ production threshold. The ARGUS collaboration has developed a pseudomass technique that can be applied to data taken far above the threshold [84]. It is based on the kinematics of the semihadronic $\tau$ decays.

If the direction of the $\tau$ neutrino were known, the $\tau$ mass could be simply calculated as the invariant mass between the hadron and the neutrino. As it is not known, ARGUS approximated it. The approximation they used has a very special feature. The mass reconstructed under the approximation matches the true mass only for some events; for all the others it is always smaller. It is called a pseudomass. The pseudomass spectrum is continuous at lower masses and has a sharp cutoff at the true mass. The position of the cutoff indicates the true value in the data.

More precisely, the $\tau$ mass can be calculated as

$$\begin{aligned} m_\tau^2 &= p_\tau^2 \\ &= (p_h + p_\nu)^2 \\ &= (E_h + E_\nu)^2 - (\boldsymbol{p}_h + \boldsymbol{p}_\nu)^2 \\ &= E_\tau^2 - (\boldsymbol{p}_h^2 + \boldsymbol{p}_\nu^2 + 2|\boldsymbol{p}_h||\boldsymbol{p}_\nu|\cos(\boldsymbol{p}_h, \boldsymbol{p}_\nu)) \\ &= E_\tau^2 - \boldsymbol{p}_h^2 - E_\nu^2 - 2|\boldsymbol{p}_h|E_\nu \cos(\boldsymbol{p}_h, \boldsymbol{p}_\nu) \\ &= 2E_h(E_\tau - E_h) + m_h^2 - 2|\boldsymbol{p}_h|(E_\tau - E_h)\cos(\boldsymbol{p}_h, \boldsymbol{p}_\nu), \end{aligned} \quad (3.6)$$

where $E_\nu = E_\tau - E_h$ is used and the neutrino mass is set to zero. The $\tau$ energy is given by the beam energy, and any other quantity in the formula can be reconstructed from the measured particles except for the angle $\angle(\boldsymbol{p}_h, \boldsymbol{p}_\nu)$. Now the direction of the neutrino is approximated by that of the hadron, i.e. the cosine of this angle is set to 1:

$$m_\tau^{*2} = 2E_h(E_\tau - E_h) + m_h^2 - 2|\boldsymbol{p}_h|(E_\tau - E_h). \quad (3.7)$$

As the angle is small owing to the boost, the cosine term contributes with a negative sign and $m_\tau^*$ is always smaller than $m_\tau$.

50   3. The Static Properties of the $\tau$

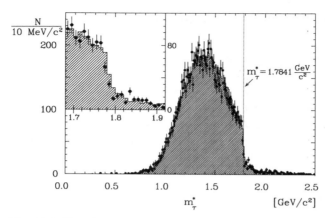

**Fig. 3.4.** Pseudomass spectrum in the $\tau \to 3\,\pi\,\nu_\tau$ channel (ARGUS collaboration [84]). *Points*, data; *histogram*, Monte Carlo simulation. The Monte Carlo distribution was generated with $m_\tau = 1784.1$ MeV, the nominal $\tau$ mass at that time. The *inset* shows an enlarged view of the threshold region

It can be shown that the events at the threshold correspond to those events where the hadron is emitted against the direction of the boost in the rest frame of the $\tau$. Then the approximation is correct.

The shape of the pseudomass spectrum strongly depends on the mass of the hadrons analyzed. For low masses, as e.g. in $\tau \to \pi\,\nu_\tau$, the method is unforgiving. For a pion emitted in the rest frame of the $\tau$ even at a very small angle to the boost, the reconstructed pseudomass falls far below the threshold. The threshold is not populated; it is not even visible in the pseudomass spectrum. The situation becomes much better for heavier hadrons. In the limit where $m_h$ approaches $m_\tau$, all events end up at the threshold at the true $\tau$ mass.

The ARGUS collaboration chose the $\tau^- \to \pi^-\,\pi^-\,\pi^+\,\nu_\tau$ channel as a compromise between reasonably heavy hadrons, a useful branching ratio, and a good reconstruction of the events. Figure 3.4 shows the measured pseudomass spectrum. From the zoom of the threshold region it is visible by eye that the threshold in the data is below the Monte Carlo position. The Monte Carlo was generated with the world average at that time ($1784.1^{+2.7}_{-3.6}$ MeV). The ARGUS measurement gives

$$m_\tau = (1776.3 \pm 2.4 \pm 1.4)\ \mathrm{MeV}, \qquad (3.8)$$

with a slightly lower error and a value almost two standard deviations lower. This measurement was confirmed soon after by CLEO (see next section) and later by the more precise BES measurement. The systematic error of the ARGUS result is dominated by the uncertainty in the momentum scale ($\pm 1.2$ MeV).

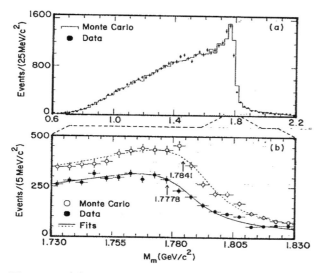

**Fig. 3.5.** (a) The spectrum of the 'minimum kinematically allowed $\tau$ mass' in the data and in the simulation (CLEO collaboration [137]). (b) An expanded view of (a), with the fit function superimposed. The vertical scale for the Monte Carlo distribution is shifted so that it can be better compared with the data points

### 3.1.4 The CLEO Method

A second pseudomass technique was invented by the CLEO collaboration [137]. Again it uses the kinematics of semihadronic decays of the $\tau$ lepton, but this method treats $\tau$ pairs as a whole, deriving only one value per event. It is based on the kinematic reconstruction of the $\tau$ direction as described in Sect. 2.2.4.

When the cones of possible $\tau$ directions around the measured hadron are calculated (see Fig. 2.7), the opening angle $\psi$ depends on the $\tau$ mass. This angle decreases with the $\tau$ mass. In general, an event correctly reconstructed with the true $\tau$ mass will have two solutions for the $\tau$ direction (see Fig. 2.8). Reducing the input $\tau$ mass in the reconstruction, the two cones shrink until at some point the two solutions degenerate. Now the cones are just touching each other in one line. If one were to reduce the mass further, there would not be a solution any more, indicating that the true $\tau$ mass cannot be that low. The mass at which the two solutions degenerate is called the 'minimum kinematically allowed $\tau$ mass'. It is a pseudomass in the sense that it is most of the time smaller than the true $\tau$ mass, but never larger. The spectrum exhibits a sharp edge right at the true $\tau$ mass and the position of this edge is used to extract the value from the data.

The CLEO collaboration used four different combinations of decay channels for $\tau^+$ and $\tau^-$: ($\pi^\pm$ versus $\pi^\mp \pi^0$), ($\pi^\pm$ versus $\pi^\mp 2\pi^0$), ($\pi^\pm \pi^0$ versus $\pi^\mp \pi^0$), and ($\pi^\pm \pi^0$ versus $\pi^\mp 2\pi^0$). The spectrum is shown in Fig. 3.5. Both the data

3. The Static Properties of the $\tau$

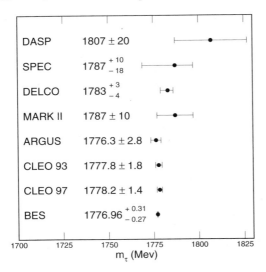

Fig. 3.6. Comparison of the different measurements of the $\tau$ mass. The current world average is $1777.05^{+0.29}_{-0.26}$ MeV [44]

and the Monte Carlo distribution have been fitted with an empirical function composed of an arctan describing the edge at $m_\tau$ multiplied by a fourth order polynomial parametrizing the slope to lower masses and a straight line to model the high-mass tail.

The parameters of the polynomials were determined from the Monte Carlo distribution, only the position of the edge and the overall normalization is varied to fit the data. The edge position fitted to the Monte Carlo distribution relative to the input value of $m_\tau$, was used as an offset to extract the measured value from the data.

There are several sources of possible bias in the measurement:

- initial/final-state radiation
- undetected or unreconstructed particles
- misidentification of leptonic decays (three-body!)
- $\pi/K$ misidentification
- background from non-$\tau$ pairs
- detector resolution
- detector energy and momentum scales
- beam energy and beam energy spread
- a nonvanishing $\tau$ neutrino mass.

The biases are corrected by the offset described above if the simulation is correct. The uncertainties in the energy ($\pm 0.30\%$) and momentum ($\pm 0.10\%$) scales dominate the systematic errors. The result is

$$m_\tau = (1777.8 \pm 0.7 \pm 1.7) \text{ MeV}. \tag{3.9}$$

Figure 3.6 summarizes the measurements of the $\tau$ mass.

## 3.2 The Lifetime

There are a large number of different methods that have been applied to determine the lifetime of the $\tau$ lepton as precisely as possible. The different methods are applied to different topologies, are sensitive to different aspects of the detector resolutions and decay kinematics, and suffer from different systematics. None of the methods is superior to the others. Thus several measurements are combined by each experiment, taking into account the statistical correlations and common systematics.

The methods can be classified by whether they exploit the information from a single $\tau$ lepton or a whole $\tau^+\tau^-$ event, and by which topologies they apply to.

**Single-$\tau$ methods**:

- **DL**. Decay length method (any three-prong $\tau$ decay) [198].
- **IP**. Impact parameter method (any one-prong $\tau$ decay) [183].

**Event methods**:

- **IPD**. Impact parameter difference method (1–1 topologies) [71].
- **IPS**. Impact parameter sum methods (1–1 topologies).
  - **IPS**. Impact parameter sum or miss distance method [157].
  - **MIPS**. Momentum-dependent impact parameter sum [58].
- **3DIP**. Three-dimensional impact parameter method (hadronic $\tau$ decays only) [53].

These methods will be described below (for more details see [317, 318]).

### 3.2.1 Decay Length (DL) Method

The decay length method can be applied to three-prong decays. The idea is straightforward: the production vertex of a $\tau$ lepton is approximated by the luminous region of the beams and the decay vertex is reconstructed from the three tracks. The distance between the two is the decay length. It shows an exponential decay law, from which the lifetime is derived.

Figure 3.7 shows the experimental conditions at SLC, LEP, and CESR on a realistic scale. The average decay length $\bar{\ell} = \beta\gamma c\tau_\tau$ is roughly 2.2 mm at $Z^0$ energies (SLC and LEP) and 0.24 mm at the $\Upsilon(4s)$ (CESR). The opening angles vary with the boost roughly like $1/\gamma$ and are therefore a factor of 8.5 larger at CESR. The linear collider (SLC) has a very small beam spot in $x$ and $y$ (perpendicular to the beam) of typically 2.5 µm. The vertical size of the beam spot is typically 10 µm at LEP and CESR, but the horizontal size is degraded owing to synchrotron radiation to about 150 µm at LEP and 350 µm at CESR. For all three accelerators the size along the direction of the beam is much larger than the decay length of the $\tau$ leptons.

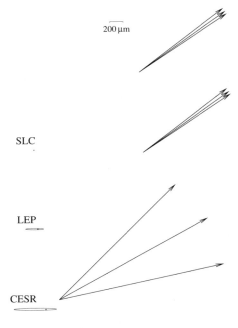

**Fig. 3.7.** Three-prong decays of a $\tau$ lepton at SLC (SLD experiment), LEP (ALEPH, DELPHI, L3, OPAL), and CESR (CLEO) [318], drawn to scale ($x$–$y$ view). The *ellipses* representing the centroids of the luminous regions include the typical uncertainty in the knowledge of the beam positions

Because the beam spot is so large in the $z$ direction, effectively only two coordinates of the decay length can be reconstructed. Hence, the decay length is measured as the displacement of the secondary vertex in the $x$–$y$ projection with respect to the center of the luminous region. An estimate of the $\tau$ direction is used to correct for the projection. This is derived from the thrust axis. Then we have

$$\bar{\ell} = \beta\gamma\, c\tau_\tau \sin\theta_\tau, \tag{3.10}$$

from which the lifetime $\tau_\tau$ can be calculated. The average decay length $\bar{\ell}$ is extracted from a fit to the data and the value of $\beta\gamma$ is taken from a sample of simulated events with radiative corrections after all selection cuts have been applied.

Although only the $x$–$y$ projection of the decay length is measured, it is nevertheless useful to have a good vertex resolution in the $z$ direction also. The $z$ measurement is correlated with and improves the resolution in the $x$–$y$ plane, especially for decays where all tracks have similar azimuthal angles.

Figure 3.8 shows the decay length distribution measured by the OPAL collaboration [222]. The exponential decay is well reproduced by the reconstructed vertices. A straight line can be seen on the logscale for positive decay lengths.

Negative values of the decay length can happen owing to measurement errors and production of $\tau$ pairs at the periphery of the luminous region. The quality of the fit for negative values is a good cross-check on the understanding

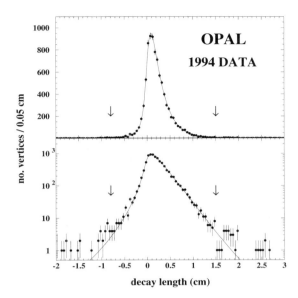

**Fig. 3.8.** Decay length distribution of vertices from three-prong $\tau$ decays selected by the OPAL collaboration [222]. The *points* are data, the *line* is the result of a fit to the data of an exponential decay convoluted with two Gaussians. The mean decay length is derived as one of the parameters of the fit. The *arrows* indicate the fit region

of these effects. From the data shown in Fig. 3.8 the following value of the lifetime is derived:

$$\tau_\tau = (289.0 \pm 3.6 \pm 1.8) \text{ fs}. \tag{3.11}$$

It can be seen from Fig. 3.8 that the smearing of the decay length distribution is small compared to the decay length itself. This indicates that the sensitivity of the measurement is close to its ideal value for a perfect detector with an infinitely small beam spot. It also means that there is no need to try any of the more elaborate methods on three-prong decays.

For the CLEO experiment at CESR, however, the horizontal width of the beam spot creates a significant smearing and dilutes the sensitivity. This problem can be avoided in 3–3 topologies, when the separation of the two secondary vertices with respect to each other is measured without reference to the primary vertex. The CLEO collaboration has done that measurement [125]. The decay length distribution is shown in Fig. 3.9. The result is

$$\tau_\tau = (309 \pm 11 \pm 9) \text{ fs}. \tag{3.12}$$

56     3. The Static Properties of the $\tau$

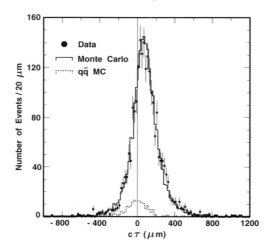

**Fig. 3.9.** Decay length distribution for events in 3–3 topology (CLEO collaboration [125]). The *points with error bars* represent the data, the *solid histogram* the Monte Carlo simulation. The *dashed histogram* shows the $q\bar{q}$ background, which exhibits no finite lifetime

### 3.2.2 Impact Parameter (IP) Method

It is not possible to reconstruct the full decay length from a one-prong decay of the $\tau$ lepton. Instead, the impact parameter of the track is measured. This is related to the decay length through the opening angle between the direction of the $\tau$ lepton and the direction of the track (see Fig. 3.10). The impact parameter is taken positive if the reconstructed track is consistent with being emitted from a forward-flying $\tau$ lepton.[1] Negative impact parameters are due to measurement errors and uncertainties in the primary vertex.

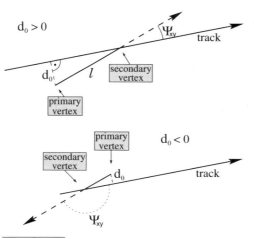

**Fig. 3.10.** Schematic view of a one-prong $\tau$ decay in the $x$–$y$ projection. The impact parameter $d_0$ is negative if $\sin\psi < 0$. The proper decay length is given by $d_0 = \ell \sin\psi$

---

[1] A different sign convention is used in the IPD and IPS methods (see [71]).

# Erratum

*Springer Tracts in Modern Physics, Vol. 160*

A. Stahl
**Physics with Tau Leptons**

ISBN 3-540-66267-7

Dear Reader

As the result of technical problems with the data conversion, some figures in this book have not been correctly reproduced. We kindly ask you to cover Figs. 3.20 (p. 67), 3.30 (p. 80), and 4.9 (p. 104) with the enclosed self-adhesive reprints.

Thank you for your understanding.

Springer-Verlag

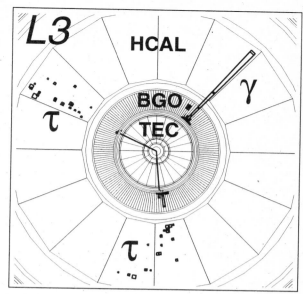

**Fig. 3.20 (p. 67)**

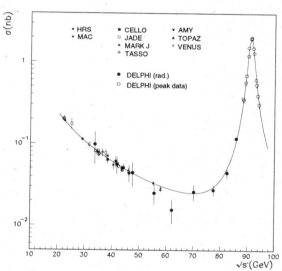

**Fig. 3.30 (p. 80)**

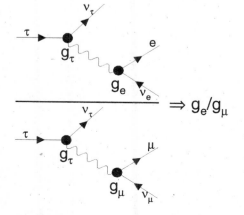

**Fig. 4.9 (p. 104)**

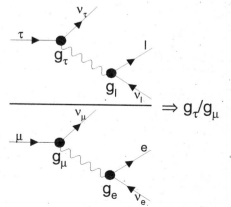

The measurement is restricted to the $x$–$y$ projection by the same argument as above, that the beam spot is so huge in the $z$ direction that no information on the lifetime can be extracted from this coordinate. Hence, for an individual event the impact parameter is related to the individual decay time of this $\tau$ lepton by

$$d_0 = \beta\gamma\, c\tau_\tau \sin\theta_\tau \sin\psi_{xy}, \qquad (3.13)$$

where $\psi_{xy}$ is the projection of $\psi$ into the $x$–$y$ plane (see Sect. 2.2.2). The individual decay time cannot be recovered, as $\psi$ is unknown. Instead, (3.13) is averaged over the event sample. The individual decay time is converted into the lifetime, but also the decay angles $\psi$ are averaged.

The $\tau$ direction is derived from the thrust axis. The difference between the thrust axis and the true $\tau$ direction smears out the lifetime information and reduces the sensitivity. This dilution has to be corrected for by Monte Carlo simulation. It depends on the dynamics of all decays involved and therefore is a potential source of systematic errors.

The measurement of the impact parameter is more sensitive to measurement errors and the size of the luminous region than the decay length in three-prong decays. The criterion here is the size of these effects with respect to a typical impact parameter, which is much smaller than a typical decay length. The resolution effects are no longer small and the sensitivity is substantially reduced compared to an ideal experiment. The reduction is by about a factor of 3 [317].

The size of the luminous region is measured from tracks in $e^+e^- \to e^+e^-$ and $e^+e^- \to \mu^+\mu^-$ events. As those tracks come directly from the primary vertex, the distribution of their impact parameters directly reflects the beam spot. Also, the resolution can be measured from such events by looking at the difference between the impact parameters of the two tracks in one event. This should be zero on average, with a width of $\sqrt{2}$ times the resolution. Figure 3.11 shows the distribution recorded by the L3 collaboration [319] in 1994.

Figure 3.12 shows the impact parameter distribution measured by the L3 collaboration from one-prong $\tau$ decays in the same data set. The finite lifetime of the $\tau$ lepton creates an asymmetry in the distribution towards positive values, which is clearly visible. The lifetime information is extracted by a fit of a physics function convoluted with a double Gaussian to the observed distribution. The physics function is a parametrization of the distribution observed from simulated events before detector effects are applied. The result from the 1994 and 1995 data set is

$$\tau_\tau = (290.5 \pm 2.7 \pm 2.9)\ \text{fs}. \qquad (3.14)$$

The largest systematic error is due to the uncertainty in the detector resolution.

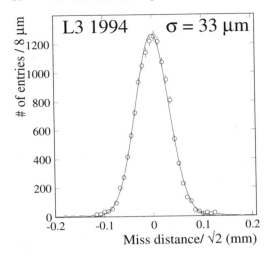

**Fig. 3.11.** The difference in impact parameter between the two tracks in Bhabha and dimuon events (L3 collaboration [319])

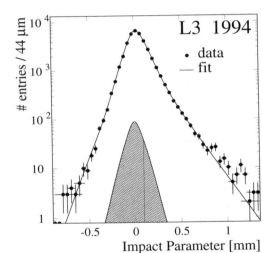

**Fig. 3.12.** Impact parameter distribution measured by the L3 collaboration from one-prong $\tau$ decays [319]. The *points with error bars* are the data; the *line* represents an unbinned maximum-likelihood fit to the data to extract the lifetime information. The *shaded area* represents the zero-lifetime background

### 3.2.3 Impact Parameter Difference (IPD) Method

In events where both $\tau$ leptons decay into a single track the difference between the impact parameters of these tracks contains information on the lifetime. It is correlated with the acoplanarity between the two tracks, i.e. the difference in their azimuthal angles $\Delta\phi = \phi_+ - \phi_- \pm \pi$ (see Fig. 3.13).

As mentioned earlier (3.13), the impact parameter of the positive track is given by

$$d_+ = \beta\gamma\, c\tau_\tau \sin\theta_\tau \sin(\phi_+ - \phi_{\tau+}), \tag{3.15}$$

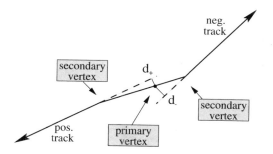

**Fig. 3.13.** Impact parameters in an event with 1–1 topology ($x$–$y$ projection)

where the replacement $\psi_{xy} = \phi_+ - \phi_{\tau+}$ has been made ($\phi_+$ is the azimuthal angle of the positive track and $\phi_{\tau+}$ that of the positive $\tau$). A similar relation holds for the negative track. The $\tau$ leptons are assumed to be produced back to back and the the approximation $\sin\alpha \approx \alpha$ is used for small angles. Then, averaging over the decay lengths, one gets

$$\overline{d_+ - d_-} = \beta\gamma\, c\tau_\tau \Delta\phi \sin\theta_\tau. \tag{3.16}$$

The average impact parameter difference is proportional to the lifetime and the acoplanarity of the tracks ($\Delta\phi$). The acoplanarity is directly measurable and a knowledge of the direction of the $\tau$ leptons (thrust axis) is only needed to determine $\sin\theta_\tau$, i.e. to correct for the projection into the $x$–$y$ plane.

To extract the lifetime, $\overline{d_+ - d_-}$ is plotted as a function of $\Delta\phi\sin\theta_\tau$. A straight line is expected with a slope that gives the lifetime.

Figure 3.14b shows the plot as measured by the SLD collaboration [320]. Each point in the plot is the average of $\overline{d_+ - d_-}$ calculated from all events in a narrow bin of $\Delta\phi\sin\theta_\tau$. A straight line is fitted through the points and the slope is compared to the slope achieved from a Monte Carlo sample with known lifetime. This implicitly corrects for any bias, e.g. from radiative corrections or the determination of $\sin\theta_\tau$. The result is

$$\tau_\tau = (287.8 \pm 7.7 \pm 3.5)\text{ fs}. \tag{3.17}$$

By construction, the method is fairly insensitive to the simulation of the dynamics of the $\tau$ decays. In the difference $d_+ - d_-$ the critical dependence of $\psi_{xy}$ on the $\tau$ direction cancels. It is also quite stable with respect to the resolution function. Only the average value of $d_+ - d_-$ enters the determination, not the width. The drawback is an enhanced dependence on the size and position of the luminous region, which enter twice.

### 3.2.4 Impact Parameter Sum (IPS) Methods

The problem of the knowledge of the luminous region is avoided by the impact parameter sum, or miss distance, method. With this method the lifetime is

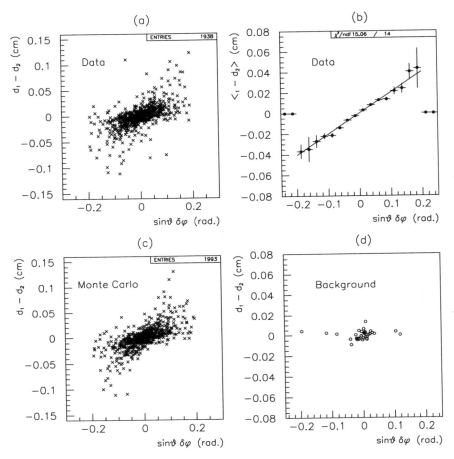

**Fig. 3.14.** The impact parameter difference method as applied by the SLD collaboration [320]. Scatter plots of $d_+ - d_-$ versus $\Delta\phi \sin\theta_\tau$ for data **(a)**, simulated $\tau$ events **(c)** and background **(d)**. Plot **(b)** shows the average $\overline{d_+ - d_-}$ for slices in $\Delta\phi \sin\theta_\tau$. The *straight line* is a fit to these points. The lifetime is extracted from this fit

determined from the sum of the impact parameters $d_+ + d_-$ in 1–1 topologies. The quantity $d_+ + d_-$ is the separation of the two tracks at the primary vertex (see Fig. 3.13). It is independent of the production point of the $\tau$ leptons to first order in the acollinearity between the two tracks.[2]

The relation to the lifetime is given by

$$d_+ + d_- = \beta\gamma\, c\tau_\tau \sin\theta_\tau \left(\sin\psi_{+,xy} + \sin\psi_{-,xy}\right), \tag{3.18}$$

---

[2] It is truly independent of the production point for two parallel tracks.

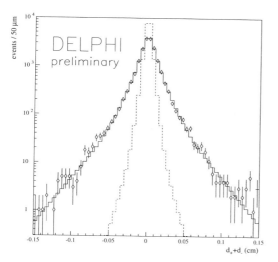

**Fig. 3.15.** Distribution of the miss distance $d_+ + d_-$ (DELPHI collaboration [321], preliminary). The *diamonds* are data, the *solid histogram* is the fit to the observed distribution, and the *dashed histogram* is a simulation of $\tau$ decays with zero lifetime showing the resolution effects only

where $\sin\psi_{+,xy}$ and $\sin\psi_{-,xy}$ are the projections onto the $x$–$y$ plane of the opening angles between the $\tau$ lepton and its daughter track, for $\tau^+$ and $\tau^-$, respectively. The distribution is symmetric around zero for these angles and therefore the average miss distance $\overline{d_+ + d_-}$ vanishes. It is the width of the distribution which contains information about the lifetime.

Figure 3.15 shows the distribution as measured by the DELPHI collaboration [321]. The lifetime was deduced from a fit to the observed distribution. The fit was performed with a physics function parametrized from the Monte Carlo simulation, convoluted by an elaborate resolution function. The resolution function depends on the momentum and the particle type of each individual track in the sample. The result is

$$\tau_\tau = (293.6 \pm 2.8 \pm 2.7) \text{ fs}. \tag{3.19}$$

The dependence of the result on the knowledge of the luminous region has been eliminated by this method. But the result now depends more crucially on the understanding of the resolution function and the dynamics of the $\tau$ decays. The dynamics enter through the angles $\sin\psi_{+,xy}$ and $\sin\psi_{-,xy}$. Their average value influences the width of the distribution.

The sensitivity of the method can be further improved by making use of some indirect information about the average size of $\sin\psi_{+,xy}$ and $\sin\psi_{-,xy}$. These angles are correlated with the momentum of the tracks. The ALEPH collaboration has exploited this information by using a momentum-dependent impact parameter sum [58], and the DELPHI collaboration has further developed the technique to include the difference in the distributions between hadronic and leptonic $\tau$ decays [322]. Figure 3.16 shows the average impact parameter for various decays.

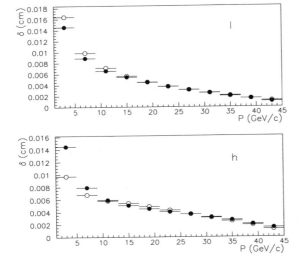

**Fig. 3.16.** The average impact parameter of one-prong $\tau$ decays in the DELPHI experiment as a function of the momentum of the daughter track [322]. The *upper plot* shows leptonic decays, the *lower* hadronic. *Open circles* are for positive helicity and *full circles* are for negative helicity. Simulation with $\tau_\tau = 300$ fs

### 3.2.5 Three-Dimensional Impact Parameter (3DIP) Method

The latest innovation is a three-dimensional impact parameter sum by the ALEPH collaboration [53]. Impact parameter sums are largely independent of the production point of the $\tau$ leptons and therefore the $z$ information is not obscured by the large beam spot in this direction. The dependence on the $\tau$ direction is removed by explicitly reconstructing it event by event by the method described in Sect. 2.2.4.

With this reconstruction the method can only be applied to events where both $\tau$ leptons decay hadronically in either 1–1 or 1–3 topologies.[3] No attempt is made to resolve the twofold ambiguity. Instead the three-dimensional impact parameters are projected onto a plane perpendicular to this ambiguity (the plane shown in Fig. 2.9). For details see [53]. The sensitivity of the method is increased, but fewer events can be used. The result is

$$\tau_\tau = (289.0 \pm 2.7 \pm 1.3) \text{ fs}. \tag{3.20}$$

### 3.2.6 Results

The results achieved with the different methods were first averaged by the collaborations themselves, taking into account the correlations between the methods. Table 3.2 and Fig. 3.17 summarize the results. The current world average is [318]

$$\tau_\tau = (290.5 \pm 1.0) \text{ fs}. \tag{3.21}$$

---

[3] There is no need to apply such a method to 3–3 topologies, as the decay length method applied separately to both hemispheres is already close to a perfect reconstruction.

**Table 3.2.** Summary of lifetime measurements [318]. The DELPHI, L3, and SLD results are preliminary. All values in fs

| Experiment | | Lifetime |
|---|---|---|
| ALEPH | [50, 53] | $290.1 \pm 1.5 \pm 1.1$ |
| CLEO | [125] | $289.0 \pm 2.8 \pm 4.0$ |
| DELPHI | [322] | $291.9 \pm 1.6 \pm 1.1$ |
| L3 | [319] | $291.7 \pm 2.0 \pm 1.8$ |
| OPAL | [222] | $289.2 \pm 1.7 \pm 1.2$ |
| SLD | [320] | $288.1 \pm 6.1 \pm 3.3$ |

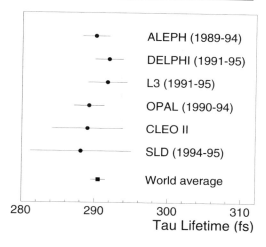

**Fig. 3.17.** Summary of lifetime measurements [318]

The world average has undergone a fairly steady decline over the years [328]. Figure 3.18 shows the world averages published by the Particle Data Group since 1986 [44, 316, 323–327].

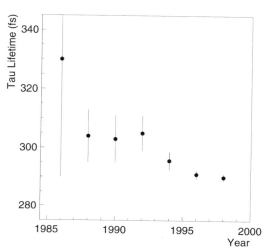

**Fig. 3.18.** World averages of the $\tau$ lifetime since 1986 from the Particle Data Group [44, 316, 323–327]

## 3.3 Form Factors of the Electromagnetic and Weak Currents

The coupling of the fermions to an external current – photon, $Z^0$, or $W^\pm$ – is point-like only in lowest order. On investigating the vertices with higher precision, $q^2$-dependent form factors develop. In the Standard Model these form factors are created by radiative corrections. The best-known is the magnetic form factor, which is the source of the anomalous magnetic moment of the electron and the muon. Also the $\tau$ carries such an anomalous magnetic moment.

In the Standard Model these effects are small and hard to measure for $\tau$ leptons. Nevertheless, these measurements are interesting, even if the sensitivity does not allow one to see the Standard Model values, as much larger effects can arise from physics beyond the Standard Model. Such effects can be introduced either by additional radiative corrections from new particles appearing virtually in loops or by a substructure of the $\tau$, when with sufficient $q^2$ one starts to resolve the constituents. In that sense the measurement of form factors is complementary to the search for excitations of the $\tau$ described in Sect. 10.4.

### 3.3.1 Definition of the Form Factors

The most general matrix element of the electromagnetic current between two fermions is given by [329, 330]:

$$\langle \Psi_\tau(p') \mid eJ_\mu^\gamma \mid \Psi_\tau(p) \rangle = -e \langle \bar{u}_\tau(p') \mid \Gamma_\mu^\gamma \mid u_\tau(p) \rangle, \tag{3.22}$$

with

$$\Gamma_\mu^\gamma = F_1^\gamma(q^2)\,\gamma_\mu \;+\; F_2^\gamma(q^2)\,\frac{i}{2\,m_\tau}\,\sigma_{\mu\nu}\,q^\nu \;+\; F_3^\gamma(q^2)\,\frac{i}{2\,m_\tau}\,q_\mu$$

$$+\; G_1^\gamma(q^2)\,\gamma_\mu\,\gamma_5 \;+\; G_2^\gamma(q^2)\,\frac{1}{2\,m_\tau}\,\sigma_{\mu\nu}\gamma_5\,q^\nu \;+\; G_3^\gamma(q^2)\,\frac{1}{2\,m_\tau}\,q_\mu\,\gamma_5, \tag{3.23}$$

where $q = p' - p$ and $\sigma_{\mu\nu} = i/2(\gamma_\mu\gamma_\nu - \gamma_\nu\gamma_\mu)$. The upper index $\gamma$ specifies the type of current, which is electromagnetic in this example. There is an identical definition for the weak neutral current. The charged weak current will not be discussed here (see [331–333]).

The form factors are in general complex functions of $q^2$. The asymptotic values of all form factors in the limit $q^2 \to 0$, however, are real. They are static properties of the fermion and are called 'moments' in the case of the electromagnetic current. For the weak current, it has become common to define the moments as the values at $q^2 = m_Z^2$.

Above some threshold for the production of intermediate particles in loops the form factors develop imaginary parts (see [330]). The real and imaginary

## 3.3 Form Factors of the Electromagnetic and Weak Currents

parts are related through a dispersion relation. At high energies the imaginary parts can become as big as or even bigger than the real parts, so that it is important to measure both in the search for new physics.

- **Charge Radii**

  The form factor $F_1^\gamma(q^2)$ describes the electric charge radius of the $\tau$ lepton. By charge renormalization $F_1^\gamma(0)$ is 1. The electromagnetic current is a pure vector current, so that the $G_1^\gamma(q^2)$ piece vanishes altogether. Similarly $F_1^{Z^0}$ and $G_1^{Z^0}$ describe the vector and axial-vector couplings to the $Z^0$, with

  $$F_1^{Z^0}(m_Z^2) = \frac{1 - 4\sin^2\theta_W}{4\sin\theta_W\cos\theta_W}, \quad (3.24)$$

  $$G_1^{Z^0}(m_Z^2) = \frac{-1}{4\sin\theta_W\cos\theta_W}. \quad (3.25)$$

- **Magnetic Dipole Moments**

  The magnetic dipole moment of the $\tau$ lepton is given in two parts: the Dirac value $F_1^\gamma(0)$ and the anomalous part $F_2^\gamma(0)$:

  $$\mu_\tau^\gamma = -\frac{e}{2\,m_\tau}\left[F_1^\gamma(0) + F_2^\gamma(0)\right], \quad (3.26)$$

  with the anomalous magnetic moment defined as

  $$a_\tau^\gamma = \frac{g_\tau^\gamma - 2}{2} = F_2^\gamma(0),$$

  $$\mu_\tau^\gamma = -\frac{g_\tau^\gamma}{2}\frac{e}{2\,m_\tau}. \quad (3.27)$$

  In the same manner the weak magnetic moment $\mu_\tau^{Z^0}$ and its anomalous part $a_\tau^{Z^0}$ are defined from $F_1^{Z^0}(m_Z^2)$ and $F_2^{Z^0}(m_Z^2)$. The Standard Model predictions are [334–336]

  $$a_\tau^\gamma(\text{SM}) = 1.1773(3) \times 10^{-3}, \quad (3.28)$$

  $$a_\tau^{Z^0}(\text{SM}) = -(2.10 + 0.61\,\text{i}) \times 10^{-6}. \quad (3.29)$$

- **Electric Dipole Moments**

  The form factor $G_2$ defines an electric dipole moment which is $\mathcal{P}$- and $\mathcal{T}$-violating:

  $$d_\tau^\gamma = \frac{e}{2\,m_\tau} G_2^\gamma(0), \quad (3.30)$$

  $$d_\tau^{Z^0} = \frac{e}{2\,m_\tau} G_2^{Z^0}(m_Z^2). \quad (3.31)$$

  It will be discussed in the section about $\mathcal{CP}$ violation (Sect. 9.1).

- **Anapole Moment**

  From the form factors $G_1$ and $G_3$ one can define yet another moment, the so-called anapole moment $\mathcal{A}_\tau$ [337, 338]. Combining their contributions to $\Gamma_\mu$ into

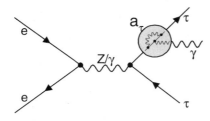

Fig. 3.19. Feynman diagram for the production of a photon through the anomalous magnetic moment of the $\tau$. The *gray bubble* symbolizes the anomalous magnetic moment. Inside the bubble the leading contribution from the Standard Model is shown. A similar bubble can be attached to the other $\tau$

$$G_4^\gamma(q^2)\left(q^2\gamma_\mu\gamma_5 - 2\,m_\tau\gamma_5 q_\mu\right), \tag{3.32}$$

one has

$$\mathcal{A}_\tau = 8\,\pi\,e\,G_4^\gamma(0). \tag{3.33}$$

The anapole moment is parity-violating, so that it is expected to vanish for the electromagnetic current.

There is an alternative way to introduce dipole moments, through couplings in an effective Lagrangian [339, 340]. If there is a new interaction at an energy scale $\Lambda$ much larger than $q^2$, the energy scale of the reaction under consideration, the effects can be parametrized by an effective Lagrangian. For example, a new $\mathcal{CP}$-violating interaction creating a weak electric dipole moment can be described by

$$\mathcal{L}_{\text{eff}} = -\frac{\text{i}}{2}\,\tilde{d}_\tau^{Z^0}\,\overline{\Psi}_\tau\sigma_{\mu\nu}\gamma_5\Psi_\tau Z^{\mu\nu} \tag{3.34}$$

($Z^{\mu\nu}$ is the field strength tensor of the $Z^0$). The real constant $\tilde{d}_\tau^{Z^0}$ is the effective weak dipole coupling. From (3.34) the dipole form factor of (3.31) can be calculated. To lowest order, one gets $d_\tau^{Z^0}(q^2) = \tilde{d}_\tau^{Z^0}$. The relation is modified by quantum corrections, which also generate the $q^2$ dependence and the imaginary part of the form factor.

The advantage of the effective-Lagrangian approach is that there are only a few universal, real constants describing all the new physics. The form factors, on the other hand, are process-dependent. Their advantage is that they are more directly related to the experimental observables. Extracting effective couplings from an experiment requires the calculation of the radiative corrections, which is largely model-dependent.

### 3.3.2 The Magnetic Moment

The electromagnetic dipole moments of the $\tau$ lepton have been investigated by the L3 [171] and OPAL [214] collaborations using the reaction $e^+e^- \to \tau^+\tau^-\gamma$ (see also [341, 342] and Fig. 3.19). Both the anomalous magnetic moment and the electric dipole moment create additional photons in this reaction. An event seen in the L3 detector is shown in Fig. 3.20.

3.3 Form Factors of the Electromagnetic and Weak Currents 67

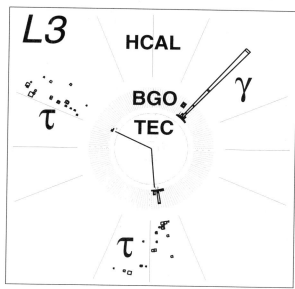

Fig. 3.20. A typical $\tau\tau\gamma$ event seen by the L3 detector [343]. The decay products of the $\tau$ leptons leave tracks in the inner tracker (TEC) and energy in the electromagnetic (BGO) and hadron calorimeters. The photon has no associated track, and leaves a considerable amount of energy in the BGO and no energy in the hadron calorimeter

These photons have high energies compared to typical bremsstrahlung photons and are emitted at larger angles with respect to the $\tau$ leptons. The differential cross sections have been calculated by several authors [342, 344–348]. A Feynman diagram is shown in Fig. 3.19. The resulting photon spectrum and angular distributions are illustrated in Fig. 3.21. There is an interference term between the amplitude from the anomalous magnetic moment and the point like coupling (see [346]). It creates a contribution to the cross section linear in $F_2^\gamma$ which is important for magnetic moments below 0.01. (The direct contribution is quadratic in $F_2^\gamma$.) Figure 3.22 shows the relative sizes of the two terms.

The experiments search for $\tau$ pair candidates with an additional photon. The standard cuts on the acollinearity between the two $\tau$ leptons have to be modified to take the radiation into account. The photon is required to have a minimum energy (3 GeV and 5 GeV for L3 and OPAL, respectively), to be well separated from the nearest $\tau$ (17° and 35°), and to be emitted at a polar angle not too close to the beam pipe ($|\cos\theta_\gamma| < 0.74$ and $0.78$). These cuts are designed to reduce the large background from initial-state radiation, nonanomalous final-state radiation, and photons from the decay of the $\tau$ leptons. Then the spectra are fitted by a superposition of the Standard Model radiation with no anomalous contribution and the expected contribution from $F_2^\gamma$. The L3 collaboration uses a two-dimensional distribution of the photon energy versus the angle between the photon and the closest $\tau$. It is shown in Fig. 3.23.

The two experiments yield the following limits (95% confidence level) on the magnetic dipole moment $a_\tau^\gamma$ and the electric dipole moment $d_\tau^\gamma$ [171, 214]:

68    3. The Static Properties of the $\tau$

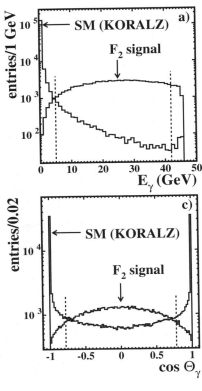

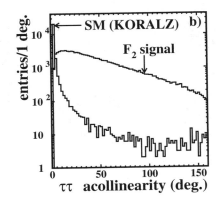

**Fig. 3.21.** Kinematics of $\tau\tau\gamma$ events from an OPAL Monte Carlo study [214]: **(a)** photon spectrum, **(b)** the reconstructed angle between the two $\tau$ leptons, and **(c)** the angle between the photon and the beam axis. The *histogram* labeled 'SM' is the Standard Model prediction without the anomalous contribution, taken from KORALZ [285]. The signal generated by $a_\tau^\gamma$ is arbitrarily normalized (no interference)

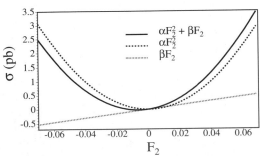

**Fig. 3.22.** Anomalous contribution to the $e^+e^- \to \tau^+\tau^-\gamma$ cross section as a function of $F_2^\gamma(0)$, showing the linear and quadratic contributions [343]

$$\begin{aligned} \text{L3}: & \quad -0.052 < a_\tau^\gamma < 0.058, \\ \text{OPAL}: & \quad -0.068 < a_\tau^\gamma < 0.065, \end{aligned} \quad (3.35)$$

and

$$\begin{aligned} \text{L3}: & \quad |d_\tau^\gamma| < 3.1 \times 10^{-16} \, e\,\text{cm}, \\ \text{OPAL}: & \quad |d_\tau^\gamma| < 3.7 \times 10^{-16} \, e\,\text{cm}. \end{aligned} \quad (3.36)$$

No sign of physics beyond the Standard Model is seen.

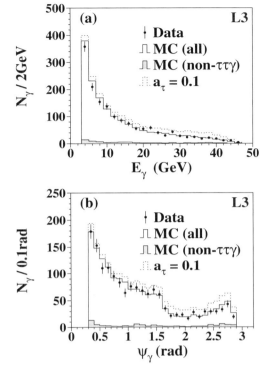

**Fig. 3.23.** The number $N_\gamma$ of photon candidates in the L3 $e^+e^- \to \tau^+\tau^-\gamma$ sample as a function of **(a)** the photon energy and **(b)** the angle relative to the closest $\tau$ [171]. The *points with error bars* denote the data and the *solid histogram* the Standard Model prediction. The *dashed histogram* shows how the distribution would change for $a_\tau^\gamma = 0.1$

### 3.3.3 The Weak Magnetic Moment

In analogy to the anomalous magnetic dipole moment of the electromagnetic current, there is the anomalous weak magnetic dipole moment of the weak neutral current. It contributes to the production of $\tau$ pairs at the $Z^0$ pole. The diagram is shown in Fig. 3.24. There is a similar contribution from the weak electric dipole moments, which violates $\mathcal{CP}$.

The additional dipole coupling changes the partial decay width of the $Z^0$ into $\tau$ leptons. This has been calculated in [341]. By comparing the measured decay width with the value expected from the Standard Model, limits on additional contributions from $a_\tau^{Z^0}$ (and $d_\tau^{Z^0}$) can be set.[4] The authors of [349] find the following limit:

$$-4 \times 10^{-3} < a_\tau^{Z^0} < 6 \times 10^{-3}. \tag{3.37}$$

The change of production rate is not the only effect of the anomalous weak magnetic moment. It has been shown that the polarization of the $\tau$ leptons produced at the Z pole is affected by $a_\tau^{Z^0}$ [350], as well as the spin correlations between the two $\tau$ leptons in an event [351].

---
[4] The sensitivity is not good enough to see the Standard Model value of $a_\tau^{Z^0}$.

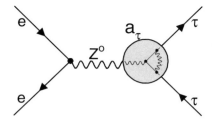

Fig. 3.24. Contribution of the anomalous weak magnetic dipole moment $a_\tau^{Z^0}$ to $\tau$ production at the $Z^0$ pole (see (3.23) and (3.27)). Inside the *gray bubble* is an example of a loop contribution

For the polarization of the $\tau$ leptons there are three spatial components: the longitudinal polarization along the direction of flight of the $\tau$ leptons and two components perpendicular to it. These are separated into the transverse polarization, in the production plane, i.e. the plane spanned by the initial beams and the $\tau$ lepton, and the normal polarization, normal to this plane (see also Sect. 4.6.3).

- **Transverse Polarization**
  There is a very small Standard Model contribution, which requires a helicityflip of one of the $\tau$ leptons. It is therefore suppressed by a factor $1/\gamma$ ($\gamma$ is the Lorentz factor $\gamma = E_\tau/m_\tau$). The leading contribution from the anomalous magnetic moment is
  $$2\gamma \left(v^2 + a^2\right) \cos\theta_\tau\, a_\tau^{Z^0}, \tag{3.38}$$
  with $v$ and $a$ being the couplings to the $Z^0$.
- **Normal Polarization**
  Contributions to this component of the polarization can only come from a $\mathcal{CP}$-violating electric dipole moment. The leading contribution is
  $$2\gamma a \left(v^2 + a^2\right) \sin\theta_\tau \cos\theta_\tau\, d_\tau^{Z^0}. \tag{3.39}$$
- **Longitudinal Polarization**
  The longitudinal polarization is different from zero owing to the parity violation of the weak interaction. It is highly sensitive to the couplings of the leptons to the $Z^0$; this fact is exploited for the tests of the Standard Model in Sect. 4.6. The contribution from the anomalous magnetic dipole moment is negligible compared to this effect.

Experimentally, the transverse and normal polarizations could be measured from asymmetries in the direction of emission of the hadrons in semi-hadronic $\tau$ decays with respect to the direction of the $\tau$. This requires the full reconstruction of the $\tau$ direction. No results have been published yet.

In [351] it has been shown how the correlations between the polarizations of the $\tau^+$ and $\tau^-$ can be used to get access to the same information with a comparable sensitivity. These correlations have been studied experimentally in [52, 151] and found to be in good agreement with the Standard Model (see also Sect. 4.6.3).

### 3.3.4 Charge Radius and Magnetic Form Factors

There have been many ideas in which leptons and quarks are considered to be composites of more fundamental constituents. There are some very strong constraints on such models, in particular the excellent agreement between experiment and present theory for the anomalous magnetic moments of the electron and the muon. Although these limits exist, there are models in which the fermions of the first two generations are elementary, whereas the third family is composite (see e.g. [352]). Thus it is necessary to obtain separate limits on the form factors of the $\tau$ lepton, even if they are much less accurate.

If the $\tau$ lepton is built from constituents, one expects to see form factors which behave as

$$F_i(q^2) \propto \frac{1}{1+q^2/\Lambda_c^2}, \tag{3.40}$$

where $\Lambda_c$ is some composite mass scale. Far below the composite scale the form factors are constant and can be replaced by their asymptotic value at $q^2 = 0$. Only as one approaches the composite scale does one see the $q^2$ dependence of the form factors themselves.

The form factors can be identified either by their contribution to the total cross section for $\tau$ production or by a modification of the angular distribution. The total cross section through the electromagnetic current is [353–355]

$$\sigma_{\tau\tau} = \frac{4\pi\alpha^2}{3q^2}\left(F_1^{\gamma\,2} + 3\,F_1^\gamma F_2^\gamma + \frac{q^2}{8m_\tau^2}\,F_2^{\gamma\,2}\right) \tag{3.41}$$

and the angular distribution is

$$\frac{d\sigma_{\tau\tau}}{d\cos\theta} \propto H_0(F_1^\gamma, F_2^\gamma) + \cos^2\theta\, H_2(F_1^\gamma, F_2^\gamma). \tag{3.42}$$

(See [353] for the functions $H_i$.) The formulas are more complicated near the $Z^0$ pole, but the general idea is the same.

The authors of [353] have fitted the data from the PETRA experiments to achieve limits on the form factors and the composite scale. They get

$$F_2^\gamma(q^2 = (7\text{--}20)^2 \text{ GeV}^2) < 0.02, \tag{3.43}$$
$$\Lambda_c > 100 \text{ GeV}. \tag{3.44}$$

More recently, the authors of [356] have investigated the cross sections and forward–backward asymmetries measured at LEP for possible signals of substructure of any fermion. For the $\tau$ lepton they find

$$F_2^{Z^0}(q^2 = m_Z^2) = 0.03^{+0.03}_{-0.06}, \tag{3.45}$$
$$\Lambda_c > 100 \text{ GeV}. \tag{3.46}$$

72     3. The Static Properties of the $\tau$

## 3.4 Branching Ratios

### 3.4.1 Topological Branching Ratios

The simplest branching ratios of the $\tau$ are the topological branching ratios, i.e. the branching ratios of the $\tau$ leptons into a particular number of charged particles. From the theoretical point of view this seems quite an arbitrary classification, but it had its importance in organizing the experimental results.

The phase space is big enough to create up to 12 pions or a maximum of 11 charged particles. This process, however, is extremely rare and has not been, and probably will never be, observed. Today one-, three-, and five-prong decays have been measured and there is an upper limit on the seven-prong branching ratio, where 'prong' stands for the signal of a charged track in a tracking chamber.

Unfortunately it is not sufficient to simply count the number of observed tracks in the events. There are some detector defects which have to be corrected. Tracks can be lost by

- the overlap of two tracks, i.e. two tracks getting so close in space that they are not separated in the tracking chamber
- tracks escaping detection, through uncovered geometrical regions (beam pipe, cracks between sectors of the chambers) or through inefficiencies of the chambers or the track-finding algorithms
- interactions with detector material, before the track enters the chambers.

Additional tracks can be generated by

- photon conversions in the detector material of photons from $\pi^0$ decays or bremsstrahlung
- tracks from Dalitz decays ($\pi^0 \to e^+e^-\gamma$), which are not considered genuine and should not be counted as prongs
- hadronic interactions of a pion or kaon in the detector material, which can lead to additional tracks
- hard multiple scattering or similar effects, which can cause a single track to be reconstructed as two.

Thus an event with a genuine $i$–$j$ topology[5] can be reconstructed with a different topology $n$–$m$. Table 3.3 shows the transfer matrix obtained from the Monte Carlo simulation of the CELLO detector from their measurement of topological branching ratios [104]. The number of events expected to be seen with topology $n$–$m$ is related to the true number of events with topology $i$–$j$ through

$$n_{nm} = \sum_{ij} \epsilon_{nm \leftarrow ij} N_{ij}, \qquad (3.47)$$

---

[5] An event with one $\tau$ decaying into $i$ prongs and the other $\tau$ into $j$ prongs is called an $i$–$j$ topology.

**Table 3.3.** Efficiency matrix of the reconstruction of topologies with the CELLO detector [104]. Each column corresponds to a generated topology, and the rows to the reconstructed topology. All entries in %

|     | 1–1   | 1–3   | 1–5   | 3–3   | 3–5   | 5–5  |
|-----|-------|-------|-------|-------|-------|------|
| 1–1 | 30.17 | 1.84  | 1.19  | 0.05  | 0.00  | 0.46 |
| 1–2 | 1.45  | 10.81 | 3.81  | 1.11  | 0.00  | 0.00 |
| 1–3 | 0.37  | 27.50 | 4.95  | 2.68  | 0.00  | 0.00 |
| 1–4 | 0.03  | 0.75  | 19.29 | 0.06  | 1.04  | 1.83 |
| 1–5 | 0.01  | 0.12  | 7.74  | 0.05  | 0.00  | 0.46 |
| 1–6 | 0.00  | 0.00  | 0.97  | 0.00  | 0.00  | 0.00 |
| 2–3 | 0.01  | 0.53  | 0.00  | 15.15 | 2.07  | 3.20 |
| 2–5 | 0.00  | 0.00  | 0.00  | 0.00  | 4.13  | 1.83 |
| 3–3 | 0.00  | 0.13  | 0.00  | 18.48 | 5.16  | 2.74 |
| 3–4 | 0.00  | 0.00  | 0.00  | 1.55  | 14.42 | 8.68 |
| 3–5 | 0.00  | 0.00  | 0.25  | 0.16  | 6.19  | 1.37 |
| 3–6 | 0.00  | 0.00  | 0.00  | 0.00  | 1.04  | 0.00 |
| 4–5 | 0.00  | 0.00  | 0.00  | 0.00  | 0.00  | 5.02 |
| 4–5 | 0.00  | 0.00  | 0.00  | 0.00  | 0.00  | 0.46 |

with $\epsilon_{nm \leftarrow ij}$ given by Table 3.3. And the number of true events $N_{ij}$ is related to the topological branching ratios $br_i$ by

$$N_{ij} = N_{\tau\tau} \times 2 \, br_i \, br_j \quad (i \neq j),$$
$$N_{ii} = N_{\tau\tau} \times br_i^2. \tag{3.48}$$

$N_{\tau\tau}$ is the total number of $\tau$ pairs produced. A likelihood function can be formed from the number of events expected from (3.47) and the number of events observed in the data. From the fit to the CELLO data they obtained the following results [104]:

$$br_1 = (84.9 \pm 0.4 \pm 0.3)\%,$$
$$br_3 = (15.0 \pm 0.4 \pm 0.3)\%,$$
$$br_5 = (0.16 \pm 0.13 \pm 0.04)\%. \tag{3.49}$$

Care has to be taken with $K^0$ mesons in the events with the decay $K^0_S \to \pi^+ \pi^-$. At lower center-of-mass energies the average decay length of the $K^0_S$ is so short that the two pions look like genuine tracks from the decay vertex of the $\tau$ and are counted as such. However, at higher energies the $K^0_S$ can have a decay length of half a meter or more, and then the pions are clearly distinct from prongs in the usual sense. This has to be taken into account when comparing the results of different experiments[6] (compare e.g. $\Gamma_{49}$ and $\Gamma_{50}$ in the Particle Data Group booklet [44]).

With the five-prong branching ratio measured, the CLEO collaboration has searched for seven-prong decays [115]. There are two problems to fight in such an analysis: the migration of three- and five-prong $\tau$ decays into the

---
[6] The CELLO result given above includes tracks from $K^0_S \to \pi^+ \pi^-$ as genuine prongs.

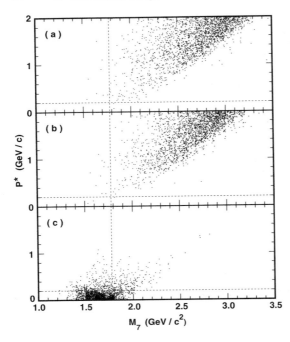

**Fig. 3.25.** The two kinematic variables used by the CLEO collaboration [115] to separate seven-prong $\tau$ decays from $e^+e^- \to$ had background. $M_7$ is the invariant mass of the seven-prong decays and $P^*$ their momentum in the $\tau$ pseudo-rest frame. **(a)** Data, **(b)** hadronic background, and **(c)** Monte Carlo simulation of $\tau^- \to 4\pi^- 3\pi^+ \nu_\tau$. The hadronic background was obtained with a high-mass tag of $M_1 > 1.8$ GeV on the recoiling $\tau$. The *dashed lines* indicate the values at which the cuts were imposed

seven-prong sample due to tracks from photon conversion and Dalitz decays, and the large background from hadronic events ($e^+e^- \to$ had).

To reject the first type of background the CLEO collaboration placed tight cuts on the impact parameters of the tracks, ensuring they came from the secondary vertex of the $\tau$ decay and not from some point of conversion in the material of the beam pipe. Furthermore, they rejected events with too many tracks consistent with an electron hypothesis.

To reduce the background from $e^+e^- \to$ had, only seven-prong decays recoiling against a one-prong with no more than two identified photons were accepted. The final suppression of this background was done by means of two kinematical quantities: the invariant mass of the seven-prong decay which has to be smaller than the $\tau$ mass, and the momentum of the seven-prong in a $\tau$ pseudorest frame, which has to be smaller than 0.2 GeV. For the Lorentz transformation back into the $\tau$ rest frame they used the thrust axis as an approximation to the $\tau$ direction. Figure 3.25 shows the distributions in these quantities for signal, background, and data. They did not find a candidate and hence set an upper limit on the branching ratio of

$$br_7 < 2.4 \times 10^{-6} \quad \text{at the 90 \% confidence level.} \tag{3.50}$$

Table 3.4 summarizes the efficiencies and numbers of expected events.

## 3.4 Branching Ratios

**Table 3.4.** Summary of the search for seven-prong $\tau$ decays by the CLEO collaboration [115]. The errors quoted are statistical only

| | |
|---|---|
| Number of events in data | 0 |
| Background from $\tau$ migration | $0.88 \pm 0.23$ |
| Background from $e^+e^- \to$ had | $1.95 \pm 1.40$ |
| Number of $\tau$ pairs analyzed | $4.21 \times 10^6$ |
| Tagging efficiency (%) | $73.0 \pm 0.3$ |
| Efficiency $4\pi^-3\pi^+$ (%) | $15.7 \pm 0.2$ |
| Efficiency $4\pi^-3\pi^+\pi^0$ (%) | $15.9 \pm 0.3$ |
| Branching ratio | $< 2.38 \times 10^{-6}$ |

### 3.4.2 Leptonic Branching Ratios

**Theoretical Expectation.** The partial decay width of the decays $\tau \to e\nu_e\nu_\tau$ and $\tau \to \mu\nu_\mu\nu_\tau$ can be calculated straightforwardly from the Feynman diagram of Fig. 3.26. It is sufficient to treat the process as an effective four-fermion interaction and add the $W^\pm$ propagator as a correction later on. The result is

$$\Gamma(\tau \to \ell\,\nu_\ell\nu_\tau) = \frac{G_F^2\, m_\tau^5}{192\,\pi^3}\,(1 + \Delta_\ell). \qquad (3.51)$$

The quantity $\Delta_\ell$ collects a number of corrections which are summarized in Table 3.5. These are

- the phase space correction due to the finite mass of the charged daughter lepton
- QED radiative corrections
- the correction due to the $W^\pm$ propagator.

The table also lists corrections which are not present in the Standard Model but will modify the decay width if new physics is present. They will be discussed at the end of this section.

The branching ratios for the two leptonic decays can be calculated from

$$br(\tau \to \ell\,\nu_\ell\nu_\tau) = \frac{\Gamma(\tau \to \ell\,\nu_\ell\nu_\tau)}{\Gamma_{\rm tot}}. \qquad (3.52)$$

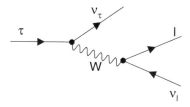

**Fig. 3.26.** Feynman diagram of a leptonic $\tau$ decay

76     3. The Static Properties of the $\tau$

**Table 3.5.** Corrections to the leptonic width of the $\tau$ lepton. The first three corrections apply to the decay in the Standard Model. The lower part of the table shows modifications due to physics beyond the Standard Model. The values in the lower part are limits independent of the leptonic branching ratios, from [44] for $m_{\nu_\tau}$ and [357] for $\eta$

|  |  | $\tau \to e\,\nu_e\nu_\tau$ | $\tau \to \mu\,\nu_\mu\nu_\tau$ |
|---|---|---|---|
| Mass of charged daughter [2] | $-8\left(\frac{m_\ell}{m_\tau}\right)^2$ | 0.00% | $-2.74\%$ |
| QED rad. corr. [358–363] | $\frac{\alpha(m_\tau^2)}{2\pi}\left(\frac{25}{4}-\pi^2\right)$ | $-0.43\%$ | $-0.43\%$ |
| $W^\pm$ boson propagator [364] | $\frac{3}{5}\left(\frac{m_\tau}{m_{W^\pm}}\right)^2$ | $+0.03\%$ | $+0.03\%$ |
| Standard Model, total |  | $-0.40\%$ | $-3.14\%$ |
| Neutrino mass [2] | $-8\left(\frac{m_{\nu_\tau}}{m_\tau}\right)^2$ | $< -0.08\%$ | $< -0.08\%$ |
| Scalar current [365] | $4\,\eta\,\frac{m_\ell}{m_\tau}$ | 0.00% | $> -0.1\%$ $< +6.1\%$ |
| Mixing with a 4th gen. $\nu$ [366] | $-\sin^2\theta$ | – | – |
| Magnetic dipole moment [331] | $\frac{\kappa}{2}+\frac{\kappa^2}{10}$ | – | – |
| Electric dipole moment [331] | $\frac{\tilde{\kappa}^2}{10}$ | – | – |

With $\Gamma_{\rm tot} = 1/\tau_\tau$ calculated from the lifetime $\tau_\tau = (290.0 \pm 1.2)$ fs and the world averages from [44], one arrives at the following predictions:

$$br(\tau \to e\,\nu_e\nu_\tau) = (17.772 \pm 0.075)\%,$$
$$br(\tau \to \mu\,\nu_\mu\nu_\tau) = (17.282 \pm 0.073)\%. \quad (3.53)$$

The errors are dominated by the uncertainty in the lifetime of the $\tau$.

**Measurements.** The leptonic branching ratios are extracted from the data by first selecting a sample of $\tau$ pairs and then counting the leptonic decays within this sample. If $N_\tau$ is the number of $\tau$ leptons selected and $N_\ell$ the number of leptonic decays (either e or $\mu$), then the naive estimate is $br(\tau \to \ell\,\nu_\ell\nu_\tau) = N_\ell/N_\tau$. This has to be corrected for background in the initial $\tau$ samples as well as in the sample of $\tau \to \ell\,\nu_\ell\nu_\tau$ decays, for the efficiency of the lepton identification, and for the bias introduced by the $\tau$ selection. The efficiency of the $\tau$ selection does not enter directly. The branching ratio is

$$br(\tau \to \ell\,\nu_\ell\nu_\tau) = \frac{1}{\varepsilon\,F_{\rm B}}\frac{N_\ell(1-f^\ell_{\rm bgd})}{N_\tau(1-f^\tau_{\rm bgd})}. \quad (3.54)$$

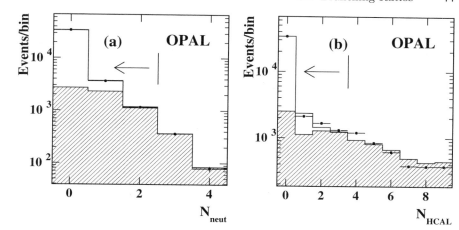

**Fig. 3.27.** Two of the variables used in the preselection of $\tau \to e\,\nu_e\nu_\tau$ candidates by the OPAL collaboration [209]: (a) the number of neutral clusters in the jet cone and (b) the penetration into the hadron calorimeter (in layers). *Points*, data; *open histogram*, Monte Carlo for the signal; *hatched histogram* for the background. The discrepancy between the data and the Monte Carlo distribution in the second variable has been carefully studied and a systematic error associated with it

The quantity $f_{\text{bgd}}^\tau$ is the fraction of background in the $\tau$ sample and $f_{\text{bgd}}^\ell$ the corresponding fraction in the $\ell$ sample. The latter has two components: background from non-$\tau$ events, e.g. electrons from Bhabha events that slipped through the $\tau$ selection in a measurement of $\tau \to e\nu_e\nu_\tau$; and background from $\tau$ decays misidentified as $\tau \to \ell\,\nu_\ell\nu_\tau$.

The quantity $\varepsilon$ is the efficiency for selecting the leptonic decay and $F_B$ a bias factor which has to be introduced because the $\tau$ selection does not in general have a uniform efficiency for selecting different $\tau$ decay channels and therefore introduces a bias into the measured value of the branching ratio. The bias factor measures the degree to which the $\tau$ selection favors or suppresses the decay $\tau \to \ell\,\nu_\ell\nu_\tau$ relative to other $\tau$ decay channels. It is defined as the ratio of the fraction of $\tau \to \ell\,\nu_\ell\nu_\tau$ decays in a sample of $\tau$ decays after the selection is applied, to the fraction before the selection.

Recently a new measurement of the electronic branching ratio was presented by the OPAL collaboration [209]. To achieve an accurate result, it is necessary to select the $\tau \to e\,\nu_e\nu_\tau$ decays with high efficiency and low background. OPAL achieved this through a two-step selection of the electrons. After a rather standard selection of the $\tau$ sample, each $\tau$ jet was first processed through a cut-based preselection, rejecting obvious non-$\tau \to e\,\nu_e\nu_\tau$ decays. The remaining candidates were then subjected to a likelihood selection. Figure 3.27 shows two out of the six variables used for the preselection. The cuts for the preselection were quite loose. The cuts associated with the two variables shown were

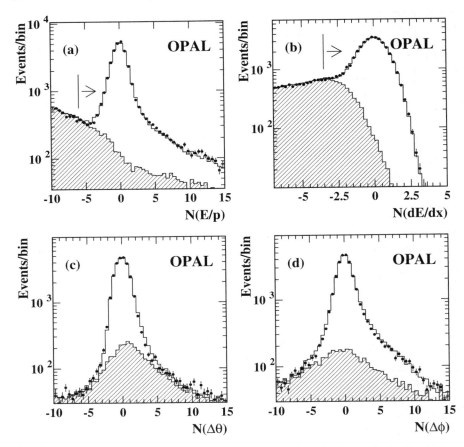

**Fig. 3.28.** Comparison of data (*points with error bars*) with Monte Carlo simulation for variables used in the likelihood selection of $\tau \to e\,\nu_e\nu_\tau$ (OPAL collaboration [209]). The *open histogram* is the signal, the *hatched* the background. The variables are (**a**) the ratio of energy deposition in the calorimeter to the momentum of the track, (**b**) the specific energy loss in the tracking chamber, and the matching of the track with the calorimeter cluster in polar angle (**c**) and azimuth (**d**). The *arrows* on the first two plots indicate preselection cuts

- no more than two neutral, electromagnetic clusters (photons) were allowed;
- the jet was rejected if a shower penetrated into the hadron calorimeter to a depth of more than three layers.

In the second step a likelihood was formed from a total of six variables; four of these are shown in Fig. 3.28. There were only two classes, $\tau \to e\,\nu_e\nu_\tau$ and the sum of all background. All variables were normalized to the expectation of the signal, which improves the performance of the likelihood:

3.4 Branching Ratios 79

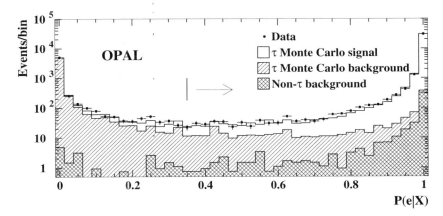

**Fig. 3.29.** The likelihood output for $\tau \to e\,\nu_e\nu_\tau$ candidates (OPAL collaboration [209])

$$N(x_i) = \frac{x_i - \mu_i(p)}{\sigma_i(p)}, \tag{3.55}$$

where $x_i$ is any of the variables and $\mu_i(p)$ and $\sigma_i(p)$ are its expectation and variance for the signal parametrized as a function of the momentum of the track. $N(x_i)$ has a distribution which is centered at zero, with a width of unity for the signal. Figure 3.29 shows the resulting likelihood distribution. A cut was placed at 0.35, a value which was optimized to give the smallest overall error on the branching ratio.

The result is

$$br(\tau \to e\,\nu_e\nu_\tau) = (17.81 \pm 0.09 \pm 0.06)\%, \tag{3.56}$$

where the first error is statistical and the second systematics.

**Lepton Universality.** In the Standard Model one expects a universal coupling of all leptons to the charged current. Replacing the Fermi constant in (3.51) by

$$\frac{G_F}{\sqrt{2}} = \frac{1}{8}\left(\frac{g_\ell}{m_{W^\pm}}\right)^2 \tag{3.57}$$

for each of the vertices, where $g_\ell$ carries the index of the flavor at that vertex, one can compare these couplings.[7] With the following current world averages as input [44]

$$m_\mu = (105.658389 \pm 0.000034)\text{ MeV},$$
$$m_\tau = (1777.05^{+0.29}_{-0.26})\text{ MeV},$$
$$\tau_\mu = (2.19703 \pm 0.00004) \times 10^{-6}\text{ s},$$

---

[7] Equation (3.51) also holds for $\mu \to e\,\nu_e\,\nu_\mu$.

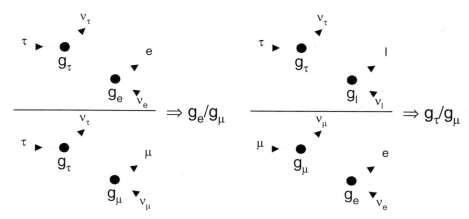

**Fig. 3.30.** Illustration of the relations between the various leptonic decays used to extract the ratios of the weak charged couplings. The $\ell$ is a fictive massless lepton with a branching ratio $\tau \to \ell \nu_\ell \nu_\tau$ calculated from the average of $\tau \to e\, \nu_e \nu_\tau$ and $\tau \to \mu\, \nu_\mu \nu_\tau$. Its coupling is assumed to be $g_e$

$$\tau_\tau = (290.0 \pm 1.2) \times 10^{-15}\ \text{s},$$
$$br(\tau \to e\, \nu_e \nu_\tau) = (17.81 \pm 0.07)\%,$$
$$br(\tau \to \mu\, \nu_\mu \nu_\tau) = (17.37 \pm 0.09)\%, \tag{3.58}$$

one obtains, from the ratio of the electronic to the muonic branching ratio of the $\tau$ lepton,

$$\frac{g_e}{g_\mu} = 0.999 \pm 0.003, \tag{3.59}$$

consistent with universality. Then, combining the two leptonic branching ratios of the $\tau$ and comparing them to the decay of the muon, one has

$$\frac{g_\tau}{g_\mu} = 1.002 \pm 0.004, \tag{3.60}$$

again consistent with universality. The ratios are illustrated in Fig. 3.30. In the second ratio the error is dominated by the uncertainty in the branching ratio and the lifetime of the $\tau$. The quantities for the muon are measured with higher precision. Hence the ratio can be interpreted as a prediction of the $\tau$ lifetime and its leptonic branching ratio from the decay of the muon. This prediction is compared to the measurement in Fig. 3.31.

These tests of lepton universality are sensitive to a large number of effects from physics beyond the Standard Model, as listed below.

A massive neutrino would change the phase space of the decays and consequently the branching ratios. The direct limits on the masses of $\nu_e$ and $\nu_\mu$ are so strict that these cannot have a measurable impact, but the $\nu_\tau$ could. Also,

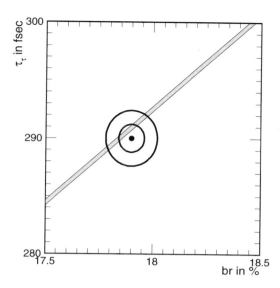

Fig. 3.31. Universality plot of the charged weak couplings. The measured values are indicated by the 1- and 2-sigma error *ellipses*. They should overlap the *diagonal band*, predicted from muon decay, if universality holds. The width of the band is dominated by the uncertainty in the mass of the $\tau$

mixing with a yet undiscovered, heavy, fourth-generation neutrino would reduce the branching ratios. The corrections are given in Table 3.5. The authors of [332, 366] derived limits from universality of

$$m_{\nu_\tau} < 38 \text{ MeV},$$
$$\sin^2 \theta < 0.008 \tag{3.61}$$

at the 95% confidence limits.

A charged Higgs boson would introduce a scalar coupling and therefore a nonvanishing Michel parameter $\eta$, which in turn modifies the branching ratios. A limit on these bosons has been found [357, 365]:

$$m_{H^\pm} > 2.3 \tan \beta \tag{3.62}$$

at the 95% confidence limit. For more details see Sect. 8.4.2.

An anomalous magnetic moment or an electric dipole moment of the charged weak current would modify the branching ratios. The correction is given in Table 3.5. Limits have been derived in [332]. They are (95% confidence level)

$$-0.015 < \kappa < 0.017,$$
$$|\tilde{\kappa}| < 0.31. \tag{3.63}$$

### 3.4.3 Hadronic Branching Ratios

$\tau \to \pi \nu_\tau$ and $\tau \to K \nu_\tau$. The branching ratio of the $\tau$ lepton into a single pion can be predicted from the analogous decay of the pion (see Fig. 3.32). The decay constant $f_\pi$ determining the strength of the coupling of the pion to the charged current is identical in both reactions. The calculation gives

$$\Gamma(\tau \to \pi \nu_\tau) = \frac{G_F^2 f_\pi^2 \cos^2 \theta_C}{16 \pi} m_\tau^3 \left(1 - \frac{m_\pi^2}{m_\tau^2}\right)^2, \tag{3.64}$$

$$\Gamma(\pi \to \mu \nu_\mu) = \frac{G_F^2 f_\pi^2 \cos^2 \theta_C}{8 \pi} m_\pi m_\mu^2 \left(1 - \frac{m_\mu^2}{m_\pi^2}\right)^2. \tag{3.65}$$

The decay $\tau \to K \nu_\tau$ can be treated in the same way, replacing $f_\pi$ by $f_K$ and the cosine of the Cabibbo angle by the sine.

The branching ratios should obey the following relations:

$$\frac{br(\tau \to \pi \nu_\tau)}{br(\pi \to \mu \nu_\mu)} = \frac{m_\tau^3 \left(1 - m_\pi^2/m_\tau^2\right)^2}{2 m_\pi m_\mu^2 \left(1 - m_\mu^2/m_\pi^2\right)^2} \frac{\tau_\tau}{\tau_\pi}, \tag{3.66}$$

$$\frac{br(\tau \to K \nu_\tau)}{br(K \to \mu \nu_\mu)} = \frac{m_\tau^3 \left(1 - m_K^2/m_\tau^2\right)^2}{2 m_K m_\mu^2 \left(1 - m_\mu^2/m_K^2\right)^2} \frac{\tau_\tau}{\tau_K}. \tag{3.67}$$

This constitutes another test of $\tau$–$\mu$ universality.

The branching ratios predicted from these equations are in reasonable agreement with the measured values [44]:

$\tau \to \pi \nu_\tau$:

predicted: $(10.87 \pm 0.05)\%$;

measured: $(11.08 \pm 0.13)\%$;

$\tau \to K \nu_\tau$:

predicted: $(7.08 \pm 0.04) \times 10^{-3}$;

measured: $(7.1 \pm 0.5) \times 10^{-3}$.

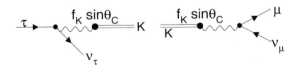

**Fig. 3.32.** Diagrams of the decays $\tau \to \pi \nu_\tau$ and $\pi \to \mu \nu_\mu$ and their Cabibbo-suppressed analogs

**Decays to an Even Number of Pions.** The decay of the $\tau$ lepton to two pions has the largest branching ratio. It is known to be dominated by the $\rho$ resonance and the subsequent decay $\rho^- \to \pi^- \pi^0$. There is also a small contribution from the $\rho'(1450)$ and maybe even from the $\rho''(1700)$.

In a simple approximation, the branching ratio has already been calculated by Tsai [23]. Assuming the decay to proceed exclusively through the $\rho(770)$ and approximating it by a narrow resonance, he gets

$$\frac{\Gamma(\tau^- \to \rho^- \nu_\tau)}{\Gamma(\rho^0 \to e^+ e^-)} = \frac{3\,G_F^2 \cos^2\theta_C\, m_\rho m_\tau^3}{32\,\pi^2\,\alpha^2} \left(1 - \frac{m_\rho^2}{m_\tau^2}\right)^2 \left(1 + 2\frac{m_\rho^2}{m_\tau^2}\right). \quad (3.68)$$

With the current world averages [44] one gets a branching ratio of $(27.0 \pm 1.3)\%$. More precise predictions can be derived using the conserved-vector-current (CVC) hypothesis (for details see Sect. 5.2.6). Any channel with an even number of pions can be calculated by this method. A recent reanalysis of the $e^+e^-$ data was given in [367]. Table 3.6 gives a comparison between the predicted and measured branching ratios. There is excellent agreement for all channels except for the $\pi^-\pi^0$. This channel shows a discrepancy of a little more than two standard deviations. The discrepancy might be explained in several ways. It might be an insufficient treatment of radiative corrections, the first sign of isospin violations, a systematic problem in the measurements, or just a statistical fluctuation after all.

**Table 3.6.** Predictions of $\tau$ branching ratios into an even number of pions from CVC compared with measurements [367] (%)

|  | Measurement | Prediction |
|---|---|---|
| $\pi^-\pi^0$ | $25.32 \pm 0.15$ | $24.52 \pm 0.33$ |
| $\pi^-3\pi^0$ | $1.11 \pm 0.14$ | $1.10 \pm 0.04$ |
| $2\pi^-\pi^+\pi^0$ | $4.22 \pm 0.10$ | $4.06 \pm 0.25$ |
| Others | $0.36 \pm 0.04$ | $0.37 \pm 0.04$ |
| Total | $31.01 \pm 0.23$ | $30.05 \pm 0.42$ |

**Decays to Three and Five Pions.** The decay of the $\tau$ lepton to three pions is dominated by the $a_1$ resonance. It decays with equal probabilities into the two charge combinations $\pi^-\pi^-\pi^+$ and $\pi^-\pi^0\pi^0$. The decays must proceed through the axial-vector current in order to conserve G parity. Thus CVC cannot be applied to predict the branching ratio. The second Weinberg sum rule [368] predicts $f_{a_1} = f_\rho$ and therefore

$$\frac{\Gamma(\tau \to a_1\,\nu_\tau)}{\Gamma(\tau \to \rho\,\nu_\tau)} = \frac{(m_\tau^2 - m_{a_1}^2)^2 (m_\tau^2 + 2m_{a_1}^2)}{(m_\tau^2 - m_\rho^2)^2 (m_\tau^2 + 2m_\rho^2)}, \quad (3.69)$$

but these kinds of relation ignore the finite width of the mesons, which is certainly not a good approximation for the $a_1$. The prediction would be $(15 \pm$

**Table 3.7.** World averages of $\tau$ decays into three and five pions [44]. All values in %; $h^\pm$ is either a charged pion or a kaon

| Channel | Branching ratio |
|---|---|
| $\pi^-\pi^+\pi^-$ | $9.23 \pm 0.11$ |
| $\pi^-\pi^0\pi^0$ | $9.15 \pm 0.15$ |
| $3\,h^-\,2\,h^+$ | $0.075 \pm 0.007$ |
| $2\,h^-h^+2\,\pi^0$ | $0.11 \pm 0.04$ |
| $h^-\,4\,\pi^0$ | $0.11 \pm 0.06$ |

1)%. It is not obvious how a broad resonance like the $a_1$ should be properly parametrized (for more detailed discussions see [300, 369–372]). The authors of [300] calculate predictions for various parametrizations. The results range from 10.2% to 25.2%.

Another technique uses the soft-pion theorems and the partially conserved axial-vector current hypothesis (PCAC). A crude estimate can be given for the decay to an odd number of pions [306]. The estimates are 10% for the total of the three-pion final states and 1% for five pions. The measurements are summarized in Table 3.7.

**Strange Decays.** Apart from the $\tau \to K\,\nu_\tau$ channel, which has been discussed above, the $K^*$ gives the largest contribution to the strange decays. It is the equivalent to the $\tau \to \rho\,\nu_\tau$ decay in the nonstrange sector. If $SU(3)_f$ were not broken, one could simply replace the cosine of the Cabibbo angle by the sine in the prediction of $\tau \to \rho\,\nu_\tau$ and get a prediction of $\tau \to K^*\,\nu_\tau$ after correcting for the difference in phase space.

With $SU(3)_f$ broken one can use the Das–Mathur–Okubo sum rule [373, 374] to relate $f_\rho$ to $f_{K^*}$:

$$\frac{f_{K^*}^2}{m_{K^*}^2} = \frac{f_\rho^2}{m_\rho^2}. \tag{3.70}$$

This gives

$$\frac{\Gamma(\tau \to K^*\,\nu_\tau)}{\Gamma(\tau \to \rho\,\nu_\tau)} = \tan^2\theta_C \frac{m_\rho^2}{m_{K^*}^2} \frac{(m_\tau^2 - m_{K^*}^2)^2 (m_\tau^2 + 2m_{K^*}^2)}{(m_\tau^2 - m_\rho^2)^2 (m_\tau^2 + 2m_\rho^2)}. \tag{3.71}$$

The prediction agrees roughly with the measurement:

predicted: $(8.8 \pm 0.1) \times 10^{-3}$;

measured: $(12.8 \pm 0.8) \times 10^{-3}$.

The $K^*$ dominates the strange decays into two mesons $\tau \to K\,\pi^0\,\nu_\tau$ and $\tau \to \pi\,K^0\,\nu_\tau$. The known strange modes with three or more mesons are summarized in Table 3.8.

**Table 3.8.** Strange decay modes of the $\tau$ lepton with three or more mesons [44]. Measured numbers in units of $10^{-4}$

| Mode | Branching ratio |
|---|---|
| $K^-\pi^+\pi^-\nu_\tau$ | $18 \pm 5$ |
| $K^- 2\pi^0 \nu_\tau$ | $8.0 \pm 2.7$ |
| $\pi^-\overline{K}^0\pi^0\nu_\tau$ | $39 \pm 5$ |
| $K^-\pi^+\pi^-\pi^0\nu_\tau$ | $2.4^{+4.3}_{-1.6}$ |
| $K^- 3\pi^0 \nu_\tau$ | $4.3^{+10.0}_{-2.9}$ |
| $\pi^-\overline{K}^0 2\pi^0 \nu_\tau$ | $6 \pm 4$ |

**Intermediate Resonances.** With samples of higher statistics becoming available, especially for the CLEO experiment, it is possible to identify not only the final state, but also intermediate resonance structures. The decay channel $\tau \to a_1 \nu_\tau$ with the $\rho$ resonance in the intermediate state of the decay chain $a_1 \to \rho\pi \to 3\pi$ is an example which has long been known. Other examples are $\rho$ and $\omega$ in $\tau \to 4\pi\nu_\tau$, and $K^{*0}$ in $\tau \to K^0 \pi^\pm \pi^0 \nu_\tau$.

A new intermediate resonance in nonstrange decays has recently been identified by the CLEO collaboration [117]: the $f_1(1285)$ in the decay $\tau^- \to f_1 \pi^- \nu_\tau$, followed by $f_1 \to \eta\pi\pi$.

Starting from an initial sample of 4.27 million $\tau$ pairs, the CLEO collaboration has searched for decays into $\pi^-\pi^+\pi^-\eta$ and $\pi^-\pi^0\pi^0\eta$. The $\eta$ mesons were identified as a narrow mass peak in their decays $\eta \to \gamma\gamma$ and $\eta \to 3\pi^0$.[8] Figure 3.33 shows the invariant masses of the photon pairs and $\pi^0$ triplets. The $\eta$ signal is clearly visible above a combinatorial background from $\tau$ decays and some $\eta$ mesons from hadronic background events. The branching ratios obtained are

$$\pi^-\pi^+\pi^-\eta : (3.4^{+0.6}_{-0.5} \pm 0.6) \times 10^{-4};$$
$$\pi^-\pi^0\pi^0\eta : (1.4 \pm 0.6 \pm 0.3) \times 10^{-4}. \quad (3.72)$$

These events were then studied for subresonances. Figure 3.34 shows the invariant mass of $\pi\pi\eta$ versus $\pi\eta$. For the three-prong mode there is an ambiguity in the choice of the same-charge pion that results in four entries per event. In the case of $\tau^- \to \pi^- 2\pi^0 \eta \nu_\tau$ there is only one $f_1 \to \pi^0\pi^0\eta$ combination, and for $\pi^0\eta$ only the combination with the higher mass is plotted. An enhancement at 1285 MeV in the $\pi\pi\eta$ mass is clearly visible. This is the $f_1$ meson ($m_{f_1} = 1281.9$ MeV, $\Gamma_{f_1} = 24$ MeV). There is also an enhancement in the $\pi\eta$ mass around 980 MeV, indicating the presence of the decay chain $f_1 \to a_0\pi$, $a_0 \to \pi\eta$. The $f_1$ is expected to decay this way in roughly 2/3 of all cases [44]. From the distribution of Fig. 3.34, CLEO fits a branching ratio of

$$br(\tau \to f_1 \pi \nu_\tau) = (5.8^{+1.4}_{-1.3} \pm 1.8) \times 10^{-4}. \quad (3.73)$$

---

[8] The decay $\eta \to 3\pi^0$ was not used for the $\pi^-\pi^0\pi^0\eta$ final state. With five neutral pions the combinatorial background is too large.

86    3. The Static Properties of the $\tau$

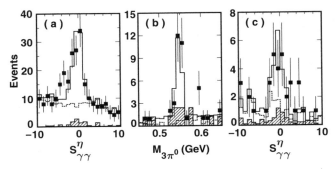

**Fig. 3.33.** The $\eta$ signal in the channels (**a**) $\tau^- \to \pi^- \pi^+ \pi^- \eta \nu_\tau$, $\eta \to \gamma\gamma$, (**b**) $\tau^- \to \pi^- \pi^+ \pi^- \eta \nu_\tau$, $\eta \to 3\pi^0$, and (**c**) $\tau^- \to \pi^- 2\pi^0 \eta \nu_\tau$, $\eta \to \gamma\gamma$ (CLEO collaboration [117]). The *solid line* is a fit to the data (*squares*) composed of the signal, the $\tau$ background (*dashed line*), and the q$\bar{\text{q}}$ background (*hatched area*). Plots (**a**) and (**c**) are binned in standard deviations from the $\eta$ mass while (**b**) has a 10 MeV bin size

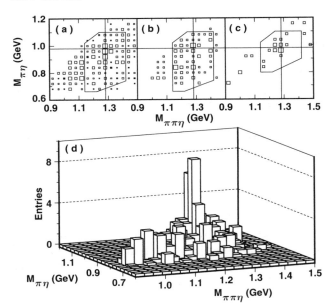

**Fig. 3.34.** Subresonances in the decay $\tau \to 3\pi\eta\nu_\tau$ (CLEO collaboration [117]). The *upper three histograms* correspond to the respective channels of Fig. 3.33. The *lower histogram* is the sum of all histograms, weighted according to the number of entries per event

### 3.4.4 Global Analysis

To avoid problems with correlated systematic errors when combining different measurements of branching ratios, several experiments have determined all branching ratios simultaneously (see e.g. [56, 72, 103]). The method is as follows. Starting from a common sample of $\tau$ pairs, each $\tau$ lepton is subjected to particle identification, attempting to identify its decay as belonging to one out of several classes. These classes are defined such that each Standard Model $\tau$ decay truly belongs to exactly one of them. Then the expected number of $\tau$ leptons identified in class $i$ is given by

$$N_i = N_\tau \sum_j \varepsilon_{ij} br_j + N_i^{\text{bgd}}, \tag{3.74}$$

where $N_\tau$ is the total number of $\tau$ leptons analyzed, $br_j$ the branching ratio into class $j$, and $\varepsilon_{ij}$ the probability that a decay of class $j$ is reconstructed in class $i$. $N_i^{\text{bgd}}$ is the non-$\tau$ background in class $i$. The branching ratios are derived by fitting $N_i$ to the number of events observed in the data.

The highest precision can be achieved if every event is assigned to the class it matches best, no matter how good the match is. This procedure, however, bears a danger: if there are events in the data which do not belong to any of the classes, they escape unrecognized. They are simply shared between the most similar classes. To avoid this problem the CELLO collaboration determined the branching ratios in several steps [103]:

1. Classification of the $\tau$ decays into seven classes which largely correspond to exclusive decay channels of the $\tau$ lepton.
2. Determination of a set of probabilities $P_i$ for each $\tau$ candidate to belong to one of the seven classes.
3. Determination of the branching ratio of each class independently.
4. Determination of the branching ratios of all classes simultaneously.
5. Investigation of systematic uncertainties.

In step 3 a decay has to have a probability $P_i$ above some cut to be assigned to class $i$, excluding decays which do not match any class from the measurement. Then in step 4 each decay is forced into one class. Unexpected decays not belonging to any of the classes would signal their presence by a difference between the results of steps 3 and 4.

The definition of the seven classes in the CELLO analysis and the results from step 4 are given in Table 3.9. Good agreement was found with the results from step 3 for a wide range of cut values.

For illustration, Fig. 3.35 shows the momentum distribution of the charged track in the first four classes in comparison with the Monte Carlo simulation. The momentum is one of the quantities used to calculate the probabilities $P_i$.

It is interesting to see how the definition of what is considered an exclusive class has changed over the years as more statistics and better detectors

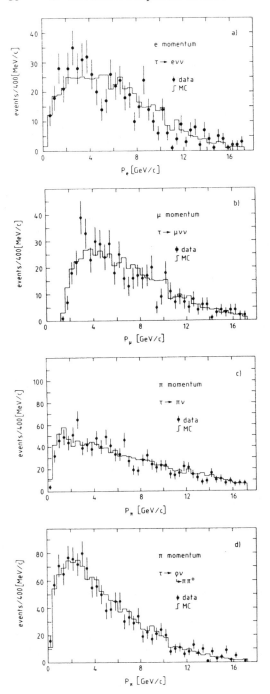

**Fig. 3.35.** Momentum distribution of the final-state charged particle in one-prong $\tau$ decays (CELLO collaboration [103]). **(a)** $\tau \to e\,\nu_e\nu_\tau$, **(b)** $\tau \to \mu\,\nu_\mu\nu_\tau$, **(c)** $\tau \to \pi\,\nu_\tau$, and **(d)** charged pions from $\tau \to \rho\,\nu_\tau$. The *points* are data; the *histogram* is the Monte Carlo simulation

**Table 3.9.** Definition of the seven classes used by the CELLO collaboration in their global analysis of the $\tau$ branching ratios [103]. All results in %

| Class | Branching ratio |
|---|---|
| $\tau \to e\,\nu_e\nu_\tau$ | $18.4 \pm 0.8 \pm 0.4$ |
| $\tau \to \mu\,\nu_\mu\nu_\tau$ | $17.7 \pm 0.8 \pm 0.4$ |
| $\tau \to$ had $\nu_\tau$ | $12.3 \pm 0.9 \pm 0.5$ |
| $\tau \to$ had $2\gamma\,\nu_\tau$ | $22.6 \pm 1.5 \pm 0.7$ |
| $\tau \to$ had $> 2\gamma\,\nu_\tau$ | $14.0 \pm 1.2 \pm 0.6$ |
| $\tau \to 3$ had $\nu_\tau$ | $9.0 \pm 0.7 \pm 0.3$ |
| $\tau \to 3(5)$had $\geq 1(0)\gamma\,\nu_\tau$ | $5.8 \pm 0.7 \pm 0.2$ |
| Sum | $99.8 \pm 2.6 \pm 1.2$ |

became available.[9] When the CELLO collaboration did its global analysis in 1990 [103] an exclusive (hadronic) class was merely a number of charged tracks (either one, or three or more) accompanied by no, two, or more than two photons. Nowadays experiments are trying to become more exclusive in several directions when they measure branching ratios:

- Photons are paired to $\pi^0$s to distinguish them from $\eta$ mesons (CELLO did already identify $\pi^0$ in some of its classes).
- Charged tracks are identified as charged pions or kaons.
- $K^0$ mesons are identified by their decay $K_S^0 \to \pi^+\pi^-$ or as $K_L^0$. No experiment has used the decay $K_S^0 \to \pi^0\pi^0$ yet. It is treated as a correction to the other multi-$\pi^0$ channels.
- Besides the $\eta$, the $\omega$ has also been explicitly identified and can be subtracted from the multi-$\pi^0$ final states.

It is easy to predict that these steps will become more and more common and new steps towards more exclusive channels will be taken in the future.

**Historical Interlude.** By the beginning of the 1980s all major decay modes of the $\tau$ lepton had been identified and the branching ratios measured. But, as pointed out by Truong [375], they did not sum to 100%. More than 5% was missing. This was called the $\tau$ puzzle, or the missing one-prong problem, as the largest discrepancy was found between the exclusive and inclusive one-prong branching ratios. There was a lot of speculation about the reasons for the deficit. It was obvious that no more major decay modes were expected from the Standard Model. Hence, the problem was either experimental or that there were new decay modes beyond the Standard Model. The deficit amounted to a little more than two standard deviations, if based purely on measurements, and increased up to seven standard deviations when some less precisely measured channels were supported by theoretical estimates (for discussions see [376–381]).

---

[9] An exclusive class is one which exclusively contains one decay channel of the $\tau$.

90     3. The Static Properties of the $\tau$

The problem was solved in 1990 when CELLO and ALEPH presented their global analysis at the $\tau$ workshop in Orsay [382, 383]. The CELLO analysis is the one described above [103], and the ALEPH results were published in [72]. Both collaborations presented measurements which were more precise in most channels than the existing results and they did not see any deficit. Mainly, their branching ratios into three pions (all charged, as well as one charged and two neutrals) were larger than the previous world averages, a tendency confirmed by later results.

Looking back, it is still interesting to identify the initial source of the problem. It was purely experimental. There are no decay modes beyond the Standard Model which contribute several percent. It can never be excluded that the whole puzzle was just a large statistical fluctuation, but it is more likely that undetected systematic errors in some of these measurements were the source of the deficit. They might have been small, compared to the statistical error of each measurement, but they caused the problem when many results were combined.

Figures 3.36 and 3.37 compare the measurements of the 1990 world averages of the three-pion channels with the current values. For both charge combinations the 1990 averages were about 2% too low, the largest part of the discrepancy.[10] In the all-charged mode the new CELLO result is already included in the average, pulling it towards higher values. Here one can also see that the scatter of the individual measurements is larger than expected from purely statistical fluctuations.

### 3.4.5 Summary of branching ratios

The 1998 Particle Data Group review [44] quotes 104 measured decay channels of the $\tau$ lepton plus upper limits on another 20 allowed modes.

Approximately 85% are one-prong modes including $K_S^0 \to \pi^+\pi^-$ (see Fig. 3.38). With 17.81% (e) and 17.37% ($\mu$), the leptonic decays make up about a third of all decays. The others are hadronic, out of which less than 3% are Cabibbo-suppressed decays. The largest resonances observed are the $\rho$ and the $a_1$. Together they make up almost half of all decays. There is a 12% branching ratio into one meson (including the kaon), 27% into two mesons, 19% into three, and then the branching ratios fall rapidly with multiplicity.

Figure 3.39 shows most of the allowed final states in nonstrange decays sorted by their mass. Figure 3.40 is the equivalent plot for the strange decays. The arrows indicate the major decay routes (those with more than a 10% branching ratio). The branching ratios of the $\tau$ into a final state are given in %, if measured. The numbers with an asterisk are branching ratios which include kaons as charged hadrons.

The figures do not include all possible states, to keep them readable.

---

[10] The three-prong topological branching ratio was also somewhat low, so that the deficit was associated with the one-prong channels only.

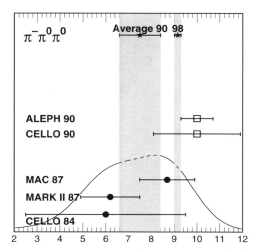

**Fig. 3.36.** The branching ratio $\tau \to \pi^{\pm} \pi^0 \pi^0 \nu_\tau$. The *filled circles* are the measurements included in the 1990 world average [384] together with their ideogram (*line*). The *open squares* are the CELLO and ALEPH results presented in Orsay [382, 383]. The *gray bars* represent the 1990 and 1998 world averages [44, 384]. All values in %

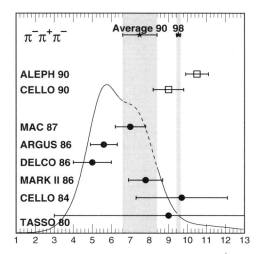

**Fig. 3.37.** The branching ratio $\tau \to 3\pi^{\pm} \nu_\tau$. The *filled circles* are the measurements included in the 1990 world average [384] together with their ideogram (*line*). The *open squares* are the CELLO and ALEPH results presented in Orsay [382, 383]. The *gray bars* represent the 1990 and 1998 world averages [44, 384]. The 1990 world average, but not the ideogram, already includes the CELLO 90 result. All values in %

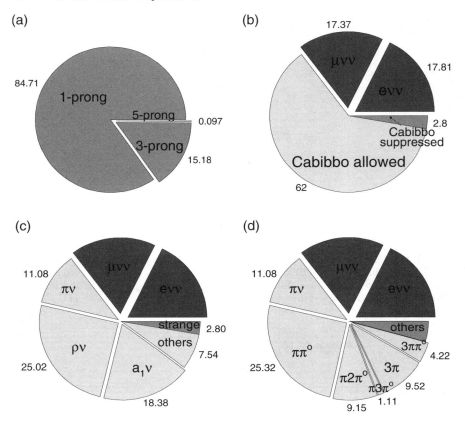

**Fig. 3.38.** Summary of the branching ratios of the $\tau$ lepton. (**a**) Topological branching ratios. (**b**) Leptonic and hadronic channels. (**c**) The major resonances observed and (**d**) the largest exclusive channels [44]. All numbers in %

- High-multiplicity states are missing, for example $\tau \to > 7\pi\,\nu_\tau$ or $\tau \to \eta > 4\pi\,\nu_\tau$. With that restriction the $\eta'\,3\,\pi$ state appears not to decay.
- All states with more than one kaon are missing. Therefore the $\Phi$ and the $f_1(1420)$ have no decay route either.
- The plots do not show the $\eta \to \gamma\gamma$ decay, but show its decay to pions.
- Some decay lines are dashed solely for readability.
- For the $f_0(1370)$ no branching ratios have been measured yet, thus no decays are entered.
- The $\sigma$ or $f_0(400-1200)$ is entered with a mass of 800 MeV.

3.4 Branching Ratios  93

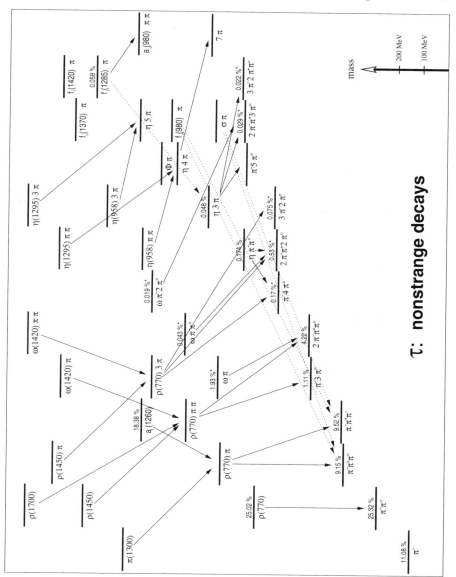

**Fig. 3.39.** Nonstrange channels of the $\tau^-$. See text

94   3. The Static Properties of the $\tau$

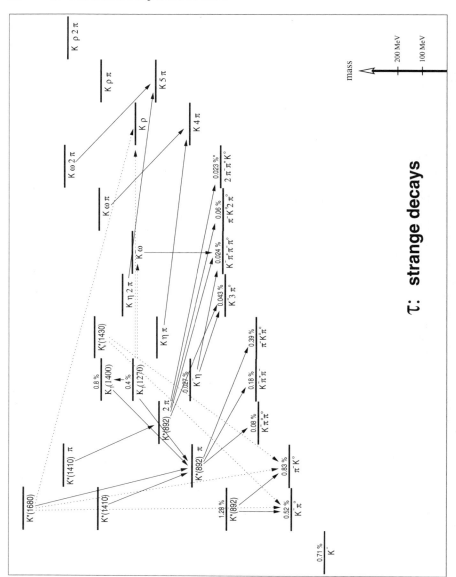

**Fig. 3.40.** Strange channels of the $\tau^-$. See text

# 4. Electroweak Physics at the $Z^0$ Pole

## 4.1 Precision Tests of the Standard Model

One may consider the precision tests of the Standard Model performed at LEP and the SLC the major achievement of these projects. The experiments confirmed the validity of the Standard Model to the level of loop corrections. The basic idea of these measurements is to determine the electroweak couplings in as many independent reactions as possible. This allows one to extract the remaining unknown parameters of the theory and still overconstrains the model. If the model under consideration appropriately describes nature, then one expects to find consistency between the different measurements, i.e. a single set of parameters is able to predict all measurements correctly. If the model is wrong, however, consistency would be a very unlikely accident. There are many excellent reviews on the topic, to which the reader is referred to for details [385–407].

The free parameters of the Standard Model are

- the coupling constants of the electromagnetic, weak, and strong interactions $\alpha_{\text{QED}}$,[1] $\alpha_{\text{weak}}$, and $\alpha_{\text{s}}$
- the masses of the fermions[2]
- the vacuum expectation value and mass of the Higgs field
- the four independent parameters of the Cabibbo–Kobayashi–Maskawa (CKM) matrix.

Some of these parameters were known prior to the LEP/SLC projects, for example $\alpha_{\text{QED}}$ and the masses of the charged fermions and lighter quarks. For some parameters, such as $\alpha_{\text{s}}$, the precision was substantially improved, and some, such as the mass of the top quark $m_{\text{t}}$, were determined for the first time.

The above list of parameters is somewhat arbitrary. It is transformed to a set of parameters more suitable for the description of the measurements. The weak coupling $\alpha_{\text{weak}}$ is replaced by the Fermi constant $G_{\text{F}}$, derived from the lifetime of the muon, and the vacuum expectation value of the Higgs

---

[1] The electromagnetic coupling constant $\alpha$ is labeled as $\alpha_{\text{QED}}$ in this section to distinguish it from $\alpha_{\text{weak}}$ and $\alpha_{\text{s}}$.
[2] Charged leptons and quarks; the neutrinos are assumed to be massless.

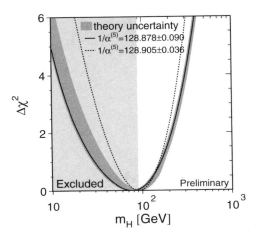

**Fig. 4.1.** Determination of the mass of the Higgs boson from the electroweak fit [385, 408]. The plot shows the variation of $\chi^2$ as a function of the Higgs mass. The preferred value is 84 GeV, with errors of $+91$ and $-51$ GeV. The *shaded area* is excluded by direct searches

field is replaced by the mass of the $Z^0$ boson $m_Z$, measured by scanning the lineshape of the $Z^0$ at LEP [408]. The mass of the Higgs boson $m_H$ is kept here, as well as $\alpha_{QED}$ and $\alpha_s$. The elements of the CKM matrix have little impact on these measurements and are quoted only for completeness.

A large number of measurements enter the electroweak fit, such as the mass and width of the $Z^0$ boson, the production cross sections and forward–backward asymmetries of leptons and quarks at the $Z^0$ peak, and the left–right asymmetry measured at the SLC [408]. There are also experimental inputs not obtained from LEP/SLC, for example the Weinberg angle measured in neutrino scattering [409] and the masses of the W boson and top quark from the TEVATRON experiments [410–414]. Figure 4.1 shows the result for the mass of the Higgs boson, the least-known parameter of the Standard Model, from the electroweak fit. A light Higgs is preferred. A mass below 89.8 GeV is excluded at the 95% confidence level by direct searches [415] and a mass beyond 280 GeV is excluded by the fit (95% confidence level). Figure 4.2 shows the excellent consistency of all the measurements with the predictions of the fit. The overall $\chi^2$ is 16.4 for 15 degrees of freedom.

Measurements from $\tau$ pairs contribute a significant amount to these tests despite the relatively small branching ratio of $Z^0 \to \tau^+\tau^-$ of $(3.360\pm0.015)\%$ [44]. The measurements of the production cross section and its forward–backward asymmetry provide a measurement of the coupling of the $Z^0$ to the initial electrons and final $\tau$ leptons. This is similar to the measurements performed with the other leptons and quarks. In addition the $\tau$ lepton provides the possibility to determine its polarization. This, together with the forward–backward asymmetry of the polarization, allows two more highly sensitive determinations of the couplings of the electrons and $\tau$ leptons to the $Z^0$. Then, with a polarized beam at SLC, there is the left–right asymmetry, another determination of the weak mixing angle from $\tau$ leptons (see Sect. 4.4).

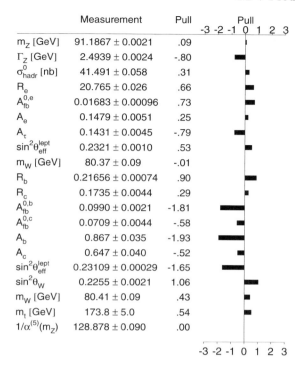

Fig. 4.2. Consistency of the overall Standard Model fit with the measurements [385, 408]. The 'pull' indicates the deviations of the measurements from the prediction in units of their standard deviations. The overall $\chi^2$ is 16.4 for 15 degrees of freedom

## 4.2 $\tau$ Production at the $Z^0$ Pole

In $e^+e^-$ collisions $\tau$ leptons are produced through the Feynman diagrams depicted in Fig. 4.3. There are two diagrams, one describing the exchange of a photon, the other the exchange of a $Z^0$ boson, plus the interference between the two. The strengths of the interactions are governed by the vector and axial-vector couplings as indicated in the figure. They are different for photons and $Z^0$ bosons, but they should be independent of the fermion type (i.e. $v_e = v_\tau$, $a_e = a_\tau$). They are given to lowest order in the Standard Model by[3]

$$v_f^\gamma = |Q_f|,$$
$$a_f^\gamma = 0,$$
$$v_f^Z = (I_f^3)_L - 2Q_f \sin^2\theta_W,$$
$$a_f^Z = (I_f^3)_L, \qquad (4.1)$$

where $Q_f$ is the electrical charge and $(I_f^3)_L$ the third component of the weak isospin of the left-handed fermion. This means that for $\tau$ leptons, $a_\tau^Z = -1/2$ and $v_\tau^Z$ is small; $v_\tau^Z$ depends on the Weinberg angle and is roughly $-0.037$.

---
[3] There are several different conventions about the normalization of the couplings to the $Z^0$.

98     4. Electroweak Physics at the $Z^0$ Pole

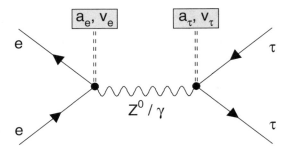

Fig. 4.3. Feynman diagram for the production of $\tau$ pairs in $e^+e^-$ collisions

Alternatively, one can use left- and right-handed couplings to describe the interaction:

$$g_L = -v_f - a_f,$$
$$g_R = +v_f - a_f. \quad (4.2)$$

In the coupling to the photon $g_L$ and $g_R$ are identical in magnitude, but for the $Z^0$ boson $g_L$ is larger (in magnitude) and therefore parity is violated.

From the Feynman diagram of Fig. 4.3 the cross section can be calculated (see e.g. [416, 417]). Unpolarized initial beams are assumed here. Beam polarization will be discussed in Sect. 4.4. The differential cross section is

$$\frac{d\sigma}{d\cos\theta_\tau} = F_0(s)\left(1+\cos^2\theta_\tau\right) + 2F_1(s)\cos\theta_\tau$$
$$- h_\tau\left[F_2(s)\left(1+\cos^2\theta_\tau\right) + 2F_3(s)\cos\theta_\tau\right]. \quad (4.3)$$

Here $\theta_\tau$ is the scattering angle of the $\tau^-$ with respect to the $e^-$ beam, $h_\tau$ the helicity of the negative $\tau$ in the event and $s$ the center-of-mass energy squared. The helicity of the positive $\tau$ is opposite in sign, ignoring the mass of the $\tau$.

The functions $F_i(s)$ describe the energy dependence of the couplings. They are given in Table 4.1. The function $\chi(s)$ determines the lineshape of the $Z^0$:

$$\chi(s) = \frac{1}{2\sin\theta_W \cos\theta_W} \frac{s}{s - m_Z^2 + is\left(\Gamma_Z/m_Z\right)}. \quad (4.4)$$

There are in general three contributions to the $F_i(s)$: from the photon diagram, the $Z^0$ diagram, and their interference. The smallness of $v_\ell$ implies that the contribution from the interference to $F_0$ and that of the $Z^0$ exchange to $F_1$ are negligible for leptonic final states. The functions $F_i(s)$ are depicted in Fig. 4.4.

From the cross section of (4.3) one can now derive the various observables. These are the total cross section $\sigma$, the forward–backward asymmetry of the cross section $A^{FB}$, the average longitudinal polarization of the negative $\tau$ lepton $P_\tau = -A_{pol}$, and the forward–backward asymmetry of the polarization $A^{FB}_{pol}$.

4.2 τ Production at the $Z^0$ Pole    99

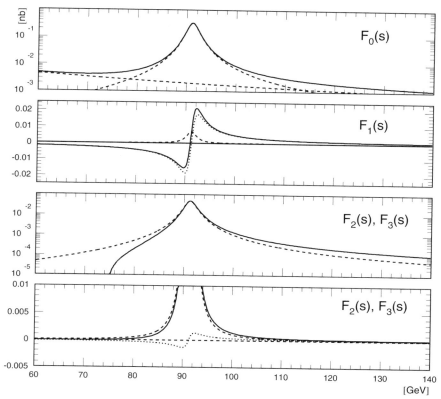

**Fig. 4.4.** The functions $F_i$ from (4.3). *Horizontal axis*: center-of-mass energy in GeV; *vertical axis*: $F_i$ in nanobarns. The *solid curve* is the total function, the *dashed lines* represent $\gamma$ and $Z^0$ exchange, and the *dotted line* the interference term. $F_2$ and $F_3$ are identical under the assumption of lepton universality and are shown twice on different scales ($m_Z = 91.1867$ GeV, $\Gamma_Z = 2.4939$ GeV, $\sin^2\theta_W = 0.23155$)

**Table 4.1.** The functions $F_i$ from (4.3) in lowest order. $v_\ell$ and $a_\ell$ are the couplings to the $Z^0$

| | $\gamma$ exchange + | interference + | $Z^0$ exchange |
|---|---|---|---|
| $F_0(s) = \frac{\pi\alpha^2}{4s}$ [ | $Q_e^2 Q_\tau^2$ | $+\ 2\mathrm{Re}\,\chi(s) Q_e Q_\tau v_e v_\tau\ +$ | $\|\chi(s)\|^2 \left(v_e^2 + a_e^2\right)\left(v_\tau^2 + a_\tau^2\right)$ ] |
| $F_1(s) = \frac{\pi\alpha^2}{4s}$ [ | | $+\ 2\mathrm{Re}\,\chi(s) Q_e Q_\tau a_e a_\tau\ +$ | $\|\chi(s)\|^2\ 2 v_e a_e\ 2 v_\tau a_\tau$ ] |
| $F_2(s) = \frac{\pi\alpha^2}{4s}$ [ | | $+\ 2\mathrm{Re}\,\chi(s) Q_e Q_\tau v_e a_\tau\ +$ | $\|\chi(s)\|^2 \left(v_e^2 + a_e^2\right) 2 v_\tau a_\tau$ ] |
| $F_3(s) = \frac{\pi\alpha^2}{4s}$ [ | | $+\ 2\mathrm{Re}\,\chi(s) Q_e Q_\tau a_e v_\tau\ +$ | $\|\chi(s)\|^2\ 2 v_e a_e \left(v_\tau^2 + a_\tau^2\right)$ ] |

$$\sigma = \sum_{h_\tau} \int \frac{\mathrm{d}\sigma}{\mathrm{d}\cos\theta_\tau} \, \mathrm{d}\cos\theta_\tau,$$

$$A^{\mathrm{FB}} = \frac{1}{\sigma}\left[\sigma(\cos\theta_\tau > 0) - \sigma(\cos\theta_\tau < 0)\right],$$

$$A_{\mathrm{pol}} = \frac{1}{\sigma}\left[\sigma(h_\tau = +1) - \sigma(h_\tau = -1)\right],$$

$$A^{\mathrm{FB}}_{\mathrm{pol}} = \frac{1}{\sigma}\left[A_{\mathrm{pol}}(\cos\theta_\tau > 0) - A_{\mathrm{pol}}(\cos\theta_\tau < 0)\right]. \tag{4.5}$$

Inserting (4.3) gives

$$\begin{aligned}
\sigma &= \tfrac{16}{3} F_0(s), \\
A^{\mathrm{FB}} &= \tfrac{3}{4}\frac{F_1(s)}{F_0(s)} \approx \tfrac{3}{4} A_e A_\tau, \\
A_{\mathrm{pol}} &= -\frac{F_2(s)}{F_0(s)} \approx -A_\tau, \\
A^{\mathrm{FB}}_{\mathrm{pol}} &= -\tfrac{3}{4}\frac{F_3(s)}{F_0(s)} \approx -\tfrac{3}{4} A_e,
\end{aligned} \tag{4.6}$$

where the last column holds only approximately on top of the $Z^0$ resonance and $A_\ell$ is defined as

$$A_\ell = \frac{2\, v_\ell\, a_\ell}{v_\ell^2 + a_\ell^2} \approx 2\, \frac{v_\ell}{a_\ell}. \tag{4.7}$$

The cross section $\sigma$ is proportional to $v_\ell^2 + a_\ell^2$ and therefore essentially measures $a_\ell$. The asymmetries, on the other hand, determine $v_\ell$. The forward–backward asymmetry $A^{\mathrm{FB}}$ is quadratic in the small quantity $A_\ell$ and thus less sensitive than the polarizations, which are linear in $A_\ell$. A study of the angular dependence of the polarization allows one to separate the couplings to the initial electrons from those to the final $\tau$ leptons.

The basic features of the asymmetries can be understood by the simple means of spin arrows as sketched in Fig. 4.5. An electron coming from the left annihilates with a positron coming from the right, making a $Z^0$ boson in the center of each figure. There are two possibilities for the orientation of the spins of the electrons. They can either both point to the right (upper row) or both point to the left. In the first case the initial electron has its spin pointing in the direction of flight, i.e. it has positive helicity, and therefore the coupling is dominantly right-handed.[4] Therefore the upper row is labeled by $g^{\mathrm{e}}_{\mathrm{R}}$. The lower row has left-handed couplings of the electron to the $Z^0$ boson and is labeled by $g^{\mathrm{e}}_{\mathrm{L}}$. Then the $Z^0$ decays into a pair of $\tau$ leptons, which can be emitted either backward (left column) or forward (right column). For simplicity of drawing, the $\tau$ leptons are shown at a slight angle to the initial

---

[4] The labels 'left' and 'right' of the couplings always refer to the particle, not the antiparticle.

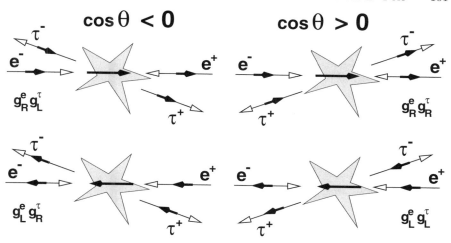

**Fig. 4.5.** Orientation of spins in $\tau$ production in forward (*right column*) and back scattering (*left column*). The *open arrow* indicates the direction of the momentum, the *solid arrow* the spin. The *large arrow* in the center of each plot indicates the spin of the $Z^0$

beam direction, but they are assumed to travel exactly in the direction of the beams. Situations at intermediate scattering angles can be extrapolated from these extreme cases. The orientation of the spins of the $\tau$ leptons is fixed by spin conservation.[5] For example, in the upper left plot the backward scattering of a $\tau$ from an initial right-handed electron leads to a left-handed $\tau$. This process is proportional to $g_R^e g_L^\tau$ as indicated in the figure.

The asymmetries can be derived from these simple arguments. For the forward–backward asymmetry $A^{\rm FB}$ one has to add the two processes of the right column (forwardscattering) and subtract the two on the left. One gets

$$A^{\rm FB} = \frac{(g_R^e g_R^\tau + g_L^e g_L^\tau) - (g_R^e g_L^\tau + g_L^e g_R^\tau)}{(g_R^e g_R^\tau + g_L^e g_L^\tau) + (g_R^e g_L^\tau + g_L^e g_R^\tau)}$$
$$= \frac{g_R^e - g_L^e}{g_R^e + g_L^e} \frac{g_R^\tau - g_L^\tau}{g_R^\tau + g_L^\tau}$$
$$\propto A_e A_\tau. \qquad (4.8)$$

For the polarization one has to add the two processes which lead to $\tau$ leptons with positive helicity (upper right and lower left) and subtract those with negative helicities:

---

[5] There is no angular momentum other than spin in these events, i.e. conservation of angular momentum implies conservation of spin.

102    4. Electroweak Physics at the $Z^0$ Pole

$$A_{\text{pol}} = \frac{(g_R^e g_R^\tau + g_L^e g_R^\tau) - (g_R^e g_R^\tau + g_L^e g_R^\tau)}{(g_R^e g_R^\tau + g_L^e g_R^\tau) + (g_R^e g_R^\tau + g_L^e g_R^\tau)}$$

$$= \frac{g_R^e + g_L^e}{g_R^e + g_L^e} \frac{g_R^\tau - g_R^\tau}{g_R^\tau + g_R^\tau}$$

$$\propto A_\tau.  \qquad (4.9)$$

For the asymmetry of the polarization,

$$A_{\text{pol}}^{\text{FB}} = \frac{(g_R^e g_R^\tau - g_L^e g_R^\tau) - (g_L^e g_R^\tau - g_R^e g_R^\tau)}{(g_R^e g_R^\tau + g_L^e g_R^\tau) + (g_L^e g_R^\tau + g_R^e g_R^\tau)}$$

$$= \frac{g_R^e - g_L^e}{g_R^e + g_L^e} \frac{g_R^\tau + g_R^\tau}{g_R^\tau + g_R^\tau}$$

$$\propto A_e.  \qquad (4.10)$$

## 4.3 Cross Sections and Asymmetries

Cross sections $\sigma$ and forward–backward asymmetries $A^{\text{FB}}$ can be measured for all three charged-lepton flavors, the electron channel $e^+e^- \to e^+e^-$ being somewhat complicated by the additional t-channel contribution. The six measurements together allow for a determination of the vector and axial-vector couplings of each lepton flavor to the $Z^0$ boson. They have been measured over a wide range of energies from the b threshold to the highest energies at LEP2 [74, 104, 165, 204, 240, 248, 254, 262, 408, 418].

The predictions from the Standard Model have been calculated with all the radiative corrections necessary to match the experimental precision [386]. Figure 4.6 shows the production cross section for $\tau$ pairs across the $Z^0$ resonance and Fig. 4.7 the forward–backward asymmetry.[6] There is a large effect visible in both plots from initial-state photon radiation. When a photon is radiated from the initial electron or positron, the effective center-of-mass energy $s'$ available for the hard process is reduced. This diminishes the cross section on top of the resonance by roughly 25% and shifts the peak position to slightly higher masses. For beam energies at or below the $Z^0$ mass the cross section at $s'$ is smaller than at $s$ and therefore initial-state radiation reduces the cross section. However, above the peak, initial-state radiation can reduce $\sqrt{s'}$ to the $Z^0$ mass and the cross section is increased. Other radiative corrections are small and would not be visible on the scale of this plot.

For the forward–backward asymmetry (Fig. 4.7) there is not much of an effect below the peak. Initial-state radiation is suppressed owing to the reduced cross section at $s'$. However, above the peak there are many events with reduced $s'$, which contribute with a smaller asymmetry.

---

[6] The plots have been produced with the program ZFITTER [419, 420], which calculates Standard Model predictions for the electroweak fits. Version 5.0, $m_Z = 91.1867$ GeV, $m_t = 171.1$ GeV, and $m_H = 100$ GeV.

4.3 Cross Sections and Asymmetries    103

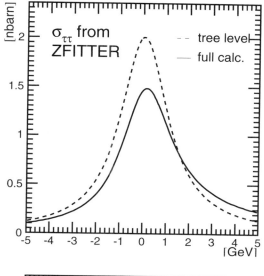

**Fig. 4.6.** Cross section for $\tau$ production across the $Z^0$ pole. *Horizontal axis*: center-of-mass energy with respect to the $Z^0$ mass. *Dashed line*: tree-level calculation. *Solid line*: full radiative corrections. Calculated with ZFITTER [419]

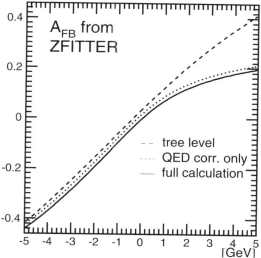

**Fig. 4.7.** Forward–backward asymmetry of $\tau$ production across the $Z^0$ pole. *Horizontal axis*: center-of-mass energy with respect to the $Z^0$ mass. *Dashed line*: tree-level calculation. *Dotted line*: QED corrections only. *Solid line*: full radiative corrections. Calculated with ZFITTER [419]

From the experimental point of view the understanding of the selection efficiency and background contamination is the key issue in the cross section measurement. The selection efficiency is determined from Monte Carlo simulated events using the KORALZ program [295]. A careful comparison of simulated and real events is necessary to test the simulation and reveal possible biases. Figures 4.8 and 4.9 show a compilation of cross section measurements at low energies by the DELPHI collaboration [421, 422] together with the measurements by DELPHI up to $\sqrt{s} = 189\,\text{GeV}$.

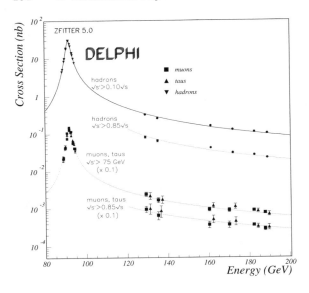

**Fig. 4.8.** Cross section for the production of $\mu$ and $\tau$ pairs (*dotted*) and hadrons (*solid*) at and above the $Z^0$ pole (DELPHI collaboration [421]). The cut on the visible energy on the upper curve suppresses the return to the $Z^0$ peak. *Symbols with error bars*, data; *lines*, Standard Model prediction

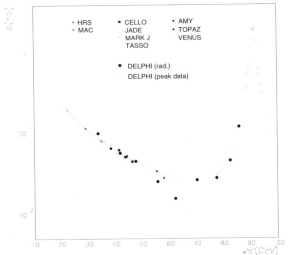

**Fig. 4.9.** Cross section for the production of muon pairs from low energies to the $Z^0$ pole (DELPHI collaboration [422]). *Symbols with error bars*, data; *line*, Standard Model prediction

For the measurement of the forward–backward asymmetry from $\tau$ pairs, most of the uncertainties in the determination of the selection efficiency cancel. Remaining biases from initial-state radiation or the asymmetry of the polarization are typically very small. The subtraction of Bhabha background is critical owing to its large intrinsic asymmetry from the t-channel contribution. The thrust axis is used to approximate the $\tau$ direction in the selected events. The events are required to have a vanishing net charge and the thrust axis is assigned an orientation in the direction of the negative $\tau$. A cut on

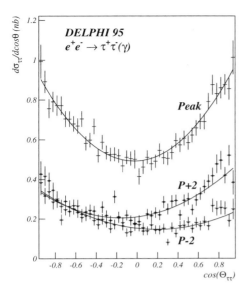

**Fig. 4.10.** Angular distribution of $\tau$ pairs produced at the $Z^0$ peak and 2 GeV above and below (DELPHI collaboration [421]). The *solid line* is the result of the fit (see text)

the acollinearity removes events with hard initial-state radiation and, after that, the angular distribution of the remaining events can be fitted by the tree-level formula

$$\frac{d\sigma}{d\cos\theta_\tau} = c\left[\frac{3}{8}\left(1 + \cos^2\theta_\tau\right) + A^{FB}\cos\theta_\tau\right]. \tag{4.11}$$

Radiative corrections to the tree-level formula introduce only a negligible bias. The fit has to be performed separately for data taken at different center-of-mass energies. Figure 4.10 shows the angular distribution of events recorded by the DELPHI experiment at the peak and 2 GeV above, and below [422]. The measurements over a wide range of $s$ are shown in Fig. 4.11. The other LEP collaborations have obtained similar results [408, 418].

## 4.4 Electroweak Physics at the SLC

At the Stanford Linear Collider (SLC) at SLAC the electron beam can be longitudinally polarized in both directions. Figure 4.12 shows the degree of polarization achieved over the years. The polarized beam allows for additional, even more sensitive electroweak measurements, compensating for the lower luminosity compared to LEP.

The cross section for $\tau$ production from (4.3) can be generalized to a polarized electron beam by making the following replacements in Table 4.1:

$$(g_R^e)^2 \to (1 + P_e)(g_R^e)^2,$$
$$(g_L^e)^2 \to (1 - P_e)(g_L^e)^2$$

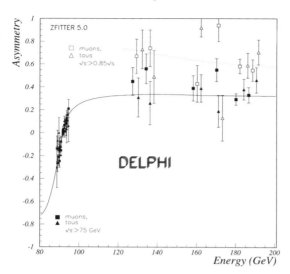

**Fig. 4.11.** Forward–backward asymmetries $A^{\rm FB}$ of $\mu$ and $\tau$ pairs at and above the $Z^0$ pole (DELPHI collaboration [421]). The cut on the visible energy on the lower curve suppresses the return to the $Z^0$ peak. *Symbols with error bars*, data; *lines*, Standard Model prediction

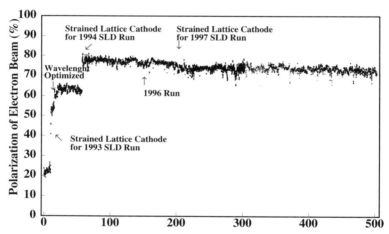

**Fig. 4.12.** Beam polarization at the SLC as a function of the number of $Z^0$s collected ($\times 1000$)

or

$$\begin{aligned} v_e^2 + a_e^2 &\to v_e^2 + a_e^2 - 2\,P_e\,v_e a_e, \\ 2\,v_e a_e &\to 2\,v_e a_e - P_e\left(v_e^2 + a_e^2\right). \end{aligned} \tag{4.12}$$

$P_e$ is the polarization of the electron beam. It is positive when the spin of the electrons is pointing preferentially in the direction of the beam and negative otherwise.

Two new observables can be defined, the left–right asymmetry $A_{\rm LR}$ and the polarized forward–backward asymmetry $A_{\rm LR}^{\rm FB}$:

$$A_{\text{LR}} = \frac{\sigma_{\text{L}} - \sigma_{\text{R}}}{\sigma_{\text{L}} + \sigma_{\text{R}}},$$
$$A_{\text{LR}}^{\text{FB}} = A_{\text{L}}^{\text{FB}} - A_{\text{R}}^{\text{FB}}, \tag{4.13}$$

where the subscripts L and R indicate that the quantities are to be measured with negative and positive beam polarization, respectively. Neglecting the contribution from the photon exchange and the interference term,[7] the differential cross section reads

$$\frac{\text{d}\sigma}{\text{d}\cos\theta_\tau} \propto (1 - P_e A_e)\left(1 + \cos^2\theta_\tau\right) + 2\left(A_e - P_e\right) A_f \cos\theta_\tau. \tag{4.14}$$

Inserting the cross section into the definition of the observables and assuming the negative and positive polarizations to be identical in magnitude gives

$$A_{\text{LR}} = |P_e| A_e,$$
$$A_{\text{LR}}^{\text{FB}} = \frac{3}{4} |P_e| A_f, \tag{4.15}$$

i.e. these quantities measure directly the couplings of the fermions to the $Z^0$.

The left–right asymmetry $A_{\text{LR}}$ is a very simple observable from the experimental point of view. If the luminosity is equal for positive and negative polarization the measurement reduces to a counting experiment:

$$A_{\text{LR}} = \frac{N_{\text{L}} - N_{\text{R}}}{N_{\text{L}} + N_{\text{R}}}, \tag{4.16}$$

where $N_{\text{L}}$ and $N_{\text{R}}$ are the numbers of events recorded with negative and positive beam polarizations, respectively. The normalization to the luminosity cancels in the ratio, as well as any systematic effects from the selection (efficiency, background, etc.) as long as they are identical for both polarizations. As $A_{\text{LR}}$ is independent of the couplings to the final state it is not even necessary to separate the different final states of $Z^0$ decays; one only has to separate $Z^0$ production from non-$Z^0$ events (two-photon background, cosmic rays, etc.) [423].

The experimental challenge is an accurate determination of the beam polarization, which is performed at SLC by a Compton polarimeter analyzing the electron beam immediately after it passed through the interaction region. The experimental setup is shown in Fig. 4.13. The relative accuracy achieved is slightly better than 1% [423–426].

The measurement of the polarized forward–backward asymmetry can be performed as a counting experiment too. Restricting the polar-angle acceptance ($|\cos\theta| < 0.8$ for SLD) reduces the sensitivity and has to be taken into account. For this measurement a separation between the decay modes of the $Z^0$ to different fermion species is desirable, as $A_{\text{LR}}^{\text{FB}}$ measures the coupling to these fermions.

---

[7] A small correction will be applied to the final result to correct for these effects.

108   4. Electroweak Physics at the $Z^0$ Pole

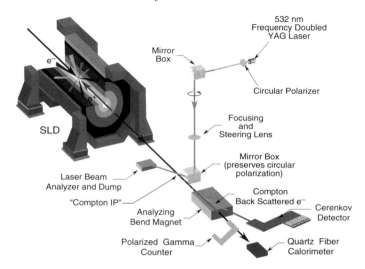

**Fig. 4.13.** The Compton polarimeter at the SLC to measure the longitudinal polarization of the electron beam. A laser beam is scattered off the electrons after they have passed through the SLD detector. Both the scattered electron and the backscattered photon are detected. The last bending magnet of the positron beam serves as an analyzing magnet for the scattered electrons. For more details see [424]

In practice the SLD collaboration selects $e^+e^- \to e^+e^-$, $\mu^+\mu^-$, $\tau^+\tau^-$, and hadrons and fits their angular distributions separately for negative and positive beam polarization, giving both $A_{\rm LR}$ and $A_{\rm LR}^{\rm FB}$. A maximum-likelihood fit is performed, taking into account photon and $Z^0$ exchange, the t-channel contribution to the $e^+e^-$ final state, and radiative corrections [427]. Figure 4.14 shows the fits to the leptons from the latest data [428]. The angular distributions are clearly distinguishable between the two beam polarizations ($A_{\rm LR}^{\rm FB} \neq 0$), and the cross section, i.e. the integral over the curves, is higher for negative (left) polarization ($A_{\rm LR} \neq 0$). The effect of the t channel is visible in the electron channel for large positive $\cos\theta$.

Figure 4.15 shows the result that the SLD collaboration has achieved for $A_\ell$ in comparison with the results from the asymmetries at LEP. There is good consistency between the couplings to the three lepton species, supporting the assumption of lepton universality, and there is good agreement between LEP and SLC at a similar level of precision.

## 4.5 Analyzing the Spin of a $\tau$ Lepton

Within the unique features of the $\tau$ is the possibility to measure the orientation of its spin. It reveals this information through the spectra and angular distributions of its decay products. This information is not obstructed by

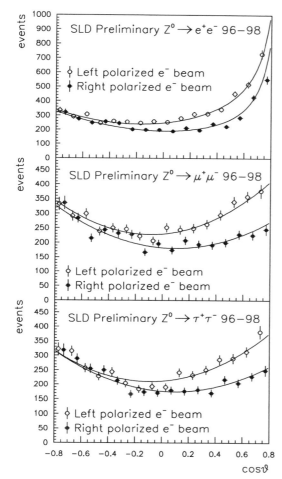

**Fig. 4.14.** Angular distribution of lepton pairs recorded by the SLD collaboration with positively and negatively polarized electron beams [428]

QCD, not even in hadronic $\tau$ decays. The spin measurement opens a variety of new possibilities to study high-energy interactions with $\tau$ leptons. One of them is the determination of the couplings of the $Z^0$ boson to $\tau$ leptons through their polarization (see next section). But first it will be described how the spin of the $\tau$ can be analyzed [23, 302, 429, 430].

### 4.5.1 The Pion Channel

The decay of the $\tau$ lepton to a single pion and a neutrino is the simplest case. (For the sake of simplicity the discussion will be restricted to the negative $\tau$; the positive is analogous.) There is no orbital angular momentum in this two-body decay and the pion carries no spin. Therefore conservation of angular momentum requires the neutrino to adopt the spin of the $\tau$. Taking

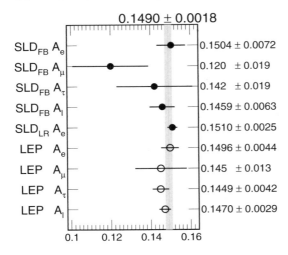

**Fig. 4.15.** Determination of $A_l$ from the asymmetry measurements at SLC and LEP [428]. The values labeled $A_l$ combine all data under the assumption of lepton universality. The first four SLD numbers are derived from $A_{LR}^{FB}$, the fifth from $A_{LR}$. The LEP numbers are derived from the cross sections, $A^{FB}$, $A_{pol}$, and $A_{pol}^{FB}$

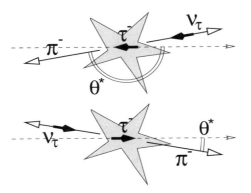

**Fig. 4.16.** The decay $\tau^- \to \pi^- \nu_\tau$ in the rest frame of the $\tau$. The *solid arrows* mark the spins of the particles, the *open arrows* their momentum. The *double arc* indicates the Gottfried–Jackson angle $\theta^*$. Boosting the decay in the direction of the *arrow on the dashed line* corresponds to the decay of a $\tau$ with negative helicity in the *upper plot* and positive helicity in the *lower plot*

further into account that the neutrino has to be left-handed, it is preferentially emitted against the direction of the $\tau$ spin as indicated in Fig. 4.16. The pion is emitted in the opposite direction (conservation of momentum). The transformation from the rest frame of the $\tau$ into the laboratory frame determines the helicity of the $\tau$ under consideration. If the dashed arrow in Fig. 4.16 indicates the direction of flight of the $\tau$ in its rest frame,[8] then boosting along this arrow gives a $\tau$ with negative helicity for the upper plot and positive for the lower. Now boosting the pion in the direction of the (dashed) arrow gives a low-energy pion in the upper plot and a high-energy one in the lower. Hence a pion with low energy in the laboratory indicates a negative helicity of the $\tau$, and a pion with high energy a positive helicity.

---

[8] To define the direction of flight of the $\tau$ in its rest frame might sound strange. It is of course not moving in this frame. More precisely, this direction means the opposite of the direction of the laboratory frame as viewed from the rest frame.

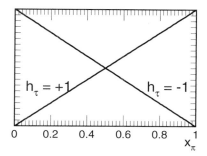

**Fig. 4.17.** Spectrum of pions from polarized $\tau$ leptons in arbitrary units; $h_\tau = +1$ corresponds to 100% positively polarized $\tau$ leptons, $h_\tau = -1$ to negative polarization. The spectrum is given in terms of the quantity $x_\pi = E_\pi/E_{\text{beam}}$

To be quantitative: the angular distribution of the neutrino in the rest frame of the $\tau$ is given by the overlap between the spin parts of the wave function of the neutrino, with negative helicity, and the wave function of the $\tau$:

$$\frac{d\Gamma}{d\tilde{\theta}} \propto \left|\langle \tau_{+/-} | \nu_L(\tilde{\theta})\rangle\right|^2, \tag{4.17}$$

where $\tau_{+/-}$ represents a spinor with helicity $+/-$ in the $z$ direction and $\nu_L(\tilde{\theta})$ one with negative eigenvalue in the direction $\tilde{\theta}$ with respect to the $z$ axis. Replacing the angle $\tilde{\theta}$ of the neutrino by the decay angle $\theta^*$ of the pion ($\theta^* = \pi - \tilde{\theta}$), one gets[9]

$$\frac{d\Gamma}{d\cos\theta^*} \propto \frac{1}{2}\left(1 + h_\tau \cos\theta^*\right). \tag{4.18}$$

Boosting the pion into the laboratory frame, ignoring terms of order $(m_\pi/m_\tau)^2$ ($\gamma_\tau = E_{\text{beam}}/m_\tau$, $\beta_\tau \approx 1$), gives the following relations:

$$E_\pi^* = \frac{m_\tau^2 + m_\pi^2}{2\,m_\tau} \approx \frac{1}{2}\,m_\tau,$$

$$p_\pi^* = \frac{m_\tau^2 - m_\pi^2}{2\,m_\tau} \approx \frac{1}{2}\,m_\tau,$$

$$\cos\theta^* \approx 2\,\frac{E_\pi}{E_{\text{beam}}} - 1 = 2\,x_\pi - 1. \tag{4.19}$$

All quantities with an asterisk refer to the $\tau$ rest frame; $E_{\text{beam}}$ is the energy of the initial beam and therefore of the $\tau$. This yields a spectrum

$$\frac{d\Gamma}{dx_\pi} \propto 1 + h_\tau\,(2x_\pi - 1). \tag{4.20}$$

The quantity $x_\pi$ is the energy of the pion in the laboratory frame scaled to the beam energy. Its range is 0 to 1. The spectrum is depicted in Fig. 4.17.

The same spectrum could also be derived straightforwardly from the Feynman diagram of Fig. 4.18. The matrix element is

---
[9] To calculate the probability, see for example [431].

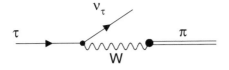

Fig. 4.18. Feynman diagram of the decay $\tau^- \to \pi^- \nu_\tau$

$$i\mathcal{M} = \frac{G_F}{\sqrt{2}} \langle \bar{u}(\nu_\tau) | \gamma^\mu (1 - \gamma_5) | u(\tau) \rangle J^\pi_\mu,$$
$$J^\pi_\mu = \cos\theta_C f_\pi \pi_\mu, \qquad (4.21)$$

where the Cabibbo angle is $\theta_C$, the decay constant of the pion is $f_\pi$, and its four-momentum is $\pi_\mu$.[10] The spectrum is

$$\frac{d\Gamma}{d\cos\theta^*} = \frac{G_F^2 f_\pi^2 \cos^2\theta_C}{32\pi} m_\tau^3 \left(1 - \frac{m_\pi^2}{m_\tau^2}\right)^2 \left(1 + h_\tau \cos\theta^*\right). \qquad (4.22)$$

The polarization of a sample of $\tau$ leptons is then defined as

$$P_\tau = \frac{N(h_\tau = +1) - N(h_\tau = -1)}{N(h_\tau = +1) + N(h_\tau = -1)}, \qquad (4.23)$$

where $N(h_\tau = +1)$ is the number of events with a $\tau^-$ lepton with positive helicity and $N(h_\tau = -1)$ with negative helicity.

### 4.5.2 More Hadronic Decays

The situation becomes slightly more complicated when the hadron produced from the $\tau$ decay carries spin, too. Figure 4.19 illustrates the situation for the decay $\tau^- \to \rho^- \nu_\tau$. The spin of the $\rho$ meson can have three different orientations with respect to its direction of flight: in, perpendicular to and against it. Adding the spin of the neutrino gives values of $+3/2$, $+1/2$, and $-1/2$, respectively, for the spin of the final state in the direction of the momentum of the $\rho$. The $+3/2$ component cannot be produced from the decay of a $\tau$; the other two are sketched in Fig. 4.19. The situation in the upper plot with the $\rho$ spin perpendicular (longitudinal $\rho$) is exactly that of the pion decay discussed above. But the situation in the lower plot (transverse $\rho$) is opposite and washes out the information about the $\tau$ spin, if only the $\rho$ meson as a whole is observed.

---

[10] When averaging over the spin directions of the $\tau$, one has to explicitly include the projector onto a certain helicity $\not{p} + m_\tau \gamma_5 \not{s}$, $s = (0; 0, 0, h_\tau)$.

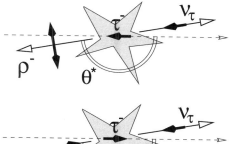
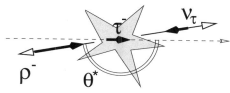

**Fig. 4.19.** The decay $\tau^- \to \rho^- \nu_\tau$ in the rest frame of the $\tau$

Quantitatively, we have [432, 433]

$$\frac{d\Gamma}{d\cos\theta^*}(\rho_T) \propto 1 - h_\tau \cos\theta^*,$$

$$\frac{d\Gamma}{d\cos\theta^*}(\rho_L) \propto \frac{m_\tau^2}{2\,m_\rho^2}\left(1 + h_\tau \cos\theta^*\right). \qquad (4.24)$$

The transverse ($\rho_T$) and longitudinal ($\rho_L$) polarization states of the $\rho$ create exactly opposite angular distributions. The relative strength of the two amplitudes depends on the mass of the meson produced. Adding the two amplitudes gives

$$\frac{d\Gamma}{d\cos\theta^*} \propto \frac{1}{2}\left(1 + \alpha_\rho h_\tau \cos\theta^*\right), \qquad (4.25)$$

with the additional factor $\alpha_\rho = 0.46$ compared to (4.18). As $\alpha$ decreases from 1, the sensitivity to the helicity of the $\tau$ lepton weakens.

When boosting the spectrum into the laboratory frame, terms of order $m_\rho^2/m_\tau^2$ cannot be neglected. The spectrum is shown in Fig. 4.20 and given by[11]

$$\frac{1}{\Gamma_\rho}\frac{d\Gamma}{dx_\rho}(\rho_T) = \frac{m_\tau^2 m_\rho^2}{(m_\tau^2 - m_\rho^2)(m_\tau^2 + 2m_\rho^2)}$$

$$\times \left[\frac{m_\tau^2}{m_\rho^2}\sin^2\omega + 1 + \cos^2\omega + P_\tau \cos\theta^*\right.$$

$$\left.\times \left(\frac{m_\tau^2}{m_\rho^2}\sin^2\omega - \frac{m_\tau}{m_\rho}\sin 2\omega \tan\theta^* - 1 - \cos^2\omega\right)\right], \quad (4.26)$$

$$\frac{1}{\Gamma_\rho}\frac{d\Gamma}{dx_\rho}(\rho_L) = \frac{m_\tau^2 m_\rho^2}{(m_\tau^2 - m_\rho^2)(m_\tau^2 + 2m_\rho^2)}$$

---

[11] For an actual data analysis $m_\rho^2$ should be replaced by $Q^2$, the invariant mass of the two pions as observed event by event.

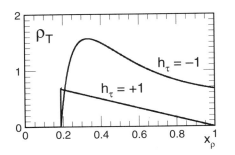 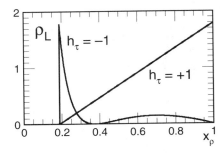

**Fig. 4.20.** Spectrum of $\rho$ mesons from polarized $\tau$ leptons; $h_\tau = +1$ corresponds to 100% positively polarized $\tau$ leptons, $h_\tau = -1$ to negative polarization. The spectra are given in the quantity $x_\rho = E_\rho/E_{\text{beam}}$. They are normalized to the total decay width of $\tau^- \to \rho^- \nu_\tau$. Left plot: transverse $\rho$. Right plot: longitudinal $\rho$ [432, 433]

$$\times \left[ \frac{m_\tau^2}{m_\rho^2} \cos^2\omega + \sin^2\omega + P_\tau \cos\theta^* \right.$$
$$\left. \times \left( \frac{m_\tau^2}{m_\rho^2} \cos^2\omega + \frac{m_\tau}{m_\rho} \sin 2\omega \tan\theta^* - \sin^2\omega \right) \right], \quad (4.27)$$

$$\cos\omega = \frac{(m_\tau^2 - m_\rho^2) + (m_\tau^2 + m_\rho^2)\cos\theta^*}{(m_\tau^2 + m_\rho^2) + (m_\tau^2 - m_\rho^2)\cos\theta^*},$$

$$\cos\theta^* = \frac{2x_\rho - 1 - m_\rho^2/m_\tau^2}{1 - m_\rho^2/m_\tau^2},$$

$$x_\rho = \frac{E_\rho}{E_{\text{beam}}}. \quad (4.28)$$

To distinguish whether the $\rho$ meson is in a transversely or longitudinally polarized state the decay of the $\rho$ must be analyzed. As the spin of the $\rho$ is transformed into orbital angular momentum in the decay, the pions are emitted preferentially perpendicular to the $\rho$ spin, affecting the splitting of the energy between them in the laboratory. An angle $\beta$ can be defined which is the angle between the charged pion and the direction of the $\rho$ in the $\rho$ rest frame, which separates the two polarization states of the $\rho$:

$$\cos\beta = \frac{m_\rho}{\sqrt{m_\rho^2 - 4m_\pi^2}} \frac{E_{\pi^-} - E_{\pi^0}}{|\boldsymbol{p}_{\pi^-} + \boldsymbol{p}_{\pi^0}|}. \quad (4.29)$$

Figure 4.21 shows the spectrum as two-dimensional distributions in $\cos\theta^*$ and $\cos\beta$. Now the two helicities are clearly distinguishable. The gain in sensitivity can be computed, defining the sensitivity as

$$S = \frac{1}{\sigma\sqrt{N}}, \quad (4.30)$$

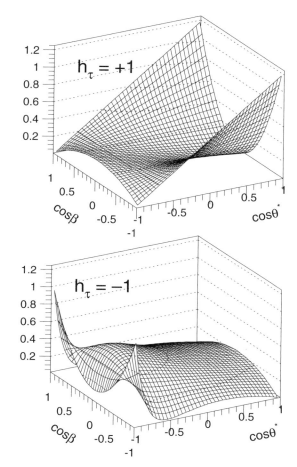

**Fig. 4.21.** Two-dimensional spectra of $\rho$ mesons from polarized $\tau$ leptons. The two angles describe the orientation of the $\rho$ in the $\tau$ rest frame ($\theta^*$) and of the charged pion in the $\rho$ rest frame ($\beta$) [432, 433]. The spectra are normalized to the total decay width

where $N$ is the number of events and $\sigma$ the expected error of the polarization measurement at $P_\tau = 0$. The numbers are given in Table 4.2 [434]. The sensitivity of the $\tau \to \rho \nu_\tau$ decays is less than half as good as for $\tau^- \to \pi^- \nu_\tau$, if only the energy (or equivalently $\cos\theta^*$) is used. Including the second angle ($\cos\beta$) improves the sensitivity to more than 80% with respect to $\tau^- \to \pi^- \nu_\tau$. For $\tau \to \rho \nu_\tau$ this is all the information available, if the direction of the $\tau$ is not reconstructed.

In terms of spin analysis the decay $\tau \to a_1 \nu_\tau$ is very similar to $\tau \to \rho \nu_\tau$ [435]. Again there are two helicity states for the $a_1$, which have to be separated. They are similar in amplitude, so that the effect of evening out the polarization information is even stronger ($\alpha_{a_1} = 0.12$ in (4.25)). The $a_1$ decays to three pions (charged or neutral). In the rest frame of the $a_1$ they form a decay plane with the orbital angular momentum normal to it. It is

**Table 4.2.** Sensitivity of the decay distributions for various channels to the $\tau$ polarization. '1-dim.': energy spectrum only. '2-dim.': distribution in $\cos\theta^*$ and $\cos\beta$. 'All but $\hat{p}_\tau$': all information, but without the knowledge of the $\tau$ direction. 'All': all information [434]

|  | Observables | | | |
|---|---|---|---|---|
|  | 1-dim. | 2-dim. | All but $\hat{p}_\tau$ | All |
| $\tau \to \ell \nu_\ell \nu_\tau$ | 0.22 | –    | 0.22 | 0.27 |
| $\tau \to \pi \nu_\tau$          | 0.58 | –    | 0.58 | 0.58 |
| $\tau \to \rho \nu_\tau$         | 0.26 | 0.49 | 0.49 | 0.58 |
| $\tau \to a_1 \nu_\tau$          | 0.10 | 0.23 | 0.45 | 0.58 |

the angle $\beta$ between this direction and the direction of flight of the $a_1$ which separates the two helicity states here [435].

From Table 4.2 one can read the sensitivities. The sensitivity is very low (0.10) if only the energy is used. It increases by more than a factor of two in the two-dimensional analysis. If all information, including the direction of the $\tau$, is used, all hadronic decay channels have the same sensitivity [436].

### 4.5.3 Leptonic Decays

The leptonic channels $\tau \to e\, \nu_e \nu_\tau$ and $\tau \to \mu\, \nu_\mu \nu_\tau$ are experimentally the cleanest, but they have low sensitivity (see Table 4.2). The low sensitivity is caused by the two unobserved neutrinos, compared to only one in hadronic decays. The spectrum can be calculated from the Feynman diagram of Fig. 4.22. The matrix element is

$$\mathcal{M} = 4 \frac{G_F}{\sqrt{2}} \, \langle \overline{\Psi}_L (\ell^-) \, | \gamma^\mu | \, \Psi_R (\overline{\nu}_\ell) \, \rangle \, \langle \overline{\Psi}_L (\nu_\tau) \, | \gamma_\mu | \, \Psi_L (\tau^-) \, \rangle, \quad (4.31)$$

with the left-handed fermions represented by $\Psi_L$ and right-handed antifermions by $\Psi_R$. This gives the spectrum in the rest frame of the $\tau$:

$$\frac{d^2 \Gamma}{dx^* d\cos\theta^*} = \frac{G_F^2 m_\tau^5}{32\,\pi^3} x^{*2} \left[ \left( \frac{1}{2} - \frac{1}{3} x^* \right) + \cos\theta^* \left( \frac{1}{6} - \frac{1}{3} x^* \right) \right]. \quad (4.32)$$

The quantity $x^*$ is the scaled lepton energy, i.e. $E_\ell^*/E_{\ell,\mathrm{max}}^*$:

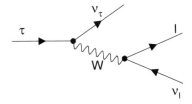

**Fig. 4.22.** Feynman diagram of a leptonic $\tau$ decay

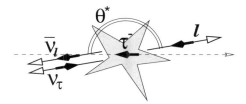

**Fig. 4.23.** Leptonic decay of a $\tau$ in its rest frame. The picture shows the creation of a high-energy lepton with both neutrinos emitted collinearly against the charged lepton. The direction of the charged lepton is preferentially against the $\tau$ spin, as shown here

$$x^*_{\min} \leq x^* \leq 1,$$
$$E^*_{\ell,\min} \leq E^*_\ell \leq E^*_{\ell,\max},$$
$$\text{with} \quad E^*_{\ell,\min} = m_\ell, \quad E^*_{\ell,\max} = \frac{m_\tau^2 + m_\ell^2}{2\,m_\tau}, \tag{4.33}$$

and $\theta^*$ is the decay angle of the charged daughter lepton with respect to the spin of the $\tau$.

The formula is more complicated than that for the hadronic decays, but some basic features can be easily understood. At the high-energy end of the $x^*$ spectrum the two neutrinos are emitted opposite to the charged lepton (see Fig. 4.23). In that case the neutrino spins combine to zero and the charged daughter lepton adopts the $\tau$ spin. In this region of the spectrum the situation is similar to that of $\tau^- \to \pi^- \nu_\tau$ and the sensitivity of these events is high. For $x^* = 1$ the decay distribution (4.32) goes to zero for $\cos\theta^* \to +1$ and reaches the maximum as $\cos\theta^* \to -1$, the preferred direction of emission. At lower energies of the charged lepton, there are many ways to combine the spins and therefore there is less sensitivity.

As the rest frame of the $\tau$ cannot be reconstructed in leptonic decays in the experiment, it is necessary to boost the spectrum into the laboratory frame[12] and to integrate over the unobservable angle $\cos\theta^*$. Depending on the helicity $h_\tau$, the charged lepton is boosted either in the direction of the $\tau$ spin or against it. Defining $x$ to be the scaled energy in the laboratory frame, one gets the relation (ignoring terms of order $(m_\ell/m_\tau)^2$ and higher)

$$x = \frac{1}{2}x^* \left(1 + h_\tau \cos\theta^*\right), \tag{4.34}$$

with

$$x = \frac{E_\ell}{E_{\ell,\max}}, \quad E_{\ell,\max} = 2\,\gamma\,E^*_{\ell,\max}. \tag{4.35}$$

Only forward emission of the lepton can create high values of $x$, and therefore the integration range in $\cos\theta^*$ has to be restricted to

$$2x - 1 < \cos\theta^* < \min\left(1,\, 2\frac{x}{x^*_{\min}} - 1\right). \tag{4.36}$$

The upper integration limit is 1, except for very small values of $x$.

---

[12] Using $\gamma = E_{\text{beam}}/m_\tau$ as the boost factor.

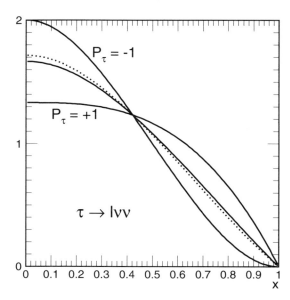

**Fig. 4.24.** Spectrum of the decay $\tau \to \ell \nu_\ell \nu_\tau$ in the scaled energy $x$ normalized to the total decay rate. The three *solid curves* represent $\tau$ polarizations of $-1$, $0$, and $1$. The *dotted curve* is the Standard Model prediction for $P_\tau = -0.147$

The resulting laboratory spectrum is

$$\frac{d\Gamma}{dx} = \frac{G_F^2 m_\tau^5}{192\,\pi^3}\left[\frac{5}{3} - 3\,x^2 + \frac{4}{3}\,x^3 - h_\tau\left(-\frac{1}{3} + 3\,x^2 - \frac{8}{3}\,x^3\right)\right]. \quad (4.37)$$

It is shown in Fig. 4.24 for various polarizations.

This concludes the discussion of the spectra of $\tau$ decays and their relation to the $\tau$ polarization. In practice one has to be more careful with phase space effects due to the masses of the final-state particles and one has to include radiative corrections in the calculation of the spectra. It is mainly the emission of real photons which disturbs the spectra. The influence can be reduced experimentally if the energy of the photons is added back to that of the charged decay products. This is usually applied to $\tau \to e\,\nu_e\nu_\tau$ decays, where the corrections are largest and the total electromagnetic energy measured already represents the sum of the electron and photons. However, in general, it is not always possible to measure all the photons. Effects of spin flips due to photon emission are small and genuine electroweak corrections are negligible.

### 4.5.4 Optimal Observables

As described in the previous section, the information about the $\tau$ spin is encoded in the momentum distributions of its decay products. In the simplest case ($\tau \to \pi\,\nu_\tau$) all the information is concentrated in only one variable: $\cos\theta^*$ or, alternatively, $x_\pi$. In more complicated situations, however, one has to study multidimensional distributions to explore the full sensitivity. For

$\tau \to a_1 \nu_\tau$, for example, there are besides $\cos\theta^*$ six more variables which carry some information about the $\tau$ spin ($Q^2$, $s_1$, $s_2$, and the Euler angles $\alpha$, $\beta$, and $\gamma$ [437]). Although there is no fundamental problem in extracting the information from a multidimensional distribution, it is in practice painful. Fortunately it has been shown that all the information can be condensed into a single variable [434, 438].

It can be shown that for any decay channel the decay distribution can be cast into the following form:

$$W(\boldsymbol{x}) = f(\boldsymbol{x}) + P_\tau g(\boldsymbol{x}), \tag{4.38}$$

where $\boldsymbol{x}$ is a vector representing all available observables. $f(\boldsymbol{x})$ is positive and normalized to 1 and $g(\boldsymbol{x})$ is normalized to zero. When a number of decays have been observed, one can extract the polarization by maximizing a likelihood function

$$\log \mathcal{L} = \sum_i \log\left[f(\boldsymbol{x}_i) + P_\tau g(\boldsymbol{x}_i)\right] \tag{4.39}$$

with respect to $P_\tau$:

$$\frac{\mathrm{d}}{\mathrm{d}P_\tau} \log \mathcal{L} = \sum_i \frac{g(\boldsymbol{x}_i)}{f(\boldsymbol{x}_i) + P_\tau g(\boldsymbol{x}_i)} = 0. \tag{4.40}$$

Rewriting this equation as

$$\frac{\mathrm{d}}{\mathrm{d}P_\tau} \log \mathcal{L} = \sum_i \frac{g(\boldsymbol{x}_i)/f(\boldsymbol{x}_i)}{1 + P_\tau g(\boldsymbol{x}_i)/f(\boldsymbol{x}_i)} = 0 \tag{4.41}$$

demonstrates that the maximum depends only on a single variable $w_i = g(\boldsymbol{x}_i)/f(\boldsymbol{x}_i)$, i.e. it is sufficient to study the one-dimensional distribution

$$W(\omega) = f(\omega)\left(1 + P_\tau \omega\right). \tag{4.42}$$

Comparing this with (4.20), one sees that for $\tau \to \pi \nu_\tau$ the scaled energy $x_\pi$ is already the optimal observable $\omega$. Also, for the leptonic channels, $x_\ell$ is the optimal observable, if the $\tau$ direction is not reconstructed. For $\tau \to \rho \nu_\tau$ and $\tau \to a_1 \nu_\tau$, however, $\omega$ is a complicated function of all kinematic variables and requires knowledge of the structure functions of the decays.

In practice, for each channel a spectrum of the optimal observable $\omega$ is extracted from Monte Carlo simulations for positive and negative helicities. The spectra are extracted with the same selection as in the data, so that they include the effects of efficiency and resolution. A linear combination of the two, plus background, is fitted to the data. The ratio of the two gives the polarization according to (4.23). The result is free of bias as long as the Monte Carlo simulation properly describes the data. Examples can be found in the next section.

## 4.6 τ Polarization

### 4.6.1 Electroweak Predictions

As mentioned in the introduction (Sect. 4.2), the $\tau$ polarization on the $Z^0$ peak measures the couplings of the $Z^0$ boson to $\tau$ leptons: $P_\tau = A_{\mathrm{pol}} \approx -A_\tau$; and the forward–backward asymmetry of the polarization determines the respective couplings to electrons: $A_{\mathrm{pol}}^{\mathrm{FB}} \approx -\frac{3}{4} A_{\mathrm{e}}$.

Figure 4.25 shows the variation of the polarization across the $Z^0$ resonance. The contributions from the $Z^0$ exchange and the $\gamma$–$Z^0$ interference are shown separately. The $\gamma$ exchange on its own creates no polarization. Assuming lepton universality, the plot for $A_{\mathrm{pol}}^{\mathrm{FB}}$ is identical to Fig. 4.25 except for the additional factor $3/4$.

The $\tau$ polarization is not constant with the scattering angle of the $\tau$, as indicated by its nonvanishing forward–backward asymmetry. It varies as (see Fig. 4.26)

$$P_\tau(\theta_\tau, s) = -\frac{(1+\cos^2\theta_\tau)F_2(s) + 2\cos\theta_\tau F_3(s)}{(1+\cos^2\theta_\tau)F_0(s) + 2\cos\theta_\tau F_1(s)}. \tag{4.43}$$

It is always zero in the exact backward direction and has its average value $A_{\mathrm{pol}}$ at $\cos\theta_\tau = 0$.

Figure 4.25 has been calculated from the tree-level formulas of Table 4.1. It does not include radiative corrections. These are shown in Figs. 4.27 and 4.28.

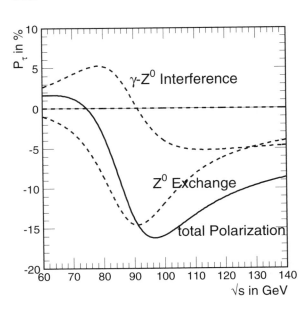

**Fig. 4.25.** The $\tau$ polarization across the $Z^0$ resonance. *Solid line*: total polarization. *Dashed lines*: contributions from $Z^0$ exchange and $\gamma$–$Z^0$ interference (tree-level calculations)

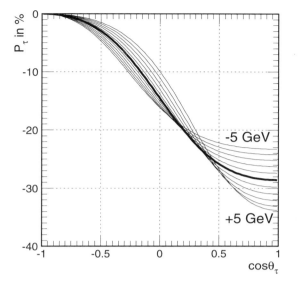

**Fig. 4.26.** Angular dependence of the $\tau$ polarization for various center-of-mass energies. The *thick line* is on peak; $\cos\theta_\tau$ is the scattering angle of the negative $\tau$ lepton

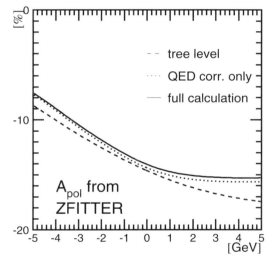

**Fig. 4.27.** Longitudinal polarization $A_{\mathrm{pol}}$ of $\tau$ leptons across the $Z^0$ peak. *Horizontal axis*: center-of-mass energy with respect to the $Z^0$ mass. *Dashed line*: tree-level calculation. *Dotted line*: QED corrections only. *Solid line*: full radiative corrections. Calculated from ZFITTER [419]

$A_{\mathrm{pol}}$ and $A_{\mathrm{pol}}^{\mathrm{FB}}$ provide a very sensitive determination of the Weinberg angle through

$$A_\ell \approx \frac{2v_\ell a_\ell}{v_\ell^2 + a_\ell^2} \approx 8\left(\frac{1}{4} - \sin^2\theta_{\mathrm{W}}\right), \tag{4.44}$$

i.e. they measure the deviation of the Weinberg angle from $1/4$. The relations with all the radiative corrections are shown in Figs. 4.29 and 4.30. There is quite a dependence on the mass of the top quark. The influence of the other

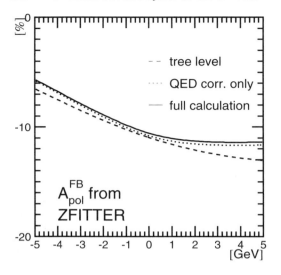

**Fig. 4.28.** Forward–backward asymmetry of the longitudinal polarization $A_{\text{pol}}^{\text{FB}}$ of $\tau$ leptons across the $Z^0$ peak. *Horizontal axis*: center-of-mass energy with respect to the $Z^0$ mass. *Dashed line*: tree-level calculation. *Dotted line*: QED corrections only. *Solid line*: full radiative corrections. Calculated from ZFITTER [419]

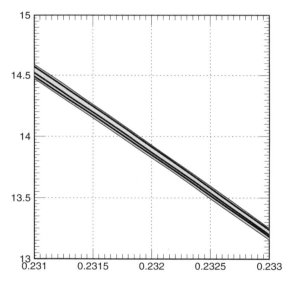

**Fig. 4.29.** Relation between the $\tau$ polarization ($-P_\tau$ is plotted in %) and the effective weak mixing angle for $\tau$ leptons at $s = m_Z^2$. The three *thick lines* correspond to three values of the Higgs mass $m_H = 100$, 200, and 300 GeV (*top* to *bottom*). The *shaded area* corresponds to $m_t = (173.8 \pm 5.0)$ GeV. Heavier top masses appear higher on the graph ($1/\alpha_{\text{QED}} = 128.896$ and $\alpha_s = 0.119$)

parameters, such as the mass of the Higgs boson, $\alpha_{\text{QED}}$, and $\alpha_s$, is small. The expected error on the Weinberg angle is

$$\Delta(\sin^2\theta_{\text{W}}) \approx \frac{1}{8}\Delta(A_\ell). \tag{4.45}$$

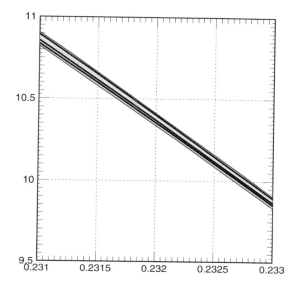

**Fig. 4.30.** Relation between the forward–backward asymmetry of the $\tau$ polarization and the effective weak mixing angle for $\tau$ leptons. The three *thick lines* correspond to three values of the Higgs mass $m_H = 100$, 200, and 300 GeV (*top* to *bottom*). The *shaded area* corresponds to $m_t = (173.8 \pm 5.0)$ GeV. Heavier top masses appear higher on the graph ($1/\alpha_{\rm QED} = 128.896$ and $\alpha_s = 0.119$)

### 4.6.2 Longitudinal Polarization

All four LEP experiments have presented polarization measurements at the $Z^0$ peak [59, 67, 155, 158, 173, 178, 179, 230, 239, 264, 439]. Measurements at lower energies can be found in [107, 440]. The most recent ALEPH results [264] (preliminary) might serve as a good example, as described in the following.

One approach is to analyze the decays of the two $\tau$ leptons in each event independently. Each one-prong hemisphere is tested separately with particle identification to identify the leptonic channels $\tau \to e\,\nu_e\nu_\tau$ and $\tau \to \mu\,\nu_\mu\nu_\tau$ and distinguish them from the hadronic decays. Then photon reconstruction is used to distinguish the hadronic decays $\tau^\pm \to \pi^\pm\,\nu_\tau$, $\tau^\pm \to \pi^\pm\,\pi^0\,\nu_\tau$, and $\tau \to \pi^\pm\,\pi^0\,\pi^0\,\nu_\tau$. ALEPH also uses the three-prong decay $\tau^\pm \to 3\,\pi^\pm\,\nu_\tau$ and performs a measurement on an inclusive hadronic class. The polarization is determined for each class and each year of data-taking independently and then the results are combined, taking into account possible correlations.

Figure 4.31 shows the spectrum of the $\tau \to \mu\,\nu_\mu\nu_\tau$ class. The spectrum has been fitted with the sum of two Monte Carlo spectra, those of $\tau$ leptons with helicity plus and minus. These are the intermediate histograms in Fig. 4.31. They correspond to the theoretical functions in Fig. 4.24, but include the efficiency function, the resolution, and the background from other $\tau$ channels. They have been derived from Monte Carlo events with the full radiative corrections and detector simulation. The non-$\tau$ background is very small, almost invisible in the plot. It comes from two-photon events $\gamma\gamma \to \mu^+\mu^-$ at low energies and muon pairs $e^+e^- \to \mu^+\mu^-$ at high energies. The result is a polarization of $P_\tau = -(13.64 \pm 2.09 \pm 0.93)\%$ from this channel only.

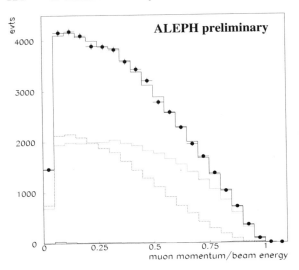

Fig. 4.31. Polarization measurement from $\tau \to \mu\, \nu_\mu \nu_\tau$ decays (ALEPH collaboration [264]). The *points* are data; the two *lower histograms* represent the Monte Carlo expectation for negative (shifted to lower momenta) and positive (higher momenta) helicities of the $\tau$. The *upper histogram* (underneath the data points) is the sum of both lower histograms plus the background (*hatched histogram* at the bottom, only just visible)

For the $2\pi$ and $3\pi$ channels ALEPH uses the optimal observables as described in Sect. 4.5.4. All the information of the decays is used except for the $\tau$ direction, which is not available with a single $\tau$ decay.

For illustration Fig. 4.32 shows the distributions in the angles $\theta^*$ and $\beta$ for the $\tau \to \pi\pi\nu_\tau$ channel. These two angles are the most important input to the optimal observable.

For the one-prong decays photons are reconstructed from the energy depositions in the electromagnetic calorimeter and combined to reconstruct the $\pi^0$s and final hadrons. Therefore a good understanding of the photon reconstruction and suppression of fake photons become essential. Also, the uncertainties in the calibration of the calorimeter contribute to the systematic errors. For the $3\pi$ modes the understanding of the hadronic structure of the decays, which enters the definition of the optimal observables, is a nonnegligible source of systematic errors, too.

In the second approach ALEPH considers each event as a whole. This makes the control of the systematics more difficult and excludes channel-specific selection procedures, but it maximizes the information from each event by making use of the reconstruction of the $\tau$ direction and the acollinearity (see next section for the acollinearity).

In events where both decays are identified as being hadronic, ALEPH tries to reconstruct the $\tau$ direction and, if successful, uses it to calculate the optimal observables. The sensitivity achieved (see (4.30) and Table 4.2) is 0.46 without and 0.51 with the reconstruction of the $\tau$ direction in the $\rho$ channel; the numbers for a perfect detector would be 0.49 and 0.58, respectively. Figure 4.33 shows the distribution of the optimal observable with and without the

4.6 τ Polarization    125

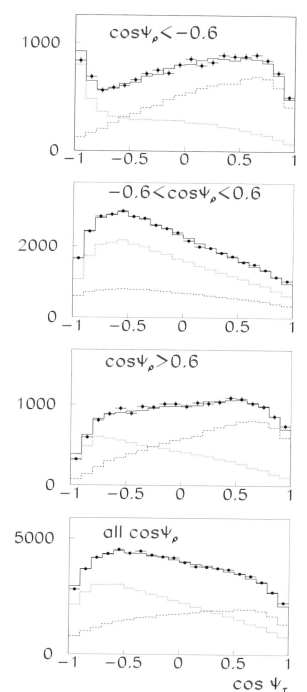

**Fig. 4.32.** Distribution of the angle $\cos\theta^* \equiv \cos\psi_\tau$ for slices in $\cos\beta \equiv \cos\psi_\rho$ (ALEPH collaboration [264], preliminary). The *dots* are the data, the *histograms* the Monte Carlo distributions: *dotted*, negative helicity; *dashed*, positive; *solid*, the sum

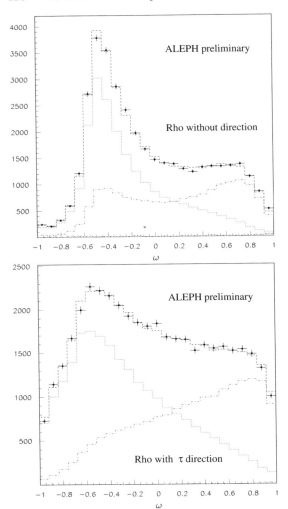

**Fig. 4.33.** The optimal observable in the decay $\tau \to \pi\pi\nu_\tau$ from ALEPH data [264]. *Upper histograms*: without reconstruction of the $\tau$ direction. *Lower histograms*: with reconstruction. The *dotted* and *dashed–dotted histograms* are the predictions for $\tau^-$ with negative and positive helicity, respectively, which are fitted to the data (*dots with error bars*) to give the result (*dashed histogram*)

$\tau$ direction reconstructed. With the $\tau$ direction the distribution looks quite similar to that of the $\tau \to \pi\nu_\tau$ channel.

With both approaches, the $\tau$ polarization is measured in nine different bins in the production angle $\theta_\tau$. The following function is fitted to the data points to extract $A_e$ and $A_\tau$ (see (4.43))

$$P_\tau(\theta_\tau) = -\frac{A_\tau(1+\cos^2\theta_\tau) + A_e(2\cos\theta_\tau)}{(1+\cos^2\theta_\tau) + 4/3\, A^{\mathrm{FB}}(2\cos\theta_\tau)}. \tag{4.46}$$

Figure 4.34 shows the measured variation of $P_\tau$ with $\theta_\tau$ (see also Fig. 4.26). The combined result from all channels and both approaches, taking

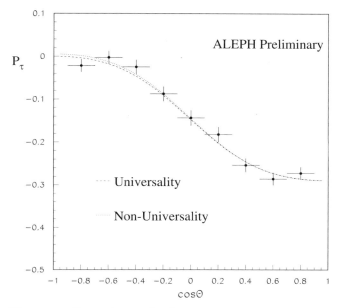

**Fig. 4.34.** Dependence of polarization on $\cos\theta_\tau$ from the ALEPH collaboration [264]. The *dashed* and *dotted* lines show the result of the fit with and without the universality constraint, respectively

into account the correlations, is

$$A_e = (15.05 \pm 0.69 \pm 0.10)\%,$$
$$A_\tau = (14.52 \pm 0.55 \pm 0.27)\%,$$
$$A_\ell = (14.75 \pm 0.43 \pm 0.16)\%.$$

The last number assumes universality between the $\tau$ lepton and the electron.

### 4.6.3 Spin Correlations

Up to now only the polarization of single $\tau$ leptons has been considered, but $\tau$ leptons are pair-produced with their spins correlated, and there is additional information in these correlations. These correlations have been ignored in (4.3). The cross section including the correlations reads [23, 441] (see also [442–444])

$$\frac{d\sigma}{d\cos\theta_\tau} = F_0(s) \frac{1 + s_z^- s_z^+}{2} (1 + \cos\theta_\tau)$$
$$+ F_1(s) \frac{1 + s_z^- s_z^+}{2} 2\cos\theta_\tau$$

$$-F_2(s)\frac{s_z^- + s_z^+}{2}(1+\cos\theta_\tau)$$

$$-F_3(s)\frac{s_z^- + s_z^+}{2}2\cos\theta_\tau$$

$$+F_4(s)\frac{s_y^- s_y^+ - s_x^- s_x^+}{2}\sin^2\theta_\tau$$

$$+F_5(s)\frac{s_y^- s_x^+ + s_x^- s_y^+}{2}\sin^2\theta_\tau, \qquad (4.47)$$

where $\boldsymbol{s}^-$ and $\boldsymbol{s}^+$ are the spin vectors of the $\tau^-$ and $\tau^+$, respectively. These are axial unit vectors defined in the rest frame of the $\tau$ leptons, with $s_z^\pm$ being the component in the direction of flight of the negative $\tau$ ($s_z^- \equiv h_\tau$). The first two terms, which correspond to the unpolarized cross section of (4.3), vanish unless the two $\tau$ spins are positively correlated longitudinally ($s_z^- = s_z^+ = +1$ or $s_z^- = s_z^+ = -1$). The next two terms differ in sign between these two cases and describe the longitudinal polarization of the $\tau$ leptons. The last two terms describe correlations between the spin components perpendicular to the direction of flight of the $\tau$ leptons. The $x$ components fall into the production plane, i.e. the plane spanned by the $\tau$ leptons and the initial beams (transverse components), and the $y$ components are perpendicular to it (normal components). Notice that there is no net polarization, only correlation, in these directions.

As the spin vectors of the $\tau$ leptons are not directly observable, the cross section of (4.47) has to be modified with the decay distributions of Sect. 4.5. The full distribution is then given by

$$\frac{\mathrm{d}\sigma}{\mathrm{d}\theta_\tau}(\boldsymbol{s}^-,\boldsymbol{s}^+)\,\frac{\mathrm{d}\Gamma^+}{\mathrm{d}\boldsymbol{x}_+}(\boldsymbol{s}^+)\,\frac{\mathrm{d}\Gamma^-}{\mathrm{d}\boldsymbol{x}_-}(\boldsymbol{s}^-). \qquad (4.48)$$

The vectors $\boldsymbol{x}_+$ and $\boldsymbol{x}_-$ are the observables of the decay of the positive and negative $\tau$, respectively, and $\theta_\tau$ is the scattering angle of the negative $\tau$, which has to be approximated by the thrust or some other appropriate axis, if not reconstructed kinematically.

The distribution (4.48) is not of much practical use, because of its high dimensionality. Depending on the application, some variables will be integrated out, leaving only the more interesting ones for the purpose of analyzing the data. The distributions used include the energy–energy correlations, where all variables are integrated except for the energies of the visible decay products of the two $\tau$ leptons, and the acollinearity distribution, where the acollinearity is the angle between the decay products of the two $\tau$ leptons. Both distributions are sensitive to the longitudinal polarizations and therefore to the electroweak couplings $A_\ell$. Also, transverse spin correlations can be studied.

**Energy–Energy Correlations.** Figure 4.35 shows the energy–energy correlation in events where both $\tau$ leptons have decayed to a single pion (Monte Carlo study [441]). In the left diagram the $\tau^-$ has positive helicity. As a

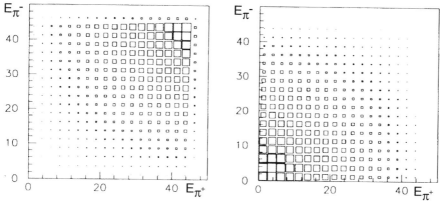

**Fig. 4.35.** Energy–energy correlation between the pions in events with $\tau^- \to \pi^- \nu_\tau$, $\tau^+ \to \pi^+ \bar{\nu}_\tau$ at LEP energies (units GeV). The helicity of the $\tau^-$ is positive in the *left plot* and negative in the *right plot* (Monte Carlo study [441]); the $\tau^+$ has opposite helicity

consequence not only does the pion from the $\tau^-$ have high energy, but also through the correlation, the pion from the $\tau^+$ has high energy. For a negative helicity of the $\tau^-$ the spectra are reversed.

These correlations are highly sensitive to the longitudinal polarization of the $\tau$ leptons, but the result is almost fully correlated with that from the measurement on single $\tau$ leptons described in the previous section. The analysis is more complicated and has lower efficiency as it requires the identification of the decay channels of both $\tau$ leptons. Therefore it is rarely in practice. Energy–energy correlations, however, have been widely used in the studies of the Lorentz structure of the $\tau$ decays (see Chap. 8).

**Acollinearity Distribution.** The acollinearity distribution has lower sensitivity than the energy–energy correlations, but it carries information independent of the single-$\tau$ spectra. It is also independent in the sense that the systematic errors associated with the measurement are different from the ones described previously. The momentum vector of the visible decay products of both $\tau$ leptons is reconstructed and the acollinearity $\eta$ is defined as $\pi$ minus the angle between the two vectors (see Sect. 2.2.3). The distribution can be written as

$$\frac{d\sigma}{d\eta} \propto Q_{11}(\eta) + \alpha_-\alpha_+ Q_{22}(\eta) + h_\tau \left[\alpha_- Q_{21}(\eta) + \alpha_+ Q_{12}(\eta)\right], \qquad (4.49)$$

with the factors $\alpha_\pm$ for the decay of $\tau^\pm$ as introduced in Sect. 4.5.2. Figure 4.36 shows the distribution for events where both $\tau$ leptons decay as $\tau \to \pi \nu_\tau$. There is good separation between the two helicity states. The $\tau$ polarization, and with it $A_\tau$, can be measured by fitting the two distributions (modified for

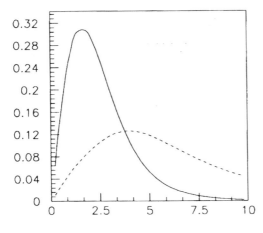

**Fig. 4.36.** Acollinearity distribution $d\sigma/d\eta$ in events where both $\tau$ leptons decay to a single pion [441]. *Solid line*, $h_\tau = +1$; *dashed line*, $h_\tau = -1$. *Horizontal axis*: $\eta$ in degrees; *vertical axis*, $d\sigma$

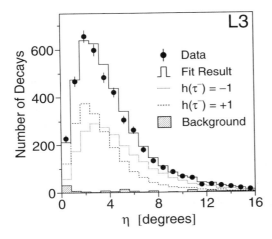

**Fig. 4.37.** Acollinearity distribution of pions from $\tau \to \pi \nu_\tau$ recoiling against a one-prong decay (L3 collaboration [173])

detector effects) to the data. There is also information on $A_e$ in the variation of the distribution with the production angle $\theta_\tau$.

In practice inclusive identifications are used for this measurement in order to keep the number of channel combinations manageable and the losses introduced by inefficiencies small. Figure 4.37 shows a measurement by the L3 collaboration of pions recoiling against any one-prong $\tau$ decay. The separation between the two helicity states is reduced compared to Fig. 4.36 owing to detector effects and the inclusive nature of the measurement, but it is still clearly visible.[13] The results from the acollinearity are $A_\tau = 0.111 \pm 0.041$ and $A_e = 0.128 \pm 0.058$ from the whole set of LEP I data. (The figure only

---

[13] For the definition of the acollinearity only the charged track of the one-prong hemisphere and the recoiling pion are used.

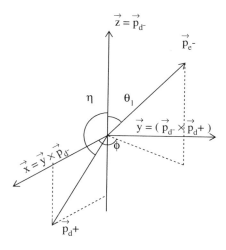

**Fig. 4.38.** Definition of the aplanarity angle $\Phi$. $\boldsymbol{p}_{d^-}$ and $\boldsymbol{p}_{d^+}$ are the momenta of the visible decay products of the $\tau^-$ and $\tau^+$, respectively. $\boldsymbol{p}_{e^-}$ is the direction of the initial electron beam [52]

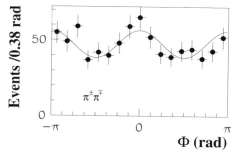

**Fig. 4.39.** Distribution of the aplanarity angle $\Phi$ in $\pi\pi$ events (ALEPH collaboration [52])

shows the 1994 data.) The correlation with the polarization measurement from the decay spectra is only 30%.

**Transverse Spin Correlations.** $\tau$ leptons produced in $e^+e^-$ collisions also exhibit correlations between the spin components perpendicular to the direction of flight of the $\tau$. The sensitivity of these correlations to the electroweak couplings is limited. Nevertheless it is interesting to verify the existence and magnitude of these correlations. DELPHI [151] and ALEPH [52] have investigated such correlations. There are two measurable correlations: the transverse-to-transverse correlation $C_{\mathrm{TT}}$ between the components of the spin in the production plane of both $\tau$ leptons, and the transverse-to-normal correlation $C_{\mathrm{TN}}$ between the component of one $\tau$ in the production plane and the component normal to it of the other $\tau$. The normal-to-normal correlation $C_{\mathrm{NN}}$ is expected to be very small. The variable carrying most sensitivity to these correlations is the aplanarity angle $\Phi$ defined in Fig. 4.38. Also, the acollinearity angle $\eta$ carries some information.

Figure 4.39 shows the distribution of the aplanarity angle measured by the ALEPH collaboration [52] in events where both $\tau$ leptons decay as $\tau \to \pi \nu_\tau$.

ALEPH has measured six combinations of decay channels altogether, but the $\pi\pi$ channel is the most sensitive. The results are

$$C_{\mathrm{TT}} = 1.06 \pm 0.13_{\mathrm{stat}} \pm 0.05_{\mathrm{sys}},$$
$$C_{\mathrm{TN}} = 0.08 \pm 0.13_{\mathrm{stat}} \pm 0.04_{\mathrm{sys}}, \qquad (4.50)$$

in agreement with the Standard Model predictions of 0.99 and $-0.01$, respectively.

## 4.7 Results

The results of the individual experiments are combined by the LEP $\tau$ polarization and electroweak working groups.[14] They check the consistency and take care of systematic errors common to all experiments. Such common systematic errors could arise from the use of the same Monte Carlo programs across the experiments or from uncertainties in the radiative corrections applied. For example, the dependence of the polarization result in the three-pion channel on the modeling of the $a_1$ decay in the Monte Carlo program TAUOLA [297] is the same for all experiments. For details see [408, 445].

Figures 4.40 and 4.41 show the combined results for $R_\ell$ and $A^{\mathrm{FB}}$ at $\sqrt{s} = m_Z$ in comparison with the Standard Model, assuming lepton universality. Instead of the production cross section for each lepton species, the ratio

$$R_\ell = \frac{\Gamma(Z^0 \to \mathrm{had})}{\Gamma(Z^0 \to \ell^+\ell^-)} \qquad (4.51)$$

is presented. Through the normalization this becomes independent of the uncertainties in the luminosity determination. It is essentially the ratio of lepton pairs over hadronic events detected, corrected for background and efficiency. The upper part of each plot shows the results experiment by experiment and the combined number. There is good consistency between the individual numbers. The lower part of each plot shows the Standard Model prediction as a function of the mass of the Higgs boson, with the input parameters quoted on the plot. The variations with $m_Z$, $m_t$, $\alpha_{\mathrm{QED}}$, and $\alpha_{\mathrm{s}}$ are also indicated. $A^{\mathrm{FB}}$ and $\Gamma(Z^0 \to \ell^+\ell^-)$ are independent of $\alpha_{\mathrm{s}}$. The high sensitivity of $R_\ell$ to $\alpha_{\mathrm{s}}$ comes in through the denominator.

---

[14] The current members of the $\tau$ polarization group are J.C. Brient, P. Garcia-Abia, J. Harton, A. Kounine, W. Lohmann, F. Matorras, D. Reid, P. Renton, J.M. Roney, M. Thomson, and H. Videau.
The current members of the electroweak working group are J. Alcaraz, D. Abbaneo, P. Antilogus, T. Behnke, B. Bertucci, D. Bloch, A. Blondel, D.G. Charlton, R. Clare (chair), P. Clarke, S. Dutta, S. Ganguli, M.W. Grünewald, A. Gurtu, K. Hamacher, M. Hildreth, R.W.L. Jones, W. Lohmann, T. Kawamoto, C. Mariotti, M. Martinez, K. Mönig, A. Nippe, A. Olshevsky, Ch. Paus, M. Pepe-Altarelli, B. Pietrzyk, G. Quast, P. Renton, D. Reid, M. Roney, D. Schlatter, R. Tenchini, F. Teubert, I. Tomalin, and P.S. Wells.

4.7 Results    133

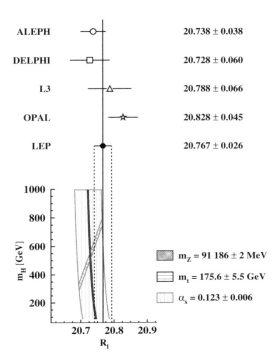

**Fig. 4.40.** LEP combined results for $R_\ell$ [408]. *Open symbols*, results from the four experiments; *filled circle*, average of the above values; *solid line*, Standard Model prediction for different Higgs masses; *hatched regions*, variation of the Standard Model prediction with the parameters as indicated on the plot

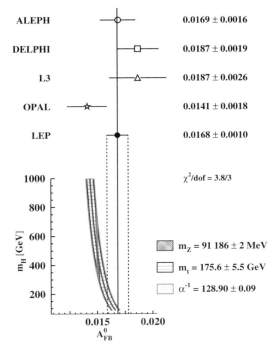

**Fig. 4.41.** LEP combined results for $A^{\mathrm{FB}}$ [408] (see Fig. 4.40 for details)

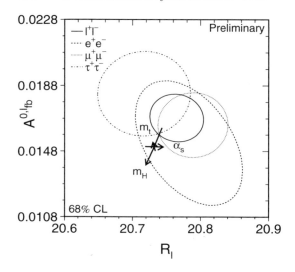

Fig. 4.42. Results for $A^{FB}$ versus $R_\ell$ for the three lepton species. The *solid line* is the average for the three species [408]

The numbers presented in Figs. 4.40 and 4.41 are numbers averaged over the three lepton species, assuming universality. Figure 4.42 shows the results for each species separately in the plane of $A^{FB}$ versus $R_\ell$. The assumption of universality is nicely fulfilled. The errors on the measurements from $\mu$ pairs are somewhat smaller compared to $\tau$ pairs, because they are experimentally easier to identify. The electron final states are complicated by the t-channel contribution, which also introduces a correlation between $A^{FB}$ and $R_\ell$.

Figures 4.43 and 4.44 show the combined results from the measurements of the $\tau$ polarization. Again there is good consistency between the experiments and agreement with the Standard Model. The dependence of the predictions on $\alpha_{QED}$ and $\alpha_s$ is only weak.

The leptonic cross sections and asymmetries allow one to extract the vector and axial-vector couplings of the $Z^0$ to the leptons individually. The results are shown by three ellipses in Fig. 4.45. The fourth, solid ellipse shows the result under the assumption of lepton universality. Also shown is the band obtained from the $A_{LR}$ measurement from the polarized beams at the SLC. The shaded area in the center is the Standard Model prediction, with the variations of $m_t$ ($\pm 5$ GeV) and $m_H$ (60 to 1000 GeV) as indicated. It once more shows the good agreement of all the data with the Standard Model and the preference for a light Higgs boson.

Finally, Fig. 4.46 shows a comparison of the Weinberg angle determined from different measurements. The first number (solid circle) is the combination of all $A^{FB}$ measurements. The solid squares, labeled $A_\tau$ and $A_e$, are the results from the $\tau$ polarization. The next three numbers (solid triangles) are from forward–backward asymmetries measured with hadronic final states (b quarks, c quarks, and inclusive hadrons). The open circle represents the average of all the above measurements. The left–right asymmetry from SLD is represented by the star, and the last number, with the shaded area, is the complete average.

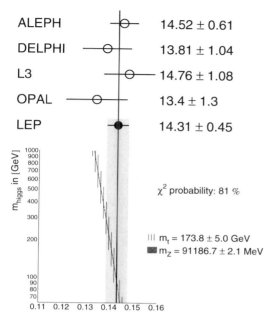

**Fig. 4.43.** LEP combined results for $A_\tau$ from the polarization measurements [445]. *Open circles*, results from the four experiments; *filled circle*, average of the above values; *central curve*, Standard Model prediction for various Higgs masses, *hatched band* variation of the Standard Model prediction with $m_t$

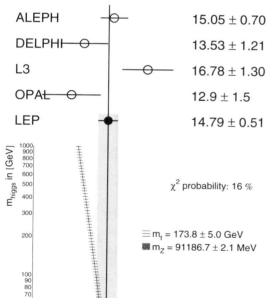

**Fig. 4.44.** LEP combined results for $A_e$ from the forward–backward asymmetry of the $\tau$ polarization [445] (see Fig. 4.40 for details)

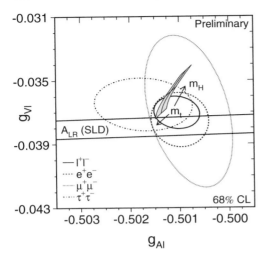

**Fig. 4.45.** Weak couplings of the $Z^0$ to the three lepton species extracted from the leptonic cross sections and asymmetries [408]

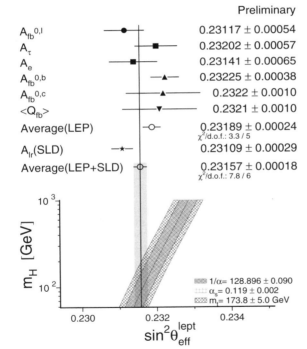

**Fig. 4.46.** Determinations of the effective weak mixing angle for leptons from LEP and SLD [408]

# 5. Strong Interactions in $\tau$ Decays

Figure 5.1 shows the basic decay of a $\tau$ into a hadronic state characterized by its mass $|Q|$, its spin, and various other quantum numbers. The intermediate vector boson $W^\pm$ couples to a quark–antiquark pair – either a $d\bar{u}$ or an $s\bar{u}$ in the case of the decay of a negative $\tau$ – which is then exposed to strong interactions.

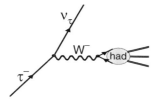

**Fig. 5.1.** The Feynman diagram of a hadronic $\tau$ decay

## 5.1 Selection Rules

This section deals with the various selection rules which apply when the $\tau$ decays into a hadronic state and a $\nu_\tau$. The aim is to explain the origin of the selection rules and classify the decays [446, 447].

### 5.1.1 Basics

Obviously, the hadronic state has to carry the same electrical charge as the decaying $\tau$.

The intermediate $W^\pm$ couples only to quarks with a total angular momentum of 0 or 1, so that $J \geq 2$ states are excluded.

The range of mesons that can be produced in $\tau$ decays is limited by the $\tau$ mass[1] and heavy mesons suffer from phase space suppression. The suppression

---

[1] Baryons would have to be pair-produced, but any baryon–antibaryon pair is heavier than the $\tau$.

138    5. Strong Interactions in τ Decays

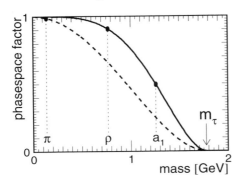

**Fig. 5.2.** Phase space suppression in τ decays for $J=1$ (*solid line*) and $J=0$ (*dashed line*) states. The *dotted lines* indicate the masses of some mesons produced in these decays

factor can be calculated from the Feynman diagram of Fig. 5.1 assuming a constant current. The result is ($J=0,1$)

$$d\Gamma \propto \left(1 - \frac{Q^2}{m_\tau^2}\right)^2 \left(1 + \frac{2Q^2}{m_\tau^2}\right)^J. \tag{5.1}$$

This is graphically depicted in Fig. 5.2.

### 5.1.2 Cabibbo Suppression

The intermediate $W^\pm$ in Fig. 5.1 couples to a $d'\bar{u}$ pair, where $d'$ is a mixture of d, s, and b quarks according to the Cabibbo–Kobayashi–Maskawa matrix [448]. The b quark is too heavy to be produced. The amplitude for the production of a $d\bar{u}$ pair is proportional to the cosine of the Cabibbo angle and that for an $s\bar{u}$ pair to the sine. More precisely [44];

$$\mathcal{M}_{d\bar{u}} \propto V_{ud} = 0.9740 \pm 0.0010,$$
$$\mathcal{M}_{s\bar{u}} \propto V_{us} = 0.2196 \pm 0.0023. \tag{5.2}$$

### 5.1.3 Helicity Suppression

In the decay of a $\tau^-$ to $d\bar{u}$, in the rest frame of the $W^\pm$, a left-handed d quark and a right-handed $\bar{u}$ quark are emitted back to back. This naturally leads to a state of total spin 1 and for massless quarks this would indeed be the only possibility. With massive quarks, however, the production of a spin-0 state is possible if one of the quarks appears with the 'wrong' helicity, i.e. the left-handed quark with positive helicity or the right-handed antiquark with negative helicity. See Fig. 5.3 for an illustration.

The amplitudes for the production of left-handed quarks and right-handed antiquarks with 'right' and 'wrong' helicities are proportional to the overlap between the eigenstates of helicity (the spinors $u_\pm(p)$) and of chirality (denoted by $u_{L/R}(p)$), i.e.[2]

---

[2] An easy way to see this is to calculate $|\frac{1}{2}(1+\gamma_5)u_+(p)|/|u_+(p)|$ in any representation.

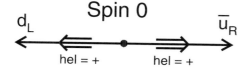

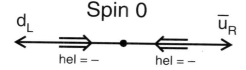

**Fig. 5.3.** The two spin configurations leading to a spin-0 hadron in $\tau$ decays

$$\langle u_+(p) | u_R(p) \rangle = \langle u_-(p) | u_L(p) \rangle = \sqrt{\tfrac{1}{2}(1+\beta)},$$
$$\langle u_+(p) | u_L(p) \rangle = \langle u_-(p) | u_R(p) \rangle = \sqrt{\tfrac{1}{2}(1-\beta)}, \qquad (5.3)$$

where $\beta$ is the velocity of the quarks. Therefore the amplitude for the production of two quarks in a spin-0 state is suppressed by

$$\sqrt{\tfrac{1}{2}(1-\beta_d)}\sqrt{\tfrac{1}{2}(1+\beta_u)} + \sqrt{\tfrac{1}{2}(1+\beta_d)}\sqrt{\tfrac{1}{2}(1-\beta_u)}$$
$$= \frac{m_d}{m_{\text{had}}} + \frac{m_u}{m_{\text{had}}} + \mathcal{O}\left(\frac{m_{d/u}^2}{m_{\text{had}}^2}\right). \qquad (5.4)$$

### 5.1.4 Isospin Suppression

The amplitude for the production of a spin-0 state can be further subdivided into scalar and pseudoscalar final states. Therefore let us define a transformation $\mathcal{P}\exp(i\pi J_{\hat{n}})$, where $\hat{n}$ is a unit vector perpendicular to the momenta of the quarks [446, 447]. The parity transformation $\mathcal{P}$ reverses the spin and momenta and $\exp(i\pi J_{\hat{n}})$ rotates the momenta back into their original directions keeping the spins reversed. Applied to the quarks, this gives[3]

$$\mathcal{P}\exp(i\pi J_{\hat{n}}) | d_\pm(p) \rangle = i | d_\mp(p) \rangle,$$
$$\mathcal{P}\exp(i\pi J_{\hat{n}}) | \bar{u}_\pm(p) \rangle = i | \bar{u}_\mp(p) \rangle. \qquad (5.5)$$

Combining two quarks into a spin-0 state in a $\tau$ decay, there are two eigenstates with opposite eigenvalues:

$$|0^+\rangle = \frac{1}{\sqrt{2}}\Big( | d_+(p), \bar{u}_+(-p) \rangle - | d_-(p), \bar{u}_-(-p) \rangle \Big),$$
$$|0^-\rangle = \frac{1}{\sqrt{2}}\Big( | d_+(p), \bar{u}_+(-p) \rangle + | d_-(p), \bar{u}_-(-p) \rangle \Big). \qquad (5.6)$$

---
[3] The phase factor i takes care of the $4\pi$ rotational symmetry of the spinors.

The amplitudes obtained from Fig. 5.3 can now be reexpressed by these eigenstates:

$$\frac{m_d}{m_{had}} \left| d_+(p), \bar{u}_+(-p) \right\rangle + \frac{m_u}{m_{had}} \left| d_-(p), \bar{u}_-(-p) \right\rangle =$$
$$\frac{m_d + m_u}{\sqrt{2}\, m_{had}} \left| 0^- \right\rangle + \frac{m_d - m_u}{\sqrt{2}\, m_{had}} \left| 0^+ \right\rangle. \quad (5.7)$$

The amplitude for the production of pseudoscalars $|0^-\rangle$ is proportional to the quark masses as expected from helicity suppression, but the amplitude for scalars $|0^+\rangle$ is further suppressed by the difference of the quark masses. In the limit of equal quark masses this amplitude would vanish.

### 5.1.5 Exotic States

Not all combinations of quantum numbers $J^{PG}$ can be produced from a quark–antiquark current. Final states with quantum numbers that cannot be produced in this way are called 'exotic'. They are linked to objects like molecules (qq$\bar{q}\bar{q}$) or hybrids (qqg).

### 5.1.6 Free Quarks versus the Hadronic Current

The last three sections described $\tau$ decays in a picture of free quarks. One might well argue that this picture is irrelevant at these energies, where quarks are strongly bound into mesons. Following this path, hadronic $\tau$ decays should be described by a hadronic current coupling to the $W^\pm$. The current comes from the vacuum and leads to a final state with one or several mesons. It is graphically depicted in Fig. 5.4. There is a vector current $V_\mu$ (first line) and an axial-vector current $A_\mu$ (second line):

$$\left\langle \Psi_{vac} \left| F_1(Q^2)\gamma_\mu + \frac{iF_2(Q^2)}{2|Q|} \sigma_{\mu\nu} q^\nu \right| \Psi_{had} \right\rangle,$$

$$\left\langle \Psi_{vac} \left| G_1(Q^2)\gamma_\mu \gamma_5 + \frac{iG_2(Q^2)}{2|Q|} \sigma_{\mu\nu}\gamma_5 q^\nu \right| \Psi_{had} \right\rangle. \quad (5.8)$$

The $\Psi$ represent the wave functions, $F_i$ and $G_i$ are form factors, $Q^2 \equiv m_{had}^2 \equiv s$ is the square of the mass of the hadronic system, and $\sigma^{\mu\nu} = i/2\,(\gamma^\mu\gamma^\nu - \gamma^\nu\gamma^\mu)$.

It is certainly not an accident that this picture gives the same pattern of branching ratios as the picture of free quarks. There is some deeper connection between the two, which we don't understand yet. The picture of currents will be exploited some more in the following sections.

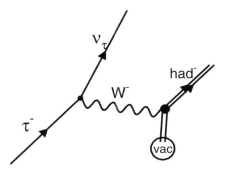

**Fig. 5.4.** The creation of mesons through a hadronic current in $\tau$ decays. In this picture the $W^{\pm}$ couples to a hadronic current coming from the vacuum

### 5.1.7 The Conserved Vector Current (CVC)

The universality of Fermi's constant $G_F$ between leptonic and hadronic weak interactions can only be understood if it is not affected by QCD corrections. There must be a weak charge which determines the strength of the coupling and which is conserved, even if a quark fluctuates into a quark plus a gluon or some other hadronic state. Mathematically this means that $\partial^\mu V_\mu = 0$ (CVC). Inserting $V_\mu$ from (5.8) and putting $\partial^\mu \Psi_{\text{had}} = iq^\mu \Psi_{\text{had}}$, one gets

$$\left\langle \Psi_{\text{vac}} \left| iF_1\left(Q^2\right) \gamma_\mu q^\mu - \frac{F_2\left(Q^2\right)}{2|Q|} \sigma_{\mu\nu} q^\nu q^\mu \right| \Psi_{\text{had}} \right\rangle. \tag{5.9}$$

$\sigma_{\mu\nu}$ contracted with two identical momenta vanishes and the first term can be converted into $F_1\left(Q^2\right) m_{\text{had}}$ by the equation of motion. This gives

$$m_{\text{had}} F_1\left(Q^2\right) \left\langle \Psi_{\text{vac}} \mid \Psi_{\text{had}} \right\rangle = 0. \tag{5.10}$$

But this is a scalar current, which means that the vector current can only be conserved if there is no scalar current present (no $J^P = 0^+$ final states).

The vector current is expected to be completely conserved in the limit of a vanishing difference between the masses of the light quarks, and this is exactly the selection rule we found in the quark picture in Sect. 5.1.4.

### 5.1.8 The Partially Conserved Axial-Vector Current

In much the same way one might compare nuclear $\beta$ decays proceeding through an axial-vector current with leptonic weak interactions. One finds a similar, but not identical coupling strength. Parametrizing the current in nuclear $\beta$ decays as $c_V \gamma_\mu - c_A \gamma_5 \gamma_\mu$, we have $c_V = 1.00$ and $c_A = 1.26$, i.e. the axial-vector current is partially conserved (PCAC).

By calculating $\partial^\mu A_\mu$ one finds a pseudoscalar current, which has to be small (suppressed) and which vanishes in the limit of vanishing quark masses, where the conservation of the axial-vector current becomes exact. This is what we called the helicity suppression in Sect. 5.1.3.

### 5.1.9 $G$ Parity and Second-Class Currents

The final states produced in $\tau$ decays can be further classified by their transformation properties under $G$ parity, a combination of charge conjugation and isospin rotation [449]. It is an experimental observation that in nature any current with quantum numbers $J^P$ appears with only one out of the two possible values of $G$. These are called the first-class currents ($J^{PG} = 0^{++}, 0^{--}, 1^{+-}, 1^{-+}, \ldots$). The second-class currents carry the opposite $G$ parity. There have been many experimental attempts to identify second-class currents, mainly in nuclear $\beta$ decays. A few experiments have reported positive results [450, 451], but none of them has been confirmed. There have been searches for second-class currents in $\tau$ decays, too[4] (see Sect. 10.1).

The classification into first- and second-class currents strictly applies to the weak vertex only, not to the subsequent fragmentation, because there $G$ can be violated owing to electromagnetic interactions or by the small mass difference of the quarks. Hence isospin-symmetry-violating interactions might introduce small admixtures of second-class currents. But in the special case of $\tau$ decays into a spin-0 state the strong interaction cannot alter $G$ parity and the separation into first- and second-class currents at the weak vertex is not washed out by fragmentation.[5]

### 5.1.10 Summary

Table 5.1 summarizes the quantum numbers of the hadronic states accessible in $\tau$ decays and the mechanisms of suppression. There are only two channels, namely $J^{PG} = 1^{-+}$ and $1^{+-}$, which do not suffer from any suppression. These are the decays $\tau \to \rho\,\nu_\tau$ and $\tau \to a_1\,\nu_\tau$ and indeed they are the two channels with the largest branching ratios, the $\rho$ being somewhat preferred by phase space. Helicity suppression or PCAC reduces the $\tau \to \pi\,\nu_\tau$ decay to number three in the list of branching ratios. All other hadronic decays are small compared to these.

All decays requiring the creation of a strange quark are Cabibbo-suppressed. They show a pattern similar to the nonstrange decays (see Table 5.2). There is no isospin symmetry in these decays, so there is neither isospin suppression nor $G$ parity classification. And, owing to the higher mass of the strange quark, helicity suppression is weaker. The $1^-$ is still the strongest

---

[4] There is one paper claiming evidence for a second-class current in $\tau$ decays, but it is in contradiction with later, more precise measurements [452].

[5] The reason is that the eigenstates of parity $\mathcal{P}$ (see (5.5)) are at the same time eigenstates of the combined transformation $G\mathcal{P}$. A flip of $G$ parity by a strong interaction would change the value of $G\mathcal{P}$ and therefore lead from one eigenstate of parity to the opposite. As QCD conserves parity, this cannot happen in the fragmentation process [446].

**Table 5.1.** Summary of the hadronic final states accessible in $\tau$ decays. The various suppression mechanisms are explained in the preceding subsections. The last column specifies the accessible resonances, but the same selection rules apply to nonresonant decays

| | Nonstrange decays | | | | |
|---|---|---|---|---|---|
| $J^{PG}$ | Helicity PCAC | Isospin CVC | Second class | Exotic | |
| $0^{++}$ | | x | | x | |
| $0^{+-}$ | | x | x | | $a_0(980)$ |
| $0^{-+}$ | x | | x | x | |
| $0^{--}$ | x | | | | $\pi$ |
| | x | | | | $\pi(1300)$ |
| $1^{++}$ | | | x | | $b_1(1235)$ |
| $1^{+-}$ | | | | | $a_1(1260)$ |
| $1^{-+}$ | | | | | $\rho(770)$ |
| | | | | | $\rho(1450)$ |
| | | | | | $\rho(1700)$ |
| $1^{--}$ | | | x | x | |

**Table 5.2.** Summary of the Cabibbo suppressed hadronic final states accessible in $\tau$ decays

| | Strange decays | | |
|---|---|---|---|
| $J^P$ | Cabibbo | Helicity | |
| $0^+$ | x | x | $K_0^*(1430)$ |
| $0^-$ | x | x | $K$ |
| $1^+$ | x | | $K_1(1270)$ |
| | x | | $K_1(1400)$ |
| $1^-$ | x | | $K^*(892)$ |
| | x | | $K^*(1410)$ |
| | x | | $K^*(1680)$ |

decay, but the $0^-$ is close now. The lightest $1^+$ and $0^+$ resonances are already close to the phase space limit and the branching ratios have not been measured yet.

## 5.2 Theoretical Description of Hadronic $\tau$ Decays

### 5.2.1 Tsai's Formula

The decay of the $\tau$ lepton into hadrons (see Fig. 5.1) was calculated by Paul Tsai [23], even before the discovery of the $\tau$. From current algebra – ignoring the propagator of the $W^\pm$ boson – the matrix element is given by the product of the leptonic and the hadronic current:

$$i\mathcal{M} = \frac{G_F}{\sqrt{2}} J^\mu_{\text{lep}} J_{\text{had}\,\mu}. \tag{5.11}$$

The leptonic current is the standard left-handed one. The hadronic current is not known a priori. We know only that it has to be a Lorentz vector. For the production of a hadron with zero spin, the only Lorentz vector available is its momentum $Q_\mu$. So the scalar part of the current has to be proportional to $Q_\mu$, and the vector part is proportional to the polarization vector $\varepsilon_\mu$:

$$J^{J=0}_{\text{had }\mu} = f_0(Q^2) Q_\mu,$$
$$J^{J=1}_{\text{had }\mu} = f_1(Q^2) \varepsilon_\mu. \tag{5.12}$$

The coefficients $f_0$ and $f_1$ are scalar functions and might depend on $Q^2$. They cannot be calculated from QCD so far, but have to be determined from experiment. They can be further decomposed according to parity and the strangeness content of the current:

$$f_0^2(Q^2) = \cos^2\theta_\text{C} \left[ \qquad a_0(Q^2) \right] + \sin^2\theta_\text{C} \left[ v_0^\text{s}(Q^2) + a_0^\text{s}(Q^2) \right],$$
$$f_1^2(Q^2)/Q^2 = \cos^2\theta_\text{C} \left[ v_1(Q^2) + a_1(Q^2) \right] + \sin^2\theta_\text{C} \left[ v_1^\text{s}(Q^2) + a_1^\text{s}(Q^2) \right]. \tag{5.13}$$

There is no interference between the different parts, so that it is good enough to decompose the $f_i^2$. The $v$ and $a$ are called spectral functions. The lower index indicates the spin, and the upper labels the strange current. The symbol $v$ stands for vector and scalar, and $a$ for axial-vector and pseudoscalar. Notice the absence of the nonstrange scalar current owing to CVC (see Sects. 5.1.4 and 5.1.7). With these currents the result of the calculation is

$$\Gamma(\tau \to \text{had } \nu_\tau) = \frac{G_\text{F}^2 m_\tau^3}{32\,\pi^2} \int_0^{m_\tau^2} dQ^2 \left(1 - \frac{Q^2}{m_\tau^2}\right)^2$$

$$\times \left\{ \cos^2\theta_\text{C} \left[ \left(1 + 2\frac{Q^2}{m_\tau^2}\right)\left(v_1(Q^2) + a_1(Q^2)\right) \qquad + a_0(Q^2) \right] \right.$$

$$\left. + \sin^2\theta_\text{C} \left[ \left(1 + 2\frac{Q^2}{m_\tau^2}\right)\left(v_1^\text{s}(Q^2) + a_1^\text{s}(Q^2)\right) + v_0^\text{s}(Q^2) + a_0^\text{s}(Q^2) \right] \right\}. \tag{5.14}$$

### 5.2.2 Parametrization of the Decays

Although Tsai's formula (5.14) is well suited to describe the mass spectrum of hadrons expected from $\tau$ decays, it does not contain any information on the dynamics of these decays. An extended parametrization including all the dynamics has been worked out by J. Kühn and his group [437, 453–455].

This again starts from the general expression for the matrix element given in (5.11) but this time the hadronic current is parametrized in terms of the mesons emerging from the decay of the hadronic resonances in Tsai's formula. This allows a description of the fully differential decay width including the angular momenta between the mesons, which, together with the spin and parity of the mesons, determine the quantum numbers of the current. However, a new parametrization is necessary now, for each decay with a given number of identical or nonidentical mesons.

**One-Meson Final States.** The decay of the $\tau$ lepton to a single meson and the $\nu_\tau$ does not contain any structure that is not included in (5.12).

**Two-Meson Final States.** For the decay $\tau^- \to \pi^- \pi^0 \nu_\tau$ the most general decomposition of the hadronic current consistent with Lorentz invariance is given by

$$J^\mu_{\text{had}} = \sqrt{2} \cos\theta_C \left[ F_1\left(Q^2\right) \left(q_1^\mu - q_2^\mu\right) + F_2\left(Q^2\right) Q^\mu \right], \tag{5.15}$$

where $q_1$ and $q_2$ are the momenta of the charged and neutral pions, respectively, and $Q = q_1 + q_2$. $F_1$ and $F_2$ are form factors depending on $Q^2$. The transformation properties of these components can be most easily determined in the hadronic rest frame, where the spatial components of $Q^\mu$ vanish: $Q^\mu = (|Q|, 0, 0, 0)$. The part proportional to $F_1\left(Q^2\right)$ transforms like a vector, the part proportional to $F_2\left(Q^2\right)$ like a scalar. Including the negative parity of the two pions, $F_1\left(Q^2\right)$ describes the vector current and $F_2\left(Q^2\right)$ the scalar current.[6] No other current can be constructed from two pions.

The same parametrization can be applied to the decays $\tau^- \to K^- \pi^0 \nu_\tau$, $\tau^- \to \pi^- K^0 \nu_\tau$, and $\tau^- \to K^- K^0 \nu_\tau$. However, this is only the simplest case of a two-meson final state, with two pseudoscalar mesons; already $\tau^- \to \pi^- \omega \nu_\tau$ has a richer structure because of the additional degree of freedom, the spin of the $\omega$ (see [454]).

**Three-Meson Final States.** The two most important decays are $\tau^- \to \pi^- \pi^- \pi^+ \nu_\tau$ and $\tau^- \to \pi^- \pi^0 \pi^0 \nu_\tau$. The hadronic current can be parametrized by

$$\begin{aligned}
J^\mu_{\text{had}} = \cos\theta_C & \\
\times \Bigg[ & F_1\left(Q^2\right) \left( q_1^\mu - q_3^\mu - Q^\mu \frac{Q\left(q_1 - q_3\right)}{Q^2} \right) \\
+ & F_2\left(Q^2\right) \left( q_2^\mu - q_3^\mu - Q^\mu \frac{Q\left(q_2 - q_3\right)}{Q^2} \right) \\
+ & \mathrm{i} F_3\left(Q^2\right) \varepsilon^{\mu\alpha\beta\gamma} q_{1\alpha}\, q_{2\beta}\, q_{3\gamma} \\
+ & F_4\left(Q^2\right) Q^\mu \Bigg],
\end{aligned} \tag{5.16}$$

which is symmetric under the exchange of $q_1$ and $q_2$, the momenta of the two like-signed pions. The momentum of the third pion is $q_3$, and $Q = q_1 + q_2 + q_3$.

---
[6] The scalar current is forbidden by CVC, i.e. $F_2 = 0$; see Sects. 5.1.4 and 5.1.7.

The form factors $F_1$ and $F_2$ together describe the axial-vector current, which dominates the decay. The component proportional to $F_4$ is a pseudoscalar current and is suppressed by helicity (see Sects. 5.1.3 and 5.1.8). A vector current proportional to $F_3$, if it exists, would have to be created by a $G$-parity-violating interaction in the fragmentation process. The three pions carry negative $G$ parity. A vector current without the $G$-parity-violating interaction would be an exotic current (see Table 5.1).

Parametrizations for three-meson decays involving kaons have been given in [455] and some information about four-meson final states can be found in [456]. These are only examples of parametrizations of the hadronic current for a few of the most important final states.

The partial decay width is then calculated from the square of the matrix element in (5.11). This turns the product of the two currents into a tensor product. The lepton tensor can be calculated perturbatively. The hadronic tensor is given by

$$H^{\mu\nu} = J^{\mu}_{\text{had}} \left(J^{\nu}_{\text{had}}\right)^{\dagger}. \tag{5.17}$$

This is a hermitian matrix with 16 independent elements, built from the momenta and spin vectors of the final-state mesons and the complex form factors $F_i\left(Q^2\right)$. All the information about the strong interactions in the decays is contained in these form factors. Everything else is kinematics and spin dynamics only. The form factors can in principle be calculated from QCD, just like the spectral functions in Tsai's formula, but they involve essentially nonperturbative phenomena so that nobody knows how to do the calculation. In the absence of such a calculation, less fundamental models are used instead and challenged by experiment.

It is convenient to perform one further manipulation on the square of the matrix element. The 16 elements $H^{\mu\nu}$ are rearranged into 16 hadronic structure functions $W_i$, thereby selecting special, more meaningful combinations. The structure function $W_A = H_{11} + H_{22}$, for example, represents the total strength of the axial-vector current. For details see [437].

### 5.2.3 Chiral Perturbation Theory

It would be an exceptionally interesting test to compare the structure functions (5.13) or the form factors from Sect. 5.2.2, calculated from pure QCD, with experimental results. Unfortunately the usual method of solving QCD by perturbative expansions in $\alpha_s$, dealing with quarks as asymptotic states, is not appropriate to calculating the resonances in exclusive $\tau$ decays. In the absence of an explicit solution of QCD, an effective field theory, based on hadrons instead of quarks as elementary fields, might be used at reasonably low energies. This is called chiral perturbation theory (ChPT) [457], emphasizing the symmetry between left- and right-handed quarks as the most relevant at these energies. (For a review of ChPT see [458] and references therein.)

This theory is based on a simple Lagrangian consistent with all symmetries of QCD, especially the chiral one. But nature does not show chiral symmetry at low energies. The lowest-lying multiplets of vector and axial-vector mesons are far from being degenerate in mass and this is even more evident for scalars versus pseudoscalars. Therefore in the theory the chiral symmetry of the Lagrangian is spontaneously broken, generating a multiplet of massless pseudoscalar Goldstone bosons. Now there is a mass gap between these bosons and the other mesons, and all the heavy states are integrated out from the Lagrangian. What is left over now describes the interaction of the Goldstone bosons at energies small compared to the masses of the other mesons. This can be expanded in powers of the momenta of the Goldstone bosons:

$$\mathcal{L} = \mathcal{L}_2 + \mathcal{L}_4 + \ldots \tag{5.18}$$

Coupling to external currents can be introduced in the usual gauge-invariant manner by covariant derivatives. These can be either the hadronic currents of the heavier mesons or a charged weak current. A small mass term explicitly breaking the chiral symmetry is added, too. Then the Lagrangian reads, to lowest order,

$$\mathcal{L}_2 = \frac{f^2}{4}\left[D_\mu U^\dagger D^\mu U + 2B_0 \mathcal{M}\left(U^\dagger + U\right)\right]. \tag{5.19}$$

Here $f$ is a constant (the pion decay constant), which might be calculated from $\alpha_s$ once a method to solve QCD at low energies is found. For the time being, its value is taken from experiment. $U = \exp\left(i\sqrt{2}/f\,\Phi\right)$ is a function of the matrix $\Phi$ built from the Goldstone fields. This matrix is

$$\Phi = \begin{pmatrix} \frac{\pi^0}{\sqrt{2}} + \frac{\eta_8}{\sqrt{6}} & \pi^+ & K^+ \\ \pi^- & -\frac{\pi^0}{\sqrt{2}} + \frac{\eta_8}{\sqrt{6}} & K^0 \\ K^- & \bar{K}^0 & -\frac{2\eta_8}{\sqrt{6}} \end{pmatrix} \tag{5.20}$$

for three flavors and

$$\Phi = \begin{pmatrix} \frac{\pi^0}{\sqrt{2}} & \pi^+ \\ \pi^- & -\frac{\pi^0}{\sqrt{2}} \end{pmatrix} \tag{5.21}$$

for two. The derivatives $D_\mu U = \partial_\mu U - i\left(v_\mu + a_\mu\right)U + iU\left(v_\mu - a_\mu\right)$ introduce the couplings to the electroweak currents, $\mathcal{M} = \text{diag}\left(m_u, m_d, m_s\right)$ is the mass matrix (omit $m_s$ for two flavors), and $B_0$ another constant. The couplings to the mesons are introduced in the same way and will be discussed in Sect. 5.2.4

Some of the vertices generated by the Lagrangian are depicted in Fig. 5.5. There are couplings to the external fields – the charged weak current in the case of $\tau$ decays – of any number of pseudoscalars and couplings between them with an even number of pseudoscalars. As an example Fig. 5.6 shows

148    5. Strong Interactions in τ Decays

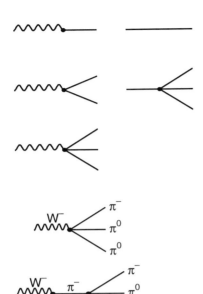

**Fig. 5.5.** Vertices generated by the chiral Lagrangian. The *wavy lines* represent the weak current, the *straight lines* pseudoscalar mesons

**Fig. 5.6.** The two lowest-order diagrams from ChPT contributing to the decay $\tau^- \to \pi^- \pi^0 \pi^0 \nu_\tau$

the two diagrams which contribute in lowest order to the decay of the $\tau$ lepton to three pions.

**Low-Energy Limits.** As mentioned earlier, ChPT can only be used at low energies. The relevant parameter of expansion is $Q^2/(4\pi f_\pi)^2$. With $4\pi f_\pi \approx$ 1.2 GeV the expansion is limited to $Q$ below 500 to 600 MeV. At these low energies there is a strong phase space suppression. It becomes overwhelming if there are kaons in the final state or more than three pions.[7] The low-energy limits of the decays $\tau^- \to \pi^- \pi^0 \nu_\tau$ and $\tau^- \to \pi^- \pi^- \pi^+ \nu_\tau$ have been calculated in [459]. Figure 5.7 shows the results for the differential decay rate of $\tau^- \to \pi^- \pi^0 \nu_\tau$. They agree well with data from the CLEO collaboration [460] and predictions from a VDM [300] up to about 500 MeV (see Sect. 5.2.4). Figure 5.8 shows the prediction from ChPT for one of the structure functions (see Sect. 5.2.2 and [437]) relevant to the decay $\tau \to 3\pi\nu_\tau$. This function is $W_E$ integrated over the Dalitz plane as a function of $Q^2$. This structure function is particularly interesting as ChPT predicts a difference in sign between the two charged modes $\pi^- \pi^- \pi^+$ and $\pi^- \pi^0 \pi^0$ which is not reproduced by VDMs [437, 453] and is not present at higher energies in the data [87, 226] (no data is available at these low energies yet).

---

[7] The decays $\tau^- \to \pi^- \nu_\tau$ and $\tau^- \to K^- \nu_\tau$ can be calculated from $f_\pi$ and $f_K$, but do not contain any dynamic structure.

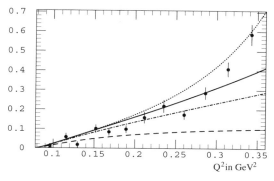

**Fig. 5.7.** Differential decay rate for $\tau^- \to \pi^- \pi^0 \nu_\tau$ normalized to $\Gamma_e$ as a function of $Q^2$ in units of GeV$^2$. *Dashed line*: ChPT prediction to $\mathcal{O}(p^2)$; *dash–dotted line*: ChPT to $\mathcal{O}(p^4)$; *solid line*: ChPT to $\mathcal{O}(p^6)$ [459]; *dotted line*: VDM; *points with error bars*: CLEO data [460]

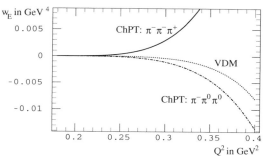

**Fig. 5.8.** The structure function $W_E$ integrated over the Dalitz plane (units GeV$^4$) as a function of $Q^2$ in units of GeV$^2$ [459]. *Solid line*: ChPT prediction to $\mathcal{O}(p^4)$ for $\pi^- \pi^- \pi^+$; *dash–dotted line*: same for $\pi^- \pi^0 \pi^0$; *dotted line*: VDM for both charge combinations

**Chiral Anomalies.** In the limit of vanishing quark masses the QCD Lagrangian is symmetric under $\gamma_5$ transformations ($\Psi(x)$ is a quark field):

$$\Psi(x) \to e^{-i\theta\gamma_5}\Psi(x). \tag{5.22}$$

Nevertheless the associated Noether current, the axial-vector current $J_\mu = \bar\Psi\gamma_\mu\gamma_5\Psi$, is not conserved[8] [461, 462]. Its divergence acquires a term in addition to the small contribution induced by the quark masses. The divergence is

$$\partial^\mu J_\mu = m\,\bar\Psi\gamma_5\Psi + \frac{\alpha}{4\pi}\,F^{\sigma\nu}F^{\rho\lambda}\epsilon_{\sigma\nu\rho\lambda}, \tag{5.23}$$

where $F$ is the field tensor of the external current and $\epsilon$ the Levi–Civita symbol. The vertices generated by this anomalous divergence can be represented at low energies by additional terms in the chiral Lagrangian [463, 464]. Some of the new vertices are depicted in Fig. 5.9. Especially interesting is the last vertex. It enables the $\pi^0$ to decay to two photons, which was excluded in ChPT until the introduction of this anomaly, which is called the Wess–Zumino anomaly, referring to the authors of [463].

---

[8] The usual Noether-type argument breaks down in perturbation theory if loop integrals with high divergences are present. In the case of QCD it is the triangular diagram, with three (axial-) vector currents coupling to a quark loop, which causes the breakdown.

150    5. Strong Interactions in $\tau$ Decays

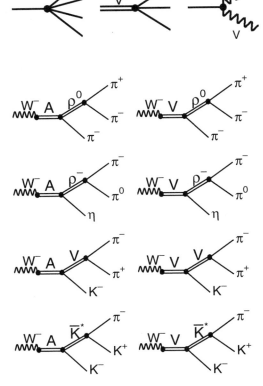

**Fig. 5.9.** Vertices generated by the Wess–Zumino anomaly. The external currents might either be of electroweak nature (*wavy lines*) or represent vector and axial-vector mesons (*double lines*)

**Fig. 5.10.** Decay diagrams for $\tau$ decays into three pseudoscalars. The diagrams in the *right column* include a vector–vector–pseudoscalar vertex, which is generated in ChPT by the Wess–Zumino anomaly

Figure 5.10 shows the production of three mesons through vector and axial-vector currents from a charged weak current in $\tau$ decays. The intermediate vertex of the diagrams on the right-hand side is generated by the Wess–Zumino anomaly. The left-hand side shows its nonanomalous counterpart. In the case of the $\tau$ lepton decaying to three pions, $G$ parity is violated at the anomalous vertex and this diagram is therefore suppressed. This is experimentally verified by the dominance of the $a_1$ meson in this decay. By the same argument the nonanomalous contribution to the decay $\tau^- \to \eta \pi^- \pi^0 \nu_\tau$ is suppressed. The fact that this decay exists at all demonstrates the presence of the Wess–Zumino anomaly in $\tau$ decays. For the other two decays, involving kaons, $G$ parity is no longer relevant and both diagrams may contribute. Similar pictures can be drawn for any number of mesons in the final state.

### 5.2.4 Vector Dominance

Most of the hadronic decays of the $\tau$ lepton are dominated by intermediate vector or axial-vector mesons, eventually decaying to pseudoscalars. This observation is the main guideline of the vector dominance models (VDMs) of $\tau$ decays. The problem, however, is that neither of these resonances in the hadronic currents can be predicted from theory. The currents are simply parametrized with Breit–Wigner form factors with the mass and width as free parameters. These and their couplings are then fixed by a fit to the experimental data. Table 5.3 summarizes the main decay channels of the $\tau$ lepton where dominating resonances have been observed.

**Table 5.3.** Resonances observed in various $\tau$ decays

| Final state | Intermediate resonances |
|---|---|
| $\tau^- \to \pi^- \pi^0 \nu_\tau$ | $\rho^- \to \pi^- \pi^0$ |
| $\tau^- \to \pi^- \pi^- \pi^+ \nu_\tau$ | $a_1^- \to \pi^- \rho^0;\ \rho^0 \to \pi^- \pi^+$ |
| $\tau^- \to \pi^- \pi^0 \pi^0 \nu_\tau$ | $a_1^- \to \pi^0 \rho^-;\ \rho^- \to \pi^- \pi^0$ |
| $\tau^- \to K^- \pi^0 \nu_\tau$ | $K^{*-} \to K^- \pi^0$ |
| $\tau^- \to \pi^- K^0 \nu_\tau$ | $K^{*-} \to \pi^- K^0$ |

From a theoretical point of view the parametrization of the hadronic current should fulfill at least the following two conditions:

- The current should be conserved.
- In the low-energy limit ($Q^2 \to 0$) the resonances vanish and the current should coincide with that of ChPT.

In the following two paragraphs the basic VDMs for $\tau^- \to \pi^- \pi^0 \nu_\tau$ and $\tau^- \to \pi^- \pi^- \pi^+ \nu_\tau$ will be presented.

**The Two-Pion Channel.** The two-pion channel is dominated by the $\rho$ resonance. The current is given by

$$J^\mu = \sqrt{2}\, F_1^{2\pi}(Q^2)\, (q_1^\mu - q_2^\mu), \tag{5.24}$$

where the $q_i$ are the momenta of the pions (see (5.15)). Or, in terms of the spectral functions (5.13),

$$v_1^{2\pi}(Q^2) = \frac{1}{12\pi} \left[F_1^{2\pi}(Q^2)\right]^2 \left(1 - \frac{4m_\pi^2}{Q^2}\right)^{\frac{3}{2}}, \tag{5.25}$$

with the same form factor $F_1^{2\pi}$. The form factor is now parametrized by a Breit–Wigner resonance. This could be

$$F_1^{2\pi} = \mathrm{BW}_\rho = \frac{m_\rho^2}{m_\rho^2 - Q^2 - \mathrm{i}\, Q\, \Gamma_\rho(Q^2)}, \tag{5.26}$$

with $m_\rho$ and $\Gamma_\rho$ the mass and width of the $\rho$ meson. This basic parametrization is the same for all VDMs. There are, however, differences in various details, such as

- the variation of the width $\Gamma_\rho$ with $Q^2$
- the inclusion of higher excitations ($\rho'$, $\rho''$)
- distortions of the Breit–Wigner resonance at high $Q^2$ due to the opening of new decay channels ($K\bar{K}$ or $4\pi$)
- the incorporation of analyticity.

For example, Kühn and Santamaria [300] use a form expected from a relativistic p-wave phase space under the assumption of a constant $\rho\pi\pi$ coupling. They use a width

$$\Gamma_\rho\left(Q^2\right) = \Gamma_\rho \frac{p^3}{Q^2} \frac{m_\rho^2}{p_\rho^3} \qquad (5.27)$$

with $\quad p = \frac{1}{2}\sqrt{Q^2 - 4m_\pi^2}, \qquad p_\rho = \frac{1}{2}\sqrt{m_\rho^2 - 4m_\pi^2}. \qquad (5.28)$

This parametrization is plotted in Fig. 5.11. Higher excitations are added through

$$F_1^{2\pi}\left(Q^2\right) = \frac{\mathrm{BW}_\rho + \beta\,\mathrm{BW}_{\rho'} + \gamma\,\mathrm{BW}_{\rho''}}{1 + \beta + \gamma}. \qquad (5.29)$$

The relative phases $\beta$ and $\gamma$ are assumed to be real. Kühn and Santamaria fix their parameters under the assumption of CVC by a fit to low-energy $e^+e^-$ data. (This requires the inclusion of the $\omega$, which is then dropped for the prediction of $\tau$ decays.) They find $m_\rho = 773$ MeV, $m_{\rho'} = 1370$ MeV, $\Gamma_\rho = 145$ MeV, $\Gamma_{\rho'} = 510$ MeV, and $\beta = -0.145$. The evidence for the presence of a $\rho''$ is inconclusive. Figure 5.12 shows the resulting partial decay width of the $\tau$. Similar models of the form factor relevant to $\tau^- \to \pi^- \pi^0 \nu_\tau$ have been given in [465–467].

**The Three-Pion Channel.** The decay of the $\tau$ lepton into three pions is dominated by the axial-vector meson $a_1$, which decays into a $\rho$ and a pion and finally to three pions. Again the form factor is parametrized by a Breit–Wigner resonance [300, 465, 468], but the functional dependence on the momenta is now given by (5.16):

$$J^\mu = -\mathrm{i}\,\frac{2\sqrt{2}}{3f_\pi}\,\mathrm{BW}_{a_1}\left(Q^2\right)$$
$$\times \left[ B_\rho(s_2)\left(q_1^\mu - q_3^\mu - Q^\mu \frac{Q(q_1 - q_3)}{Q^2}\right) \right.$$
$$\left. + B_\rho(s_1)\left(q_2^\mu - q_3^\mu - Q^\mu \frac{Q(q_2 - q_3)}{Q^2}\right) \right]. \qquad (5.30)$$

The current is symmetric under the exchange of the two identical pions in accordance with Bose symmetry. The normalization factor $-\mathrm{i}(2\sqrt{2})/(3f_\pi)$

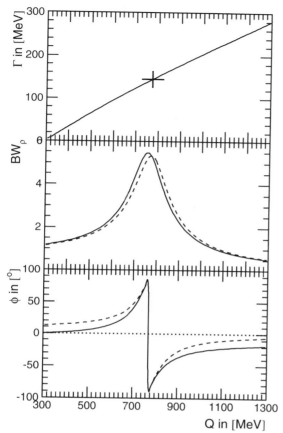

**Fig. 5.11.** Parametrization of the form factor for $\tau^- \to \pi^- \pi^0 \nu_\tau$ by a Breit–Wigner resonance. *Upper plot*: $Q^2$ dependence of the width. *Central plot*: absolute value of $BW_\rho$ for the Kühn/Santamaria model (*solid line*) and constant width (*dashed line*). *Lower plot*: phase of $BW_\rho$ (*solid/dashed lines* as above)

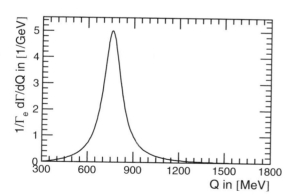

**Fig. 5.12.** Partial decay width of the decay $\tau^- \to \pi^- \pi^0 \nu_\tau$ calculated from the Kühn/Santamaria model

is chosen so that the current meets the chiral prediction in the limit $Q^2 \to 0$ ($\mathrm{BW}_{a_1} \to 1$ for $Q^2 \to 0$). A Breit–Wigner resonance as in (5.26) is used in the Kühn/Santamaria model with the parameters fixed by a fit to data from the ARGUS experiment [96] ($m_{a_1} = 1251$ MeV and $\Gamma_{a_1} = 599$ MeV). The functions $B_\rho$ ($s_1 = (q_2 + q_3)^2$, $s_2 = (q_1 + q_3)^2$) parametrize the intermediate resonances in $a_1 \to (\pi\pi)\pi \to \pi\pi\pi$. It is not necessarily true that the same vector mesons that make up the $\tau^- \to \pi^- \pi^0 \nu_\tau$ decay appear as $2\pi$ subresonances here, but assuming $B_\rho \equiv F_1^{2\pi}$ gives a good description of the data. Figure 5.13 shows the resulting decay distribution for the decay $\tau^- \to \pi^- \pi^- \pi^+ \nu_\tau$ and Fig. 5.14 gives an example of a decay distribution in the Dalitz plane. Notice that there are two possible combinations of $\pi^+ \pi^-$ that can make up the $\rho$ and there is constructive interference between them. This creates the peak in the Dalitz plane where the two $\rho$ bands cross.

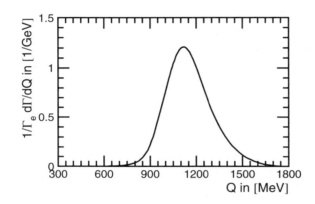

**Fig. 5.13.** Partial decay width of the decay $\tau^- \to \pi^- \pi^- \pi^+ \nu_\tau$ calculated from the Kühn/Santamaria model

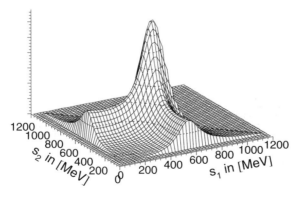

**Fig. 5.14.** Decay distribution in the Dalitz plane of the decay $\tau^- \to \pi^- \pi^- \pi^+ \nu_\tau$ calculated from the Kühn/Santamaria model at $Q^2 = 1.25$ GeV$^2$

VDMs for $\tau$ decays involving kaons can be found in [455], for $\tau^- \to \pi^- \omega \nu_\tau$ in [454], for the decay into three pseudoscalars with predictions of the Wess–Zumino contributions in [303], and for $\tau$ decays to four pions in [456].

**Other models.** Models of vector dominance in $\tau$ decays are very popular, but they are not the only type of models available. It is not intented to give a complete discussion of all models here, but a few more will be mentioned:

- Isgur, Morningstar, and Reader [372] have presented an alternative way of describing the decay chain $\tau \to a_1 \nu_\tau$; $a_1 \to \rho \pi$; $\rho \to \pi \pi$. They couple the hadrons at the vertices using the most general forms of their currents and then propagate them to the next vertex by propagators introducing the parameters of these mesons. To avoid problems in their calculations with nonresonant contributions, they restrict themselves to amplitudes which only produce transversely polarized vector mesons.[9] A sufficiently rich parametrization of these nonresonant amplitudes is added incoherently later on. (They choose a third-order polynomial in $m_\pi^2$.) The couplings of the currents and their $Q^2$ dependence are calculated from a flux-tube-breaking model on a lattice [469].
- There have been attempts to include the vector mesons in the effective Lagrangian of ChPT. Braaten, Oakes and Tse [470] used the most general Lagrangian including these fields consistent with chiral symmetry and the electroweak gauge symmetry, fixed the additional coupling constants from vector meson decays, and predicted several branching ratios of the $\tau$. And recently Davoudiasl and Wise [471] have calculated the dynamics of the decays $\tau \to \rho \pi \nu_\tau$, $\tau \to K^* \pi \nu_\tau$, and $\tau \to \omega \pi \nu_\tau$ in regions of the phase space where the pion is slow in the rest frame of the vector meson.
- Special quark models have been used in $\tau$ decays, for example [472].
- Truong and collaborators have applied current algebra to $\tau$ decays [306, 473–476]. Here again the form factors of the currents have been parametrized by Breit–Wigner-type functions representing the dominating vector and axial-vector mesons.

### 5.2.5 Isospin Relations

Multipion states like those created in $\tau$ decays can be characterized by their isospin or similar quantum numbers. But Pais pointed out in 1960 [477] that it might be more important to classify the different states according to their transformation properties under the exchange of individual pions. If one class dominates a reaction, this becomes a powerful, predictive tool, if not, it is still a convenient way to organize the wave functions of the final states. So he grouped the particles of an $N \pi$ state into singlets, doublets, and triplets, where the wave function describing the configuration should be antisymmetric under the exchange of two pions within a multiplet and symmetric between

---
[9] This is strictly true only if the $a_1$ and $\rho$ are completely on-shell.

multiplets. Notice that no more than three pions can be antisymmetrized, because with four at least two are identical and Bose symmetry requires a symmetric wave function for these.

A partition is identified by three numbers $(N_1\ N_2\ N_3)$ which define the numbers of triplets $(N_3)$, doublets $(N_2 - N_3)$, and singlets $(N_1 - N_2)$. It is a set of configurations with different pions making up the triplets, doublets, and singlets. The interesting point is that for $I < 2$ and no more than eight pions, the partition numbers $(N_1\ N_2\ N_3)$ determine uniquely the isospin of the partition and, by construction, the transformation properties. If one further specifies the charge and labels the configuration, i.e. one specifies which pions $\pi_i$ with momenta $\boldsymbol{p}_i$ make up the triplets, doublets, and singlets, the wave function is uniquely determined.

Table 5.4 shows the simplest example, the two partitions of the 2 $\pi$ states. The partitions can be graphically displayed by Young tableaux: each square represents a pion, and pions in the same column belong to the same multiplet and have to be antisymmetrized. Along a row the wave function has to be symmetric.

The numbers label the different configurations. The 2 $\pi$ state has only one configuration per partition. Table 5.5 shows the labeled Young tableaux for the 3 $\pi$ state, which is more illuminating. The rule is that the labels always have to increase within a row from left to right, and in a column from top to bottom, otherwise one reproduces configurations already accounted for. (Table 5.4 also gives the isospin of the partitions and the wave functions which go with the different charges.)

First of all, the partitions are a convenient way of organizing the wave functions of multipion states. But they also have predictive power, especially if an $N\ \pi$ state is known to be dominated by a single partition. In that case it is sufficient to measure a single charge combination of the $N$ pions. The dynamics will be identical between all charge combinations and the branching ratios can be calculated from the partition. Tables 5.5–5.8 present all the partitions which contribute to $\tau$ decays into three to six pions, display the Young tableaux, and give the branching into the different channels.

**Applications.**

- From Table 5.5 one reads that the $(2\,1\,0)$ partition would create equal branching ratios of the $\tau$ into $\pi^-\ \pi^0\ \pi^0$ and $\pi^-\ \pi^+\ \pi^-$, whereas the $(3\,0\,0)$ partition would make the all-charged mode four times as big as the neutral mode. Measurements [44] show almost equal branching ratios, $(9.15 \pm 0.15)\,\%$ for $\tau^- \to \pi^-\ \pi^0\ \pi^0\ \nu_\tau$ and $(9.23 \pm 0.11)\,\%$ for $\tau^- \to \pi^-\ \pi^-\ \pi^+\ \nu_\tau$, identifying the 3 $\pi$ decay as being dominated by the $(2\,1\,0)$ partition. Quantitatively, defining $\alpha$ as the contribution of the $(3\,0\,0)$ partition to the total $\tau \to 3\,\pi\,\nu_\tau$ decay, we have

$$\frac{br\left(\pi^-\ \pi^0\ \pi^0\right)}{br\left(\pi^-\ \pi^+\ \pi^-\right)} = \frac{\frac{1}{5}\alpha + \frac{1}{2}(1-\alpha)}{\frac{4}{5}\alpha + \frac{1}{2}(1-\alpha)}, \tag{5.31}$$

**Table 5.4.** Partitions, Young tableaux, and wave functions for a two-pion final state

| | | | |
|---|---|---|---|
| (2 0 0) | $\boxed{1\,2}$ | | |
| $I = 0$ | $Q = 0$ | | $\frac{1}{\sqrt{3}} \left( \pi_1^+ \pi_2^- - \pi_1^0 \pi_2^0 + \pi_1^- \pi_2^+ \right)$ |
| (1 1 0) | $\begin{array}{c}\boxed{1}\\\boxed{2}\end{array}$ | | |
| $I = 1$ | $Q = +1$ | | $\frac{1}{\sqrt{2}} \left( \pi_1^+ \pi_2^0 - \pi_1^0 \pi_2^+ \right)$ |
| | $Q = 0$ | | $\frac{1}{\sqrt{2}} \left( \pi_1^+ \pi_2^- - \pi_1^- \pi_2^+ \right)$ |
| | $Q = -1$ | | $\frac{1}{\sqrt{2}} \left( \pi_1^0 \pi_2^- - \pi_1^- \pi_2^0 \right)$ |

**Table 5.5.** Partitions and labeled Young tableaux of a three-pion final state. The second line in each row gives the branching fractions for the $Q = -1$ states ($\tau$ decays)

| | | | |
|---|---|---|---|
| (3 0 0) | $\boxed{1\,2\,3}$ | | $I = 1$ |
| | $\pi^- \, 2\pi^0$ : | $\frac{1}{5}$ | |
| | $\pi^- \pi^+ \pi^-$ : | $\frac{4}{5}$ | |
| (2 1 0) | $\begin{array}{c}\boxed{1\,2}\\\boxed{3}\end{array}$ | $\begin{array}{c}\boxed{1\,3}\\\boxed{2}\end{array}$ | $I = 1$ |
| | $\pi^- \, 2\pi^0$ : | $\frac{1}{2}$ | |
| | $\pi^- \pi^+ \pi^-$ : | $\frac{1}{2}$ | |
| (1 1 1) | $\begin{array}{c}\boxed{1}\\\boxed{2}\\\boxed{3}\end{array}$ | | $I = 0$ |

**Table 5.6.** Partitions, Young tableaux, and branching ratios for $\tau \to 4\pi\,\nu_\tau$

---

(4 0 0)

$I = 0$

---

(3 1 0)

$I = 1$   $\pi^- 3\pi^0$ :   $\frac{2}{5}$

$2\pi^- \pi^+ \pi^0$ :   $\frac{3}{5}$

---

(2 2 0)

$I = 0$

---

(2 1 1)

$I = 1$   $\pi^- 3\pi^0$ :   $0$

$2\pi^- \pi^+ \pi^0$ :   $1$

---

which results in $\alpha = (0.7 \pm 1.7)\,\%$, consistent with zero. Isospin symmetry then predicts the structures of the two decays to be identical, i.e. they have the same $Q^2$, Dalitz plane, and angular distributions.

- A similar calculation can be applied to the $4\pi$ decays of the $\tau$. The (3 1 0) partition contributes to the decays to $\pi^- 3\pi^0$ and $2\pi^- \pi^+ \pi^0$ in the ratio 2:3, whereas the (2 1 1) partition only feeds the second channel. We have

$$\frac{br\left(\pi^- 3\pi^0\right)}{br\left(2\pi^- \pi^+ \pi^0\right)} = \frac{\frac{2}{5}\alpha}{\frac{3}{5}\alpha + (1-\alpha)} \qquad (5.32)$$

with $\alpha$ quantifying the (3 1 0) contribution. The branching ratios are [44] $(1.11 \pm 0.14)\,\%$ for $\tau^- \to \pi^- 3\pi^0 \nu_\tau$ and $(2.49 \pm 0.10)\,\%$ for $\tau^- \to \pi^- \pi^- \pi^+ \pi^0 \nu_\tau$. These numbers result in $\alpha = (77 \pm 7)\,\%$. Both partitions contribute with similar strength and no further prediction of the structures of the decays can be made.

- Even if there is not enough information from the branching ratios to disentangle the contributions from the different partitions, isospin symmetry can place useful constraints on the branching ratios. In the case of $\tau$ decays to five pions, for example, there are four partitions but only three branching ratios. It is impossible to solve for the four unknowns. Nevertheless it can be seen from Table 5.7 that the branching ratio to one charged and four neutral pions can never be more than 30% of the total five pion branching ratio. In that limiting case the $5\pi$ decays would be dominated

**Table 5.7.** Partitions, Young tableaux, and branching ratios for the decay of the $\tau$ lepton into five pions

---

**(5 0 0)**

$I = 1$  $\pi^- 4\pi^0$ :  $\frac{3}{35}$

$2\pi^- \pi^+ 2\pi^0$ :  $\frac{8}{35}$

$3\pi^- 2\pi^+$ :  $\frac{24}{35}$

---

**(4 1 0)**

$I = 1$  $\pi^- 4\pi^0$ :  $\frac{3}{10}$

$2\pi^- \pi^+ 2\pi^0$ :  $\frac{3}{10}$

$3\pi^- 2\pi^+$ :  $\frac{2}{5}$

---

**(3 2 0)**

$I = 1$  $\pi^- 4\pi^0$ :  $0$

$2\pi^- \pi^+ 2\pi^0$ :  $\frac{3}{5}$

$3\pi^- 2\pi^+$ :  $\frac{2}{5}$

---

**(3 1 1)**

$I = 0$

---

**(2 2 1)**

$I = 1$  $\pi^- 4\pi^0$ :  $0$

$2\pi^- \pi^+ 2\pi^0$ :  $1$

$3\pi^- 2\pi^+$ :  $0$

---

160     5. Strong Interactions in $\tau$ Decays

**Table 5.8.** Partitions, Young tableaux, and branching ratios for $\tau \to 6\pi\nu_\tau$

---

**(6 0 0)**

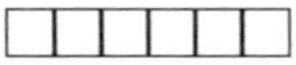

$I = 0$

---

**(5 1 0)**

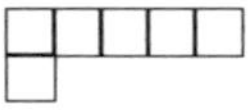

| $I = 1$ | $\pi^- 5\pi^0$ : | $\frac{9}{35}$ |
|---|---|---|
| | $2\pi^- \pi^+ 3\pi^0$ : | $\frac{2}{7}$ |
| | $3\pi^- 2\pi^+ \pi^0$ : | $\frac{16}{35}$ |

---

**(4 2 0)**

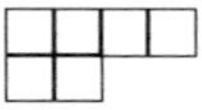

$I = 0$

---

**(3 3 0)**

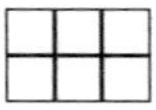

| $I = 1$ | $\pi^- 5\pi^0$ : | 0 |
|---|---|---|
| | $2\pi^- \pi^+ 3\pi^0$ : | $\frac{4}{5}$ |
| | $3\pi^- 2\pi^+ \pi^0$ : | $\frac{1}{5}$ |

---

**(4 1 1)**

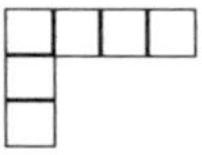

| $I = 1$ | $\pi^- 5\pi^0$ : | 0 |
|---|---|---|
| | $2\pi^- \pi^+ 3\pi^0$ : | $\frac{1}{5}$ |
| | $3\pi^- 2\pi^+ \pi^0$ : | $\frac{4}{5}$ |

---

**(3 2 1)**

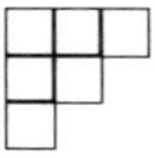

| $I = 1$ | $\pi^- 5\pi^0$ : | 0 |
|---|---|---|
| | $2\pi^- \pi^+ 3\pi^0$ : | $\frac{1}{2}$ |
| | $3\pi^- 2\pi^+ \pi^0$ : | $\frac{1}{2}$ |

---

**(2 2 2)**

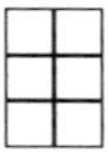

$I = 0$

---

by the (4 1 0) partition. Overall, the constraints on the five pion-channel are the following [478]:

$$
\begin{aligned}
0 &\leq \frac{\pi^- \, 4\pi^0}{\text{all}\,(5\,\pi)^-} \leq \frac{3}{10}, \\
\frac{8}{35} &\leq \frac{2\pi^- \, \pi^+ \, 2\pi^0}{\text{all}\,(5\,\pi)^-} \leq 1, \\
0 &\leq \frac{3\pi^- \, 2\pi^+}{\text{all}\,(5\,\pi)^-} \leq \frac{24}{35}.
\end{aligned}
\quad (5.33)
$$

These constraints apply to the total branching ratios as well as to the differential at any point in phase space. Similar constraints can be set on many other channels (see [478, 479]).

- The CVC hypothesis can be used to relate $\tau$ decays to hadron production in $e^+e^-$ annihilation for the channels with an even number of pions. The relative contributions from the different partitions should be equal in both reactions. As an example, Table 5.6 gives the splitting of the two $I=1$ partitions for four pions into the different final states. The following ratio turns out to be independent of the relative strength of the two partitions [480]:

$$
\frac{\tau^- \to \pi^- \, \pi^+ \, \pi^- \, \pi^0 \, \nu_\tau}{\tau^- \to \pi^- \, \pi^0 \, \pi^0 \, \pi^0 \, \nu_\tau} = 1 + 2 \, \frac{e^+e^- \to \pi^+ \, \pi^- \, 2\pi^0}{e^+e^- \to 2\pi^+ \, 2\pi^-}. \quad (5.34)
$$

This is valid for any value of $Q^2$, as well as for the integral from 0 to $m_\tau^2$ which gives the total branching ratios on the left-hand side.

### 5.2.6 Predictions from CVC

On the basis of the experimental observation of a universal coupling strength in all reactions involving the vector part of the charged weak current, Feynman and Gell-Mann [481] incorporated the hypothesis of a conserved vector current into their theory of weak interactions. Under this assumption the weak charge is not altered by quantum corrections and the coupling constant becomes a process-independent quantity dependent only on $Q^2$. In addition CVC includes the assumption that the charged weak current forms an isospin triplet with the $I=1$ part of the electromagnetic current, with the same structure in all three $I_3$ components. Together with lepton universality CVC relates hadronic $\tau$ decays to hadron production in low-energy $e^+e^-$ collisions.

The sequence of transformations from $e^+e^-$ annihilation to $\tau$ decays is shown in Fig. 5.15. The first diagram (A) shows the $e^+e^-$ annihilation. When an isospin rotation is applied, A is transformed into an electron–neutrino collision annihilating into a $W^-$ (diagram B). At the same time the total charge of the created hadrons changes from zero to one and the electromagnetic couplings are replaced by weak couplings. It is in this step that CVC ensures that the strong interactions which create the hadrons are not altered under the transformation. Then, going from diagram B to C, the $e^-$ and $\bar\nu_e$ are

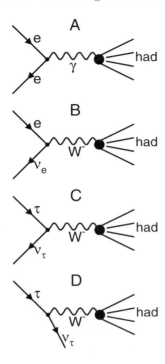

**Fig. 5.15.** The sequence of transformations relating low-energy hadron production to $\tau$ decays

replaced by $\tau^-$ and $\bar{\nu}_\tau$ assuming lepton universality. Finally, the direction of the $\bar{\nu}_\tau$ is reversed in time, changing the $\tau^-$–$\bar{\nu}_\tau$ collision into a $\tau^- \to \nu_\tau$ decay. The essential point is that none of these transformations changes the structure of the hadronic current, so that the same spectral functions (5.13) apply to $e^+e^-$ annihilation and $\tau$ decays. Quantitatively, the relation is

$$\Gamma(\tau \to \mathrm{had}\, \nu_\tau) = \frac{\cos^2 \theta_\mathrm{C}\, G_\mathrm{F}^2\, m_\tau^3}{32\, \pi^2}$$

$$\times\, 2 \int_0^{m_\tau^2} ds \left(1 - \frac{s}{m_\tau^2}\right)^2 \left(1 + 2\frac{s}{m_\tau^2}\right) \frac{s}{8\pi^2 \alpha^2}\, \sigma_{e^+e^- \to \mathrm{had}}^{I=1}(s)\,. \quad (5.35)$$

Remember that this relation only applies to the nonstrange vector final states, and the proper charge combinations of the hadrons still have to be selected. These are listed in Table 5.9.

Several authors have used these relations to predict branching ratios of the $\tau$. In the older papers (e.g. [478, 480]) the authors used to parametrize the measured $e^+e^-$ cross section and integrate the parametrization. This introduces an unnecessary model dependence which is avoided in more recent publications by a point-to-point summation of the experimental values [367, 482–484]. The different approaches differ in the treatment of the uncertainties of the measurements, which are largely of systematic nature. Ta-

**Table 5.9.** Isospin relations between hadronic $\tau$ decays and $e^+e^-$ annihilation into hadrons. The table gives the $\sigma\left(e^+e^- \to \text{had}\right)$ to be inserted into (5.35)

| $\Gamma\left(\tau \to \text{had}\, \nu_\tau\right)$ | $\sigma^{I=1}_{e^+e^- \to \text{had}}$ |
|---|---|
| $\Gamma\left(\tau^- \to \pi^-\,\pi^0\,\nu_\tau\right)$ | $\sigma\left(e^+e^- \to \pi^+\,\pi^-\right)$ |
| $\Gamma\left(\tau^- \to \pi^-\,3\,\pi^0\,\nu_\tau\right)$ | $\frac{1}{2}\sigma\left(e^+e^- \to 2\,\pi^+\,2\,\pi^-\right)$ |
| $\Gamma\left(\tau^- \to 2\pi^-\,\pi^+\pi^0\nu_\tau\right)$ | $\frac{1}{2}\sigma\left(e^+e^- \to 2\,\pi^+\,2\,\pi^-\right)$ |
|  | $+\sigma\left(e^+e^- \to \pi^+\,\pi^-\,2\,\pi^0\right)$ |
| $\Gamma\left(\tau^- \to \pi^-\,\omega\,\nu_\tau\right)$ | $\sigma\left(e^+e^- \to \pi^0\,\omega\right)$ |
| $\Gamma\left(\tau^- \to \pi^-\,\pi^0\,\eta\,\nu_\tau\right)$ | $\sigma\left(e^+e^- \to \pi^+\,\pi^-\,\eta\right)$ |

**Table 5.10.** Comparison between branching ratios predicted from CVC [482] and the experimental values. The measurement on the $\pi^-\,\omega$ channel includes $K^-\,\omega$

| Channel | CVC prediction | Measurement |
|---|---|---|
| $\pi^-\,\pi^0$ | $(23.4 \pm 0.8)\%$ | $(25.32 \pm 0.15)\%$ |
| $\pi^-\,3\,\pi^0$ | $(1.07 \pm 0.06)\%$ | $(1.11 \pm 0.14)\%$ |
| $2\,\pi^-\,\pi^+\,\pi^0$ | $(4.3 \pm 0.3)\%$ | $(4.22 \pm 0.10)\%$ |
| $\pi^-\,\omega$ | $(2.2 \pm 0.3)\%$ | $(1.93 \pm 0.06)\%$ |
| $\pi^-\,\pi^0\,\eta$ | $(0.13 \pm 0.02)\%$ | $(0.17 \pm 0.03)\%$ |

ble 5.10 compares the predictions from [482] with the experimental values [44]. There is in general good agreement. See also Sect. 3.4.3.

Not only does CVC relate the integral over the $e^+e^-$ cross section to $\tau$ branching ratios, but it is also possible to extract the spectral functions from both processes with their $Q^2$ dependence and compare them directly. This will be shown in Sect. 5.4.3 (Fig. 5.37).

## 5.3 Experimental Studies

This section presents studies of the hadronic structure in some exclusive decay channels of the $\tau$ lepton. It is not intended to be complete.

### 5.3.1 The Two-Pion Channel

The decay $\tau^- \to \pi^-\,\pi^0\,\nu_\tau$ is known to be dominated by the $\rho$ resonance ($\tau^- \to \rho^-\,\nu_\tau$ followed by $\rho^- \to \pi^-\,\pi^0$). The presence of the $\rho'\,(1450)$ has also been clearly identified. The evidence for a second excitation – the $\rho''\,(1700)$ – is still inconclusive. Many of the earlier papers [300, 369, 465, 485–487] used $e^+e^-$ data to study the resonances and then CVC to convert the results to $\tau$ decays. Figure 5.16 shows a fit of the Gounaris/Sakurai model [465] to the data collected in [485]. The step in the data at about 780 MeV is due to the interference with the narrow $\omega$ resonance, which is not present

164    5. Strong Interactions in $\tau$ Decays

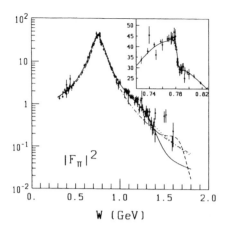

**Fig. 5.16.** Fits of the Gounaris/Sakurai parametrization of the pion form factor to the mass spectrum of two pions from $e^+e^-$ collisions (taken from [300]). $W$ is the invariant mass of the two pions

in $\tau$ decays ($I = 0$). Various modifications to describe the high-mass tail of the spectrum have been applied. The dash–dotted curve simulates only the $\rho$ and does not describe the data well ($\chi^2$/d.o.f. $= 1255/136$). ('d.o.f.' $=$ 'degrees of freedom'). The solid curve includes the $\rho'$ ($\chi^2$/d.o.f. $= 159/133$) and the dashed curve the $\rho''$ in addition ($\chi^2$/d.o.f. $= 151/132$). Both give a reasonable description of the data. The fit with the $\rho'$ only is still used today in TAUOLA, the Monte Carlo generator for $\tau$ decays [295, 297].

An extensive study of the two-pion channel in $\tau$ decays has been presented by the ALEPH collaboration [54]. ALEPH also reanalyzed the $e^+e^-$ data. Figure 5.17 shows the comparison of their data with the $e^+e^-$ data. There is perfect agreement between the two. Throughout most of the region accessible by $\tau$ decays the $\tau$ data is more precise. The normalization is better understood. The $e^+e^-$ cross sections suffer from uncertainties introduced through the luminosity measurement. The mass resolution, however, is better in the $e^+e^-$ data. It is defined by the energies of the beams there. In $\tau$ decays the energy resolution of the reconstructed neutral pions dominates the mass resolution. It is only of the order of 100 MeV$^2$ and makes a complicated unfolding procedure necessary. The results from the fits are presented in Table 5.11. The major obstacle in the identification and reconstruction of the decays is fake photons. These are fragments or fluctuations of the hadronic shower of the charged pion misidentified as photons. Their impact is especially large at very low masses, where the neutral pion is expected to be close to the charged pion in angle and low in energy.

The theoretical problem with the parametrization of these resonances is their large width. There are several things which might affect such a broad resonance (see also Sect. 5.2.4):

- The angular momentum of the two pions requires the introduction of a mass-dependent width (a so-called p-wave Breit–Wigner). There are differences on how to do that (see [486]).

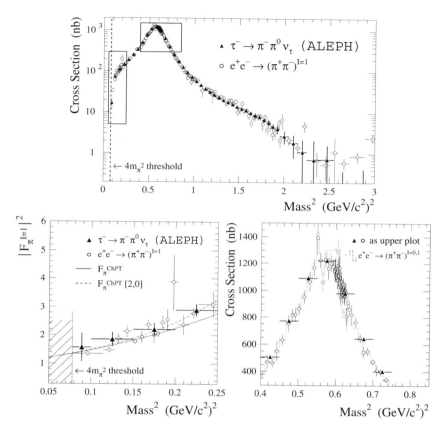

**Fig. 5.17.** The two-pion data from $\tau$ decays compared to the corresponding isovector $e^+e^-$ cross section [54]. Both distributions are shown with statistical and systematic errors. The two *rectangles* indicate the regions that are expanded in the *lower plots*

- Threshold effects of new channels opening across the resonance distort the resonance.
- The coupling constants are not necessarily constant over the wide $Q^2$ range of a broad resonance (see [369]).
- The various $\rho$ resonances interfere, which creates strong correlations between their parameters.

These problems lead to strong model dependences of the parameters extracted from the fits. This is particularly true for the $\rho'$, sitting on the still strong shoulder of the $\rho(770)$. This might also explain the unexpectedly large scatter of results for the mass of the $\rho'$ and its width extracted from $\tau$ decays

**Table 5.11.** Fit results of the pion form factor in $\tau \to \pi\pi\nu_\tau$ from ALEPH [54]. The two models fitted are those of Kühn and Santamaria [300] and Gounaris and Sakurai [465]. Values without errors were taken from Particle Data Group '96 [327] and were fixed in the fit

| Parameter | Kühn and Santamaria | Gounaris and Sakurai |
|---|---|---|
| $m_{\rho^\pm(770)}$ | $774.9 \pm 0.9$ | $776.4 \pm 0.9$ |
| $\Gamma_{\rho^\pm(770)}$ | $114.2 \pm 1.5$ | $150.5 \pm 1.6$ |
| $\beta$ | $-0.094 \pm 0.007$ | $-0.077 \pm 0.008$ |
| $m_{\rho^\pm(1450)}$ | $1363 \pm 15$ | $1400 \pm 16$ |
| $\Gamma_{\rho(1450)}$ | $\equiv 310$ | $\equiv 310$ |
| $\gamma$ | $-0.015 \pm 0.008$ | $0.001 \pm 0.009$ |
| $m_{\rho^\pm(1700)}$ | $\equiv 1700$ | $\equiv 1700$ |
| $\Gamma_{\rho(1700)}$ | $\equiv 235$ | $\equiv 235$ |
| $\chi^2/$d.o.f. | $81/65$ | $54/65$ |

and other measurements. Figure 5.18 shows all of the mass measurements quoted by the Particle Data Group [44].

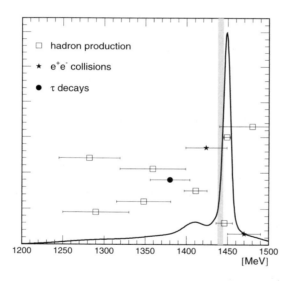

**Fig. 5.18.** Measurements of the mass of the $\rho'(1450)$ from Particle Data Group [44] (*points with error bars*). The *line* is the superposition of Gaussian probability density functions for each measurement weighted by $1/\sigma$ (ideogram)

### 5.3.2 The Three-Pion Channel

The $\tau$ decay to three pions and a neutrino is dominated by the $a_1$ resonance with the quantum numbers $J^{PG} = 1^{+-}$. It decays through the intermediate state of $(\rho\pi)$, with mostly $\rho(770)$ and an admixture of $\rho(1450)$ at higher masses, followed by $\rho \to \pi\pi$. One expects the two charge combinations $\tau^- \to \pi^-\pi^-\pi^+\nu_\tau$ and $\tau^- \to \pi^-\pi^0\pi^0\nu_\tau$ to have a similar structure.

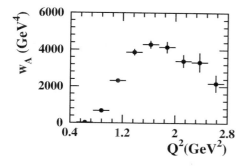

Fig. 5.19. The structure function of the axial-vector current in the decay $\tau \to 3\pi\nu_\tau$ integrated over the Dalitz plane as measured by the OPAL collaboration [217]

Most experimental studies use the all-charged mode. It is easier to identify and the resolution is usually better.

Kühn and his coworkers have developed a method to analyze the hadronic structure of $\tau$ decays in a completely model-independent way (see Sect. 5.2.2 and, for the $3\pi$ channel, [437]). The hadronic structure can be characterized by the quantum numbers $J^{PG}$ of the final states and their distributions in the invariant mass $Q^2$ and the two Dalitz plane variables $s_1 = (q_2 + q_3)^2$ and $s_2 = (q_1 + q_3)^2$. (The two like-signed pions have momenta $q_1$ and $q_2$. The third pion has momentum $q_3$.) For a given point $(Q^2, s_1, s_2)$ in the spectrum the orientations of the angular momenta and therefore the angular distributions of the pions in the decay are fixed by their quantum numbers $J^{PG}$. To extract the hadronic structure, the data is binned in $(Q^2, s_1, s_2)$ and within each bin the structure functions (see Sect. 5.2.2) are assumed to be constant and are disentangled by a fit to the angular distribution.

OPAL [217, 226] has applied the method to its three-pion decays. Figure 5.19 shows the measured structure function $W_A$ integrated over the Dalitz plane, from [217]. $W_A$ describes the total strength of the dominating axial-vector current and shows the $a_1$ resonance unfolded from the impact of the phase space of the $\tau$.

The structure functions are extracted from the data by an extended maximum-likelihood fit. For each bin in $(Q^2, s_1, s_2)$ the following function is maximized:

$$\sum_{\substack{\text{events} \\ \text{in bin}}} \ln \frac{\mathrm{d}\Gamma(\boldsymbol{x}, W_i)}{\mathrm{d}\boldsymbol{x}} \frac{N_{\text{tot}}}{\Gamma_{3\pi}} - \int_{\text{bin}} \frac{\mathrm{d}\Gamma(\boldsymbol{x}, W_i)}{\mathrm{d}\boldsymbol{x}} \frac{N_{\text{tot}}}{\Gamma_{3\pi}} \mathrm{d}\boldsymbol{x}. \qquad (5.36)$$

The first term is a standard log likelihood normalized to the total number of events in the $\tau \to 3\pi\nu_\tau$ decay, and the second term takes care of the normalization, which depends on $W_i$. The quantity $\boldsymbol{x}$ is an abbreviation for a set of kinematic quantities describing completely each event. The likelihood of (5.36) cannot be applied as it is, it has to be corrected for the experimental shortcomings: efficiency, resolution, and background. This is done by the following replacements:

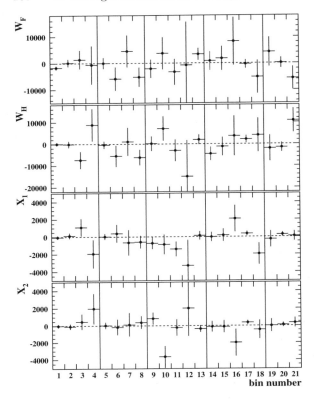

**Fig. 5.20.** The nonaxial-vector structure functions of the decay $\tau \to 3\pi\nu_\tau$ [217]. The results are consistent with zero and demonstrate the dominance of the axial-vector. The *vertical lines* indicate five bins in $\sqrt{Q^2}$: $0.8 < 1.0 < 1.1 < 1.2 < 1.4 < 1.7$ in GeV (*left to right*); each point within a bin represents a different region in the Dalitz plane

$$\frac{d\Gamma(x, W_i)}{dx} \to \frac{d\Gamma(x, W_i)}{dx} \epsilon(x),$$
$$\frac{d\Gamma(x, W_i)}{dx} \to \int \frac{d\Gamma(x', W_i)}{dx} R(x' - x) dx',$$
$$\frac{d\Gamma(x, W_i)}{dx} \to \frac{d\Gamma(x, W_i)}{dx} + \frac{d\Gamma_{\text{bgd}}(x)}{dx}. \qquad (5.37)$$

The first replacement multiplies the theoretical prediction with an efficiency function depending on the kinematic quantities $x$. The second correction folds the prediction with the resolution function. It gives the probability that an event with the true kinematic quantities $x'$ is reconstructed with $x$. The last step adds the background distribution. All three correction functions $\epsilon$, $R$, and $\Gamma_{\text{bgd}}$ have to be generated from Monte Carlo events and carefully checked. The major systematic uncertainties are associated with these corrections, especially the uncertainties in the hadronic structure of the $\tau \to 4\pi\nu_\tau$ decays entering through $\Gamma_{\text{bgd}}$.

With this method is was possible to demonstrate modelindependently the dominance of the axial-vector current in $\tau \to 3\pi\nu_\tau$. The result is displayed in Fig. 5.20, which shows the measurement of the structure functions describing the interference of the axial-vector current with the vector and pseudoscalar.[10] These interference terms are linear in the nonaxial-vector current form factors and therefore more sensitive than the pure vector spectral function $W_B$ or the pure pseudoscalar $W_{SA}$. $W_F$, $W_G$, $W_H$, and $W_I$ are connected to the vector current interference and $W_{SB}$ and $W_{SD}$ to the pseudoscalar. For details see [217]. All measurements are consistent with zero, i.e. with a pure axial-vector current. For comparison the measurement of the pure axial-vector structure function $W_A$ gives typical values between 10 000 and 20 000 in these units.

Figure 5.21 shows the measurement of the axial-vector structure functions in the Dalitz plane for one of the bins in $Q^2$ (1.1 GeV $< \sqrt{Q^2} <$ 1.2 GeV) in comparison with the model of Kühn and Santamaria [300].

The DELPHI collaboration compared their data [150] in detail to the models of Kühn and Santamaria [300], Isgur, Morningstar, and Reader [372], and Feindt [468]. They found good agreement in general, but discrepancies between the data and all three models in the Dalitz plane projection for the highest values of $Q^2$ (see the left plot of Fig. 5.22) and also a slight excess in the mass spectrum at the high end. The discrepancy could be explained by a new resonance close to the $\tau$ mass. The $a_1'$, a radial excitation of the $a_1$, would be a good candidate. It has probably already been seen by the VES collaboration [488] and is expected theoretically (see [150] for discussion). The situation is still inconclusive, but an admixture of the $a_1'$ according to

$$\mathrm{BW}_{a_1} \to \mathrm{BW}_{a_1} + \kappa e^{i\Phi} \mathrm{BW}_{a_1'} \tag{5.38}$$

improves the description of the data. A fit gives an admixture with a strength $\kappa$ between 1/2 and 3/4. This is shown in the right plot of Fig. 5.22.

The CLEO collaboration has presented a preliminary result at the 1998 $\tau$ workshop [489, 490] which demonstrates how much more structure one can see with more data. They analyzed about 80 000 identified events in $\tau^- \to \pi^-\pi^-\pi^+\nu_\tau$ as well as in $\tau^- \to \pi^-\pi^0\pi^0\nu_\tau$, a good order of magnitude more than the LEP experiments. They fitted a total of seven amplitudes of the $a_1$ with different intermediate resonances and found that all seven are necessary to give a good description of their data. Table 5.12 specifies the amplitudes and their contributions to the neutral mode. The CLEO collaboration also studied the mass spectrum for contributions different from the $a_1$. They could exclude branching ratios into the $\pi(1300)$ (pseudoscalar current) at the level of $10^{-4}$ and also found that an admixture of an $a_1'$ improves the fits, at a significantly lower branching ratio than the DELPHI result [150], though.

### 5.3.3 The Four-Pion Channel

There is no dominating resonance in the $\tau \to 4\pi\nu_\tau$ channel. Only the charged mode $\tau^- \to \pi^-\pi^-\pi^+\pi^0\nu_\tau$ has been studied in detail up to now. There is

---

[10] $X_1$ is a combination of $W_G$ and $W_{SD}$ and $X_2$ a combination of $W_I$ and $W_{SB}$.

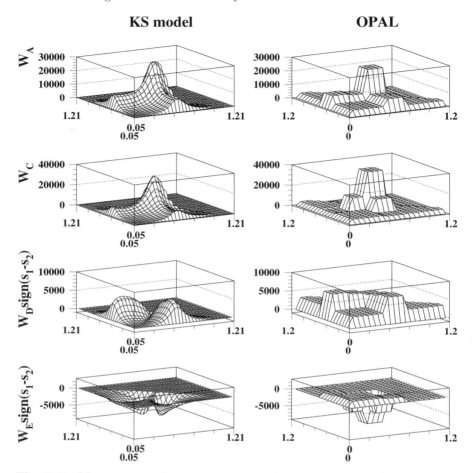

**Fig. 5.21.** Measurement of the axial-vector structure functions of the decay $\tau \to 3\pi\nu_\tau$ from OPAL [217]. $W_A$ gives the total strength of the current, $W_C$ and $W_D$ two possible polarizations of the hadronic system, and their difference $W_E$ is a parity-violating quantity. The *right column* shows the data and the *left column* the result of the model of Kühn and Santamaria for comparison [300]

a contribution from $\tau \to \omega \pi \nu_\tau$ followed by $\omega \to \pi^+ \pi^- \pi^0$, and also $(\rho\pi\pi)$ in various charge combinations has been seen as intermediate states. These $(\rho\pi\pi)$ states probably come from the decay of the $\rho(1450)$, but this has not been proven and not much is known about the origin of the $(\omega\pi)$ states (the $\rho(1450)$ does not decay to $(\omega\pi)$ [44]).

The ALEPH collaboration studied $\omega$ production in $\tau$ decays [55]. They found that roughly half of the $4\pi$ events are due to $\tau \to \omega\pi\nu_\tau$; more precisely, the branching ratio for $\tau \to 3h^\pm \pi^0 \nu_\tau$ is $(4.30\pm0.13)\%$ [56] and for $\tau \to \omega\pi\nu_\tau$ they find $(1.91 \pm 0.09)\%$. Figure 5.23 shows their $(\pi^+\pi^-\pi^0)$ mass spectrum.

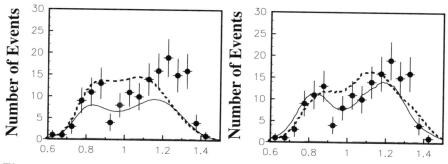

**Fig. 5.22.** Distribution of $\sqrt{s_1}$ in GeV for $Q^2$ in the bin $2.3\text{ GeV}^2 < Q^2 < 3.2\text{ GeV}^2$ (DELPHI collaboration [150]). The *points* represent the data, the *dashed line* the Kühn and Santamaria model [300], and the *solid line* the Isgur et al. model [372]. *Left plot*, without and *right* with an $a_1'$

**Table 5.12.** Results from a fit of the $a_1$ with different intermediate resonances to $\tau^- \to \pi^- \pi^0 \pi^0 \nu_\tau$ events by the CLEO collaboration [489–491]. The first column gives the intermediate state, the second specifies the angular momentum between the two particles, the third is the significance of the amplitude in standard deviations from zero, and the last column is the fraction of events going through that intermediate state

|  | L | Significance | Fraction |
|---|---|---|---|
| $\rho(770)\pi$ | s |  | 68.11 |
| $\rho(1450)\pi$ | s | 1.4 $\sigma$ | 0.30 ± 0.66 |
| $\rho(770)\pi$ | d | 5.0 $\sigma$ | 0.36 ± 0.18 |
| $\rho(1450)\pi$ | d | 3.1 $\sigma$ | 0.43 ± 0.29 |
| $f_2(1270)\pi$ | p | 4.2 $\sigma$ | 0.14 ± 0.06 |
| $\sigma\pi$ | p | 8.2 $\sigma$ | 16.18 ± 4.06 |
| $f_0(1370)\pi$ | p | 5.4 $\sigma$ | 4.29 ± 2.40 |

The peak of the $\omega$ at 782 MeV is clearly visible. The width of the peak is dominated by the experimental resolution.

Most of the non-$\omega\pi$ events show a $\rho$ mass in one of the charge combinations (see Fig. 5.24). Fitting an incoherent mixture of the three charge states of the $\rho$, taking into account kinematic reflections and Bose symmetrization, ALEPH finds them consistent with 100% $\rho\pi\pi$ production. The individual contributions to the decay of the negative $\tau$ are

$$\begin{aligned} \rho^+\pi^-\pi^- &: \quad 42 \pm 2\%; \\ \rho^-\pi^+\pi^- &: \quad 38 \pm 2\%; \\ \rho^0\pi^-\pi^0 &: \quad 20 \pm 2\%; \end{aligned} \quad (5.39)$$

the errors quoted are purely statistical and the quality of the fit indicates that the systematic errors might be substantial (see also [86]).

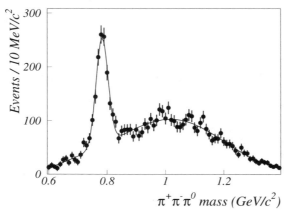

Fig. 5.23. Distribution of the invariant mass of $\pi^+\pi^-\pi^0$ in $\tau \to 4\pi\nu_\tau$ from the ALEPH collaboration [55]. There are two entries per event

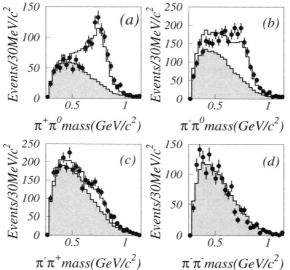

Fig. 5.24. Distribution of the invariant mass of two pions in $\tau \to 4\pi\nu_\tau$ for events with no $\omega$ (ALEPH collaboration [55]). The *points* are data, with two entries per event in (c) and (d). The *open histogram* is the result of the fit. The *shaded histogram* shows the kinematic reflections from the other charge combinations

### 5.3.4 Strange Decays

Not much is known about the hadronic structure of the strange decays of the $\tau$. Besides the simple $\tau \to K\nu_\tau$ mode there is the $\tau \to K^*\nu_\tau$, which dominates both charge combinations of the $\tau \to K\pi\nu_\tau$ final state [176]. $K_1$ mesons have been observed in the $K_S^0\pi^-\pi^0$ final state [48]. Probably both the $K_1(1270)$ and the $K_1(1400)$ are present, with a slight preference for the first. And that is about all that is known to date.

## 5.4 Inclusive Decays

Although QCD at its current state of understanding is unable to describe any of the exclusive decay channels of the $\tau$ lepton from first principles, it can by virtue of the global quark–hadron duality predict the inclusive decay rate [492, 493]. Here global quark–hadron duality states that theoretical predictions, calculated in terms of quark fields, equal experimental measurements on real hadrons, if only the process in question averages over a sufficient number of these states.

The strong coupling constant $\alpha_s$ at $Q^2 = m_\tau^2$ is as big as 0.35 and nonperturbative effects might not be negligible. In fact the hadronic decays of the $\tau$ might well be the lowest-energy process from which the running coupling constant can be extracted cleanly [494, 495]. The $\tau$ offers a unique situation with several advantages:

- The perturbative corrections are large enough to give a good experimental sensitivity; nonperturbative corrections are predicted to be small and can be verified by experiment.
- The perturbative corrections have been calculated to third order in $\alpha_s$.
- The predictions of partially inclusive spectra for strange/nonstrange decays and vector/axial-vector currents can be calculated with the same techniques. They are more sensitive to the shortcomings of these calculations and hence allow for stringent tests.

(For reviews on this topic see [496–500].)

### 5.4.1 Theoretical Framework

**The Definition of $R_\tau$.** The $R$ ratio, defined in $e^+e^-$ annihilation to be the ratio of the total hadronic cross section ($e^+e^- \to$ had) to the leptonic cross section ($e^+e^- \to \mu^+\mu^-$), can be generalized to $\tau$ decays:

$$R_\tau := \frac{\Gamma(\tau \to \nu_\tau \text{had})}{\Gamma(\tau \to e\,\nu_e\nu_\tau)} = \frac{\Gamma_\text{had}}{\Gamma_e}. \tag{5.40}$$

The theoretical prediction for $R_\tau$ is obtained through an integration of the hadron spectrum over the invariant mass squared:

$$R_\tau = \frac{1}{\Gamma_e} \int_0^{m_\tau^2} \frac{d\Gamma_\text{had}}{ds}\,ds. \tag{5.41}$$

The naive prediction for $R_\tau$ is 3, as the quark pairs come in three different colors. It is then modified by QCD corrections.

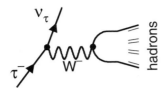

**Fig. 5.25.** The optical theorem relates the production of hadrons from $\tau$ decays (*upper diagram*) to the imaginary part of the forward scattering amplitude of the *lower diagram*

**The Optical Theorem.** From the upper diagram in Fig. 5.25, one calculates $\Gamma_{\text{had}}$ as [494, 495]

$$d\Gamma_{\text{had}} = \frac{1}{2\,m_\tau}\,(2\pi)^4\,\Delta^4\,(\ldots)\,\frac{G_F}{\sqrt{2}}\,\left(|V_{ud}|^2 + |V_{us}|^2\right)$$
$$\times L^{\mu\nu} \sum_{\text{had}} \langle\,0\,|J_\mu(q)|\text{had}\rangle\langle\text{had}|J_\nu^\dagger(q)|\,0\,\rangle\,d\Phi_{\text{had}}\,d\Phi_{\nu_\tau}, \quad (5.42)$$

where $d\Phi$ denotes the invariant phase space elements of the hadrons and the neutrino, and $L^{\mu\nu}$ is the leptonic tensor of the $\tau$. The total 4-momentum of the hadronic system is written as $q$ $(q^2 \equiv s)$. Now the optical theorem can be used to reexpress the matrix element for the production of the hadrons by the imaginary part of the forward scattering amplitude:

$$d\Gamma_{\text{had}} = \frac{1}{2\,m_\tau}\,\frac{G_F}{\sqrt{2}}\,\left(|V_{ud}|^2 + |V_{us}|^2\right) L^{\mu\nu}\,2\,\text{Im}\,\langle\,0\,|J_\mu(q)\,J_\nu^\dagger(q)|\,0\,\rangle\,d\Phi_{\nu_\tau}.$$
$$(5.43)$$

This is quite an important step. While (5.42) requires the calculation of matrix elements of exclusive final states (and their summation), the new (5.43) contains a loop integral which implicitly includes all final states. The first one cannot be handled by perturbative QCD, but the second one can.

**The Correlators.** The dynamics of the amplitude in (5.43) can now be made explicit by decomposing the hadron tensor into all possible Lorentz covariants. One gets

$$\langle\,0\,|J_\mu(q)\,J_\nu^\dagger(q)|\,0\,\rangle = \left(q_\mu q_\nu - g_{\mu\nu}q^2\right)\Pi^{(1)}(q^2) + q_\mu q_\nu\,\Pi^{(0)}(q^2). \quad (5.44)$$

The two functions $\Pi^{(J)}$ introduced are called the two-point correlators of the quark currents. They describe the creation of hadronic states with total angular momentum $J$ from the vacuum by means of QCD. By convention the couplings of the quark currents to the $W^\pm$ are not included in $\Pi^{(J)}$. The integration over the phase space of the neutrino can now be performed. The result is $(|V_{ud}|^2 + |V_{us}|^2 = 1)$ [11]

---

[11] $|V_{ud}|^2 + |V_{us}|^2 = 1 - |V_{ub}|^2 \approx 1$ to a very good approximation.

$$\Gamma_{\text{had}} = \frac{G_F^2 m_\tau^5}{16\pi^2} \int_0^{m_\tau^2} \frac{ds}{m_\tau^2} \left(1 - \frac{s}{m_\tau^2}\right)^2 \left[\left(1 + \frac{2s}{m_\tau^2}\right) \operatorname{Im} \Pi^{(1)}(s) + \operatorname{Im} \Pi^{(0)}(s)\right]$$

(5.45)

and, after normalization to $\Gamma_e$,

$$R_\tau = 12\pi \int_0^{m_\tau^2} \frac{ds}{m_\tau^2} \left(1 - \frac{s}{m_\tau^2}\right)^2 \left[\left(1 + \frac{2s}{m_\tau^2}\right) \operatorname{Im} \Pi^{(1)}(s) + \operatorname{Im} \Pi^{(0)}(s)\right].$$

(5.46)

The correlators can be decomposed according to the quantum numbers of the states involved, if one so wishes:

$$\Pi^{(J)}(q^2) = |V_{ud}|^2 \left(\Pi^{(J)}_{ud,V}(q^2) + \Pi^{(J)}_{ud,A}(q^2)\right)$$
$$+ |V_{us}|^2 \left(\Pi^{(J)}_{us,V}(q^2) + \Pi^{(J)}_{us,A}(q^2)\right).$$

(5.47)

The lower indices on $\Pi^{(J)}$ indicate the quark flavors and type of current (vector or axial-vector) in question. In the limit of massless quarks the vector and axial-vector currents are conserved and the contribution of $\Pi^{(0)}$ to $R_\tau$ vanishes. Furthermore, $\Pi^{(1)}_{ij,V}$ and $\Pi^{(1)}_{ij,A}$ become identical.

**Extension into the Complex Plane.** We shall now use the analyticity of the amplitudes to solve (5.46) as a contour integral in the plane of complex $s$ (see Fig. 5.26). One has to be careful as $s$ approaches the positive real axis. There are singularities expected there! The correlators can be expressed as a power series in $\ln(-s)$ with real coefficients and then the imaginary part of $\Pi^{(J)}$ can be calculated as[12]

$$\operatorname{Im} \Pi^{(J)} = -\frac{i}{2}\left[\Pi^{(J)}(s + i\varepsilon) - \Pi^{(J)}(s - i\varepsilon)\right].$$

(5.48)

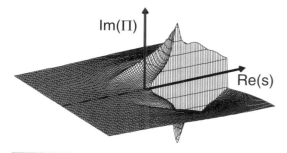

Fig. 5.26. This sketch (not a calculation!) illustrates the behavior of the spin-1 correlator in the complex plane. The correlator vanishes at the origin and develops a structured cut as one moves in the direction of the positive real axis

---

[12] $\operatorname{Im}[\ln(-s)] = -i/2 \left[\ln(-s - i\varepsilon) - \ln(-s + i\varepsilon)\right] = -i/2 \left[\ln(s) + i\pi - \ln(s) + i\pi\right] = \pi.$

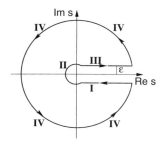

**Fig. 5.27.** The contour of integration of $R_\tau$ in the complex plane. The radius of the outer circle is $m_\tau^2$

Now consider the path depicted in Fig. 5.27. An integral of a correlator integrated over the full path vanishes owing to analyticity. There are no singularities enclosed by the path. However, splitting the path into the four sections labeled in Fig. 5.27 gives the following contributions:

I,III After changing the direction of integration on section I, the contribution of this section can be combined with section III (see (5.48)). The result is the integral to be computed: $R_\tau$.
II The contribution of section II becomes negligible as $\varepsilon$ goes to zero.
IV Section IV converges to a full circle in the limit of $\varepsilon \to 0$.[13]

Thus the result is (symbolically: III $-$ ($-$I) $= -$IV)

$$R_\tau = 6\pi i \oint_{|s|=m_\tau^2} \frac{ds}{m_\tau^2} \left(1 - \frac{s}{m_\tau^2}\right)^2 \left(1 + \frac{2s}{m_\tau^2}\right) \Pi^{(1)}(s). \qquad (5.49)$$

Again, this is quite an important step. Equation (5.46) involves the calculation of processes at all values of $s$ between $m_\pi^2$ and $m_\tau^2$, which cannot be handled by perturbative QCD. But the new integral (5.46) moves away from the resonances along the real axis. It can be calculated reliably.

**The Adler Function.** For technical reasons it is preferred to work with the logarithmic derivative of the correlators, the so-called Adler function [461]. It is defined as

$$D(s) = -s \frac{d\Pi^{(1)}(s)}{ds} \qquad (5.50)$$

and can be expressed as a power series in the strong coupling constant $\alpha_s$:

$$D(s) = \frac{1}{4\pi^2} \sum_{n=0}^{\infty} K_n \left(\frac{\alpha_s(-s)}{\pi}\right)^n. \qquad (5.51)$$

---

[13] It now crosses the real axis and one has to worry about poles there.

It is introduced into (5.49) by partial integration

$$R_\tau = -6\pi i \oint_{|s|=m_\tau^2} \frac{ds}{s} \left(G\left(m_\tau^2\right) - G(s)\right) D(s) \tag{5.52}$$

with

$$G\left(m_\tau^2\right) - G(s) = 1 - 2\frac{s}{m_\tau^2} + 2\frac{s^3}{m_\tau^6} - \frac{s^4}{m_\tau^8}. \tag{5.53}$$

**The Final Calculation of $R_\tau$.** Assuming that the coefficients of the Adler function have been calculated (see the next section), there are two more technical steps to be done to obtain $R_\tau(\alpha_s)$:

- As the Adler function is expanded in powers of $\alpha_s(-s)$ (5.51), one has to use a renormalization group equation (RGE) to calculate the running of $\alpha_s(-s)$ along the circle in the plane of complex $s$ (see Fig. 5.27) in terms of $\alpha_s(m_\tau^2)$:

$$\frac{d\alpha_s(-s)}{d\phi} = \left[i(\phi - \pi) + \ln m_\tau^2\right] \sum_{i=0}^{\infty} \beta_i \left(\frac{\alpha_s}{4\pi}\right)^{i+2}, \tag{5.54}$$

where $-s = m_\tau^2 e^{i(\phi - \pi)}$.

- Finally, the integration (5.52) has to be performed, yielding $R_\tau$ as a function of $\alpha_s(m_\tau^2)$ with the coefficients $K_n$ of the Adler function and $\beta_i$ from the RGE as parameters.

Unfortunately there is no analytic solution for these steps, in general. An exact solution can be obtained only if one is willing to truncate the $\beta$ function already after the first term; an approximate solution is available including the $\beta_1$ term. Beyond that the RGE and the integral have to be solved numerically (see e.g. [501]).

Although the $\beta$ function of the RGE (5.54) itself is a polynomial in $\alpha_s$, this is not the case for its solution, the running $\alpha_s$. For example, the approximate two-loop solution reads

$$\alpha_s(-s) = \left[i(\phi - \pi)\left(\frac{\beta_0}{4\pi} + \frac{\beta_1}{16\pi^2}\alpha_s\left(m_\tau^2\right)\right) + \frac{1}{\alpha_s(m_\tau^2)}\right]^{-1}. \tag{5.55}$$

A natural approach is to expand $\alpha_s(-s)$ in a power series in $\alpha_s(m_\tau^2)$ and truncate it where the first unknown $\beta_i$ coefficient comes into play. After performing the integration this gives $R_\tau$ as a power series in $\alpha_s(m_\tau^2)$, where the coefficients of all terms included are exact and no higher ones are present. This is called 'fixed-order perturbation theory' (FOPT).

A different approach would be to keep the full solution of the RGE and perform a numerical integration. Now the result includes all the terms from FOPT and in addition some higher orders in $\alpha_s(m_\tau^2)$ which are generated by

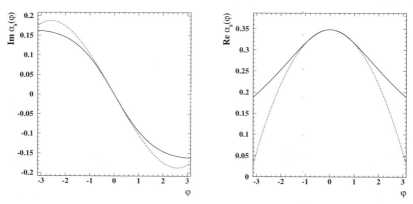

**Fig. 5.28.** The running coupling constant $\alpha_s$ along a circle of radius $m_\tau^2$ in the plane of complex $s$. Exact solution of the RGE to fourth order (CIPT) shown as *solid line* and its Taylor expansion to fourth order as the *dashed line* (RCPT)

the running. This is called 'contour-improved perturbation theory' (CIPT) [502]. The additional terms included are of the form $\ln(-s/m_\tau^2)$, which are large in some parts of the integration range (see Fig. 5.28). There are theoretical indications that this procedure should be preferred compared to FOPT. However, there is no rigorous proof that the partial inclusion of higher orders actually improves the result.

A completely different approach has been motivated by the observation that the largest contributions to the result are of the form $\beta_0^{n-1}\alpha_s^n$. Both FOPT and CIPT include only the first few of these terms. This can be improved by resumming these terms to all orders, i.e. including the renormalon chains (see Fig. 5.29). This approach is called 'renormalon chain perturbation theory' (RCPT). It has to be mentioned however, that there are different versions of RCPT, which show a nonnegligible scatter in their results.[14]

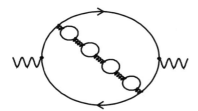

**Fig. 5.29.** Graphical representation of a renormalon chain contribution to $R_\tau$

---

[14] In the following, RCPT will be used as described in [211].

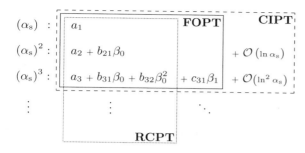

**Fig. 5.30.** Contributions to $R_\tau$ in powers of $\alpha_s$ and $\beta_i$. The *three boxes* indicate the terms included in the three methods described in the text

Figure 5.30 summarizes the various terms included in the three approaches. The RCPT result can be further extended by adding some of the missing terms from FOPT.

### 5.4.2 Calculating $R_\tau$

$R_\tau$ will now be written as the tree-level result modified by various corrections, which will be discussed in the following sections:

$$R_\tau = 3\left(|V_{ud}|^2 + |V_{us}|^2\right) S_{ew} \left(1 + \Delta_{pert} + \Delta_{nonpert} + \Delta_{ew}\right). \tag{5.56}$$

The perturbative correction $\Delta_{pert}$ will turn out to be by far the largest correction (about 20%). The electroweak corrections $S_{ew}$ and $\Delta_{ew}$ and the non-perturbative correction $\Delta_{nonpert}$ are of the order of 1% or below.

**Perturbative QCD Corrections.** The Adler function has been calculated to third order in $\alpha_s$ and the $\beta$ function is even known to fourth order. The Adler function in the $\overline{MS}$ scheme reads [503–507]

$$D(s) = \frac{1}{4\pi^2} \sum_{n=0}^{\infty} K_n \left(\frac{\alpha_s(-s)}{\pi}\right)^n, \tag{5.57}$$

$$K_0 = 1,$$
$$K_1 = 1,$$
$$K_2 = 1.63982,$$
$$K_3 = 6.37101.$$

Figure 5.31 depicts the diagrams contributing to the Adler function in first and second order.

The $\beta$ function for three quark flavors is [508–513]

$$\frac{1}{\pi}\frac{d\alpha_s(s)}{d\ln s} = -\frac{\alpha_s(s)}{\pi} \sum_{n=0}^{\infty} \beta_n \left(\frac{\alpha_s(s)}{\pi}\right)^{n+1}, \tag{5.58}$$

$$\beta_0 = 2.25,$$
$$\beta_1 = 4,$$
$$\beta_2 = 10.0599,$$
$$\beta_3 = 47.2306.$$

The resulting perturbative correction to $R_\tau$ is shown in Fig. 5.32.

180    5. Strong Interactions in $\tau$ Decays

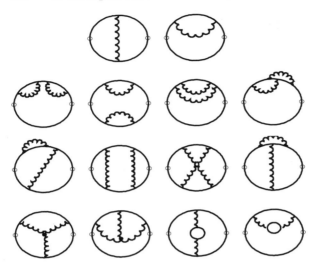

**Fig. 5.31.** The perturbative corrections to $R_\tau$ in leading (*first row*) and next-to-leading order. The two gluons in the third diagram of the third row are not coupled to each other

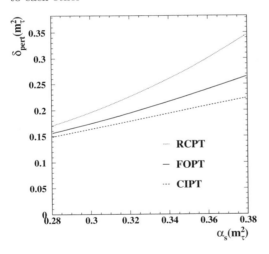

**Fig. 5.32.** The perturbative correction to $R_\tau$ calculated from the three approaches FOPT, CIPT, and RCPT

The perturbative series is usually calculated in the $\overline{\text{MS}}$ scheme assuming the appropriate scale $\mu^2$ to be equal to $m_\tau^2$. In an ideal world, where the full perturbative series was known, the result would be independent of these choices. However, as the series is truncated, the resulting $R_\tau$ shows an artificial dependence on the scheme and the scale, which has to be included in the result as a systematic uncertainty. Furthermore, the missing knowledge of the higher-order coefficients introduces a theoretical uncertainty, which is usually estimated by varying $K_4$ within reasonable bounds. (The impact of $\beta_4$

is most probably negligible, as the $\beta_3$ contribution is already very small.) The variation of the scale, the scheme, and the $K_4$ coefficient all give an estimate of the unknown higher-order contributions [514, 515]. There is some double counting between the three effects, but as the amount of double counting cannot be estimated quantitatively, the effects are added together.

- The scale uncertainty is estimated by varying $\mu^2$ somewhere between $0.3\, m_\tau^2$ and $2.0\, m_\tau^2$.
- The coefficients up to and including $K_2$ and $\beta_1$ are scheme-independent. Not so, however, the higher orders. The scheme dependence is estimated by varying $\beta_2$ between 0 and twice its $\overline{\text{MS}}$ value and at the same time changing $K_3$ accordingly. Also, $\beta_3$ should be adjusted simultaneously, to compensate for the variation in $\beta_2$, but unfortunately its scheme dependence is not explicitly known [502, 515].
- There are several estimates of the size of the $K_4$ coefficient:
  - Geometrical scaling: $K_3/K_2 = K_4/K_3$    25
  - Principle of minimal sensitivity    27.5    [516, 517]
  - Effective charge approach    27.5    [517, 518]
  - Naive nonabelianization prescription    24.8    [519, 520]
  - Estimate from data    29    [521].

Overall, a value of $K_4$ of 25 or 50 is assumed with an error of $\pm 50$.

**Nonperturbative Corrections.** The operator product expansion (OPE) [522–524] is used in the spirit of Shifman, Vainshtein, and Zakharov [525–527] to organize the nonperturbative corrections. The basic idea behind this approach is that the power corrections, not the higher orders in the $\alpha_s$ series, limit the applicability of the perturbative series towards lower and lower $q^2$. First of all, by including these power corrections, one extends the range of validity of the calculation.[15] And second, one gets an indicator of the reliability of the calculation: The result should be reliable, as long as the power corrections are small.[16] The OPE also elegantly combines the perturbative and power corrections into a single framework.

The corrections are organized as a series of nonlocal operators constructed from quark and gluon fields with increasing dimensions of mass:

$$\Pi^{(J)} = \sum_{D=0,2,4,\ldots} \frac{1}{m_\tau^D} \sum_{\dim \mathcal{O}=D} C^{(J)}(s,\mu) \langle \mathcal{O}(\mu) \rangle. \qquad (5.59)$$

For example, the operator built from two gluon fields[17] $\langle (\alpha_s/\pi)GG \rangle$, the so-called gluon condensate, is of dimensions GeV$^4$. It describes the nonperturbative creation of two gluons (at different points in space–time) from the

---
[15] Shifman, Vainshtein, and Zakharov themselves claim that this should work down to $q^2 \approx m_\rho^2$.
[16] These corrections to $R_\tau$ are about 1%, compared to 20% for the perturbative correction.
[17] More precisely, $G \equiv G_{\mu\nu}$ is the field strength tensor of the gluon field.

Fig. 5.33. Illustration of a vacuum polarization diagram involving the gluon condensate

QCD vacuum, which are then perturbatively coupled to the $\tau$ decay. The latter is described by the Wilson coefficient $C^{(J)}(s,\mu)$ [528]. The scale $\mu$ separates long-distance effects included in the condensates from short-distance effects accounted for by the Wilson coefficients.[18] Figure 5.33 illustrates a process involving the gluon condensate.

The following list summarizes the operators of various dimensions [495]. Quark mass corrections have been calculated in [529–532]. The Wilson coefficients have been calculated for dimensions 4 [533], 6 [534–536], and 8 [537]. A summary of all ingredients can be found in [495, 538].

$D = 0$. This is the perturbative correction $\Delta_{\text{pert}}$ (see (5.56)) for massless quarks. The corresponding operator is the unit operator.

$D = 2$. In dimension 2 we find the leading quark mass corrections to the perturbative series [529–532, 538]. They are proportional to $m_i m_j/m_\tau^2$, which is negligible for u and d quarks, but has to be taken into account for the Cabibbo-suppressed decays. Again the corresponding operator is the unit operator.

It is important to realize that it is impossible to build a dynamical operator of dimension 2 from the quark and/or gluon fields. That means that the leading nonperturbative correction is absent!

$D = 4$. Apart from another perturbative quark mass correction in fourth order of $m_i/m_\tau$ [529–532, 538], there are two truly nonperturbative power corrections: the quark condensates $\langle m_j \bar{\Psi}_i \Psi_i \rangle$ and the already mentioned gluon condensate. The quark condensates are reasonably well understood. They can be parametrized as [495]

$$\langle m_j \bar{\Psi}_i \Psi_i \rangle = -\hat{m}_j \hat{\mu}_i^3, \tag{5.60}$$

with [539]

$$\hat{m}_u = (8.7 \pm 1.5) \text{ MeV},$$
$$\hat{m}_d = (15.4 \pm 1.5) \text{ MeV},$$
$$\hat{m}_s = (270 \pm 30) \text{ MeV},$$
$$\hat{\mu}_u = \hat{\mu}_d = (189 \pm 7) \text{ MeV},$$
$$\hat{\mu}_s = (160 \pm 10) \text{ MeV}.$$

---

[18] The typical choice is $\mu^2 = m_\tau^2$ [495].

Less is known about the gluon condensate. A rough estimate gives [539]

$$\left\langle \frac{\alpha_s}{\pi} GG \right\rangle = (0.02 \pm 0.01) \text{ GeV}^4. \tag{5.61}$$

The contribution of the dimension-4 operators to $R_\tau$ is suppressed. It actually vanishes if the weak s-dependence of the Wilson coefficients (and the gluon condensate through $\alpha_s$) is ignored. This can be traced back to the absence of an $s^2/m_\tau^4$ term in (5.53).

$D = 6.$  With the $\dim(\mathcal{O}) = 2$ corrections absent and the $\dim(\mathcal{O}) = 4$ suppressed, the $\dim(\mathcal{O}) = 6$ is the largest nonperturbative correction to $R_\tau$. It comes from four-quark operators of the form $\langle \bar{\Psi}_i \Gamma_\nu \Psi_j \bar{\Psi}_k \Gamma^\nu \Psi_l \rangle$. Other contributions, from the triple gluon condensate and from dimension-4 operators multiplied by quark masses, are negligible.

$D = 8.$  There are a number of operators (17) of dimension 8 (they can be found in [519]). Overall, their contribution to $R_\tau$ is expected to be small. The largest probably comes from the square of the gluon condensate.

$D \geq 10.$  The contributions from dimensions 10 and higher are expected to be even smaller. Furthermore, they are suppressed by the same mechanism as for the $\dim(\mathcal{O}) = 4$ corrections.

Although there are theoretical estimates for all the nonperturbative power corrections which are accurate enough not to spoil the $\alpha_s$ measurement from $R_\tau$, the experiments take a different approach. The corrections are left as free parameters in a fit to the data. The details will be explained in Sect. 5.4.3. This places the $\alpha_s$ measurement on solid ground concerning the power corrections, their uncertainty now being included in the statistical error.

**Electroweak Corrections.** Apart from the QCD corrections, the process $\tau^- \to \nu_\tau \, d' \, \bar{u}$ is affected by QED corrections caused by the emission of real and virtual photons,[19] just as $\tau^- \to e^- \bar{\nu}_e \nu_\tau$ is modified by such diagrams. To the extent that the corrections are identical for both processes, they cancel in the ratio $R_\tau$. But this is not the case for all corrections. For more details see [363, 540–543].

The first nonvanishing contribution to $R_\tau$ comes from the UV behavior of corrections with virtual photons. Figure 5.34 displays the relevant diagrams for $\tau^- \to e^- \bar{\nu}_e \nu_\tau$. All three of them are divergent. However, in the Fermi limit, i.e. neglecting the $W^\pm$ propagator, they cancel each other exactly. But this is no longer the case for $\tau^- \to \nu_\tau d' \bar{u}$. There are now two charged particles in the final state with charges $-1/3$ and $-2/3$. There is a total of six diagrams, all with different couplings, which cancel only partially. The corrections become finite once they are embedded in the full electroweak theory. The three graphs for $\tau^- \to e^- \bar{\nu}_e \nu_\tau$ still vanish, but a finite correction remains for $\tau^- \to \nu_\tau d' \bar{u}$:

---

[19] d' is understood to represent the Cabibbo mixture of d and s quarks.

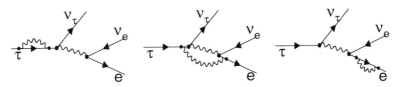

**Fig. 5.34.** Radiative corrections to $\tau^- \to e^- \bar{\nu}_e \nu_\tau$ with virtual photons

$$S_{\text{ew}} = 1 + 4\,\frac{\alpha}{2\pi}\,\ln\frac{m_Z}{m_\tau}. \qquad (5.62)$$

It is a short-distance correction. Processes with gluons exchanged between the final-state quarks or other QCD corrections will be modified in exactly the same way. Therefore $S_{\text{ew}}$ is taken as a multiplicative factor to the tree-level result for $R_\tau$ and its QCD corrections (see (5.56)). Numerically, the correction is $S_{\text{ew}} = 1.018$. It increases to 1.019 after resummation of the leading logarithms.

The second correction which does not cancel completely in $R_\tau$ comes from the radiation of real photons. Again the difference comes from the different charges of (d'u) and (e$^-\nu_e$). The corrections are:

$$\begin{aligned}
\tau^- \to \nu_\tau\, d'\, \bar{u} &: \quad \frac{\alpha}{2\pi}\left(\frac{85}{12} - \pi^2\right); \\
-\quad \tau^- \to e^- \bar{\nu}_e \nu_\tau &: \quad \frac{\alpha}{2\pi}\left(\frac{75}{12} - \pi^2\right); \\
\hline
\Delta_{\text{ew}} &: \quad \frac{\alpha}{2\pi}\,\frac{5}{6}\,.
\end{aligned} \qquad (5.63)$$

Numerically, the correction is very small ($\Delta_{\text{ew}} = 0.0010$).

### 5.4.3 Spectral Functions

Although it is possible to determine $R_\tau$ exclusively from the leptonic branching ratios and use theoretical estimates of the nonperturbative corrections to determine $\alpha_s$, this is not the preferred method. The nonperturbative corrections can be derived directly from the data from the shape of the mass spectra of hadronic $\tau$ decays, leading to a more robust result. Its reliability can be further supported by several tests of various aspects of the QCD calculations performed on the same spectra (see the following sections).

The experimental input to all of these tests and measurements are the spectral functions defined in (5.13). The spectral functions are related to the correlators (5.47) simply through the following (strange decays are analogous):

$$\begin{aligned}
2\pi\,\text{Im}\,\Pi^{(J)}_{\text{ud},V}(s) &= v_J(s), \\
2\pi\,\text{Im}\,\Pi^{(J)}_{\text{ud},A}(s) &= a_J(s).
\end{aligned} \qquad (5.64)$$

The experimental determination of the spectral functions is quite a challenging task. So far only nonstrange spectral functions have been published. Measurements of the strange spectral functions are just evolving [544, 545]. The spectral functions are derived from the normalized distributions $1/N_i\, dN_i/ds$ of the invariant mass of the hadrons in the final state. The distribution is divided by

$$\frac{6 S_{\rm ew}|V_{ij}|^2}{m_\tau^2}\left(1-\frac{s}{m_\tau^2}\right)^2\left(1+2\frac{s}{m_\tau^2}\right) \qquad (5.65)$$

to unfold if from the kinematics of the decay; the last factor is omitted for scalars and pseudoscalars. Each major hadronic decay channel of the $\tau$ lepton has to be identified exclusively. Strange decays with an odd number of kaons ($K^\pm, K_S^0$ or $K_L^0$) are rejected to extract the nonstrange spectral functions. The spin-0 part is simple: there is no scalar contribution (see Sect. 5.1) and the pseudoscalar is completely dominated by one channel: $\tau \to \pi\,\nu_\tau$. The remaining events have to be separated into vector channels with an even number of pions and axial-vector channels with an odd number. A highly efficient $\pi^0$ reconstruction is necessary to minimize the crosstalk between the channels.

There are two exceptions where this simple assignment does not work: channels with two kaons and electromagnetic decays. In the absence of detailed studies of the decays with two kaons, these decays are assigned with a fraction of $(50 \pm 50)\,\%$ to both spectral functions. Channels involving isospin-violating, electromagnetic decays of $\omega$ or $\eta$ mesons are assigned to the wrong channel by this procedure and have to be corrected afterwards.

Another demand on the $\pi^0$ reconstruction is a good resolution of the momentum of the $\pi^0$, which is needed for an accurate measurement of $s$. In the end the measured spectra have to be corrected for efficiency, background, crosstalk between the channels, and the finite resolution by an unfolding procedure. These corrections are substantial, especially for channels with high multiplicity. In addition, statistical errors become large for high values of $s$ owing to the phase space suppression. Finally, each decay channel $i$ is weighted by

$$\frac{br\,(\tau \to {\rm had}_i\,\nu_\tau)}{br\,(\tau \to {\rm e}\,\nu_{\rm e}\nu_\tau)} \qquad (5.66)$$

and the appropriate channels are added to the vector and axial-vector spectral function. Details of the experimental procedures can be found in [54, 211]. Table 5.13 summarizes the channels and their weights as used in a measurement by the ALEPH collaboration [47]. The resulting spectral functions are shown in Fig. 5.35 and 5.36. Similar results have also been obtained by OPAL [211].

The vector spectral function from $\tau$ decays can be compared to the isospin $I = 1$ part of the $e^+e^-$ annihilation cross section, which is related to $v_1$ by

186   5. Strong Interactions in $\tau$ Decays

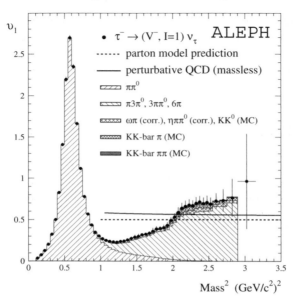

**Fig. 5.35.** Vector spectral function as measured by the ALEPH collaboration [47]. The *lines* show the predictions from the naive parton model and from massless perturbative QCD using $\alpha_s(m_Z^2) = 0.120$

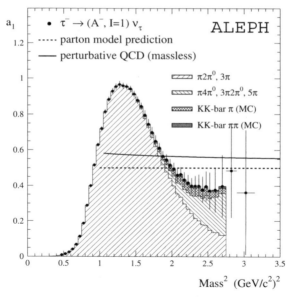

**Fig. 5.36.** Axial-vector spectral function as measured by the ALEPH collaboration [47]. The *lines* show the predictions from the naive parton model and from massless perturbative QCD using $\alpha_s(m_Z^2) = 0.120$

### 5.4 Inclusive Decays

**Table 5.13.** Vector and axial-vector hadronic $\tau$ decay modes with their contributing branching fractions. ALEPH collaboration [47]

| Vector | Branching ratio (%) | Axial-vector | Branching ratio (%) |
|---|---|---|---|
| $\pi^-\pi^0\nu_\tau$ | $25.34 \pm 0.19$ | $\pi^-\nu_\tau$ | $11.23 \pm 0.16$ |
| $\pi^-3\pi^0\nu_\tau$ | $1.18 \pm 0.14$ | $\pi^-2\pi^0\nu_\tau$ | $9.23 \pm 0.17$ |
| $2\pi^-\pi^+\pi^0\nu_\tau$ | $2.42 \pm 0.09$ | $2\pi^-\pi^+\nu_\tau$ | $9.15 \pm 0.15$ |
| $\pi^-5\pi^0\nu_\tau$ |  | $\pi^-4\pi^0\nu_\tau$ | $0.03 \pm 0.03$ |
| $2\pi^-\pi^+3\pi^0\nu_\tau$ | $\left.\right\}0.04 \pm 0.02$ | $2\pi^-\pi^+2\pi^0\nu_\tau$ | $0.10 \pm 0.02$ |
| $3\pi^-2\pi^+\pi^0\nu_\tau$ |  | $3\pi^-2\pi^+\nu_\tau$ | $0.07 \pm 0.01$ |
| $\omega\pi^-\nu_\tau$ | $1.93 \pm 0.10$ | $\omega\pi^-\pi^0\nu_\tau$ | $0.39 \pm 0.11$ |
| $\eta\pi^-\pi^0\nu_\tau$ | $0.17 \pm 0.03$ | $\eta 2\pi^-\pi^+\nu_\tau$ | $0.04 \pm 0.01$ |
| — | — | $\eta\pi^-2\pi^0\nu_\tau$ | $0.02 \pm 0.01$ |
| $K^-K^0\nu_\tau$ | $0.19 \pm 0.04$ | — | — |
| $K^-K^+\pi^-\nu_\tau$ | $0.08 \pm 0.08$ | $K^-K^+\pi^-\nu_\tau$ | $0.08 \pm 0.08$ |
| $K^0\bar{K}^0\pi^-\nu_\tau$ | $0.08 \pm 0.08$ | $K^0\bar{K}^0\pi^-\nu_\tau$ | $0.08 \pm 0.08$ |
| $K^-K^0\pi^0\nu_\tau$ | $0.05 \pm 0.05$ | $K^-K^0\pi^0\nu_\tau$ | $0.05 \pm 0.05$ |
| $K\bar{K}\pi\pi\nu_\tau$ | $0.08 \pm 0.08$ | $K\bar{K}\pi\pi\nu_\tau$ | $0.08 \pm 0.08$ |
| Total | $31.58 \pm 0.29$ | Total | $30.56 \pm 0.30$ |

$$\sigma^{I=1}_{e^+e^-\to\text{had}} = \frac{4\pi\alpha^2}{s} f_x v_1(s) \tag{5.67}$$

($f_x$ is a channel-dependent constant which can be derived from Table 5.9). According to the CVC hypothesis the vector structure functions extracted from both processes should be identical. The comparison is shown in Fig. 5.37. There is good agreement and the $\tau$ data are more accurate in a large fraction of the area accessible through $\tau$ decays.

#### 5.4.4 The Nonperturbative Contributions and $\alpha_s$

It has been pointed out by Le Diberder and Pich [538] that the nonperturbative corrections in the various dimensions can be determined from data through weighted integrals of the spectra, the spectral moments

$$R^{kl}_\tau = \int_0^{m_\tau^2} ds \left(1 - \frac{s}{m_\tau^2}\right)^k \left(\frac{s}{m_\tau^2}\right)^l \frac{dR_\tau}{ds}. \tag{5.68}$$

These can be calculated in the same manner as the original $R_\tau \equiv R^{00}_\tau$ (5.40). Depending on the values of the integers $k$, $l$, some of the QCD corrections are enhanced and others are suppressed. So, by taking enough of these moments one can disentangle the corrections of different dimensions. To illustrate these features Table 5.14 shows the relative weights of the corrections to five different moments, neglecting their small $s$ dependence.

ALEPH [47], CLEO [130], and OPAL [211] have determined the power corrections from the vector (V) and axial-vector (A) spectra simultaneously

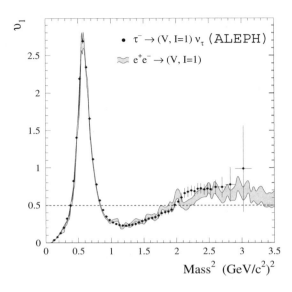

**Fig. 5.37.** Hadronic vector current spectral function from $\tau$ decays (data points, from [54]) and the corresponding distribution calculated from $e^+e^-$ isovector states. The *shaded band* includes statistical and systematic errors. The *dashed line* corresponds to the naive quark–parton prediction

**Table 5.14.** Relative weights of QCD corrections of dimension $D$ to the $kl$ moments. For simplification, the corrections are assumed to be $s$-independent

| $kl$ | $D=2$ | $D=4$ | $D=6$ | $D=8$ | $D=10$ |
|------|-------|-------|-------|-------|--------|
| 00   | +1    | 0     | −3    | −2    | 0      |
| 10   | +1    | +1    | −3    | −5    | −2     |
| 11   | 0     | −1    | −1    | +3    | +5     |
| 12   | 0     | 0     | +1    | +1    | −3     |
| 13   | 0     | 0     | 0     | −1    | −1     |

with the strong coupling constant $\alpha_s$. Table 5.15 shows the results from OPAL [211]. The total nonperturbative correction is about 2% for V and A individually and compatible with 0 in the sum. The measurements confirm the smallness of the nonperturbative corrections to $R_\tau$. The suppression of the dimension-4 correction is clearly visible. It is interesting to note that the corrections of dimension 6 and 8 for V and A are about equal in size, but opposite in sign. This indicates that they almost cancel in the sum V+A, which is expected from global quark–hadron duality: the more inclusive the measurement, the less sensitive it should be to nonperturbative effects.

The results for $\alpha_s$ from ALEPH [47], CLEO [130], and OPAL [211] are summarized in Fig. 5.38; $\alpha_s$ has a value of 0.34 with an error of 0.02 at the $\tau$ mass. The error is dominated by the theoretical uncertainties due to the unknown higher orders in the perturbative series. Varying $K_4$ by ±50 changes $\alpha_s$ by ±0.012, changing the renormalization scheme has about the same impact, and a variation of the renormalization scale gives ±0.006. (For a detailed study of the theoretical uncertainties see [546].) The experimental

**Table 5.15.** Power corrections to $R_\tau$ and $\alpha_s$ determined from the vector and axial-vector spectra [211] using the CIPT approach to calculate the perturbative correction. Theoretical estimates from [495]. The errors on $\Delta^4$ come from the uncertainty of the gluon condensate only; the uncertainties from the quark masses and condensates have to be added and are important in the case of $\Delta_A^4$

| Observable | OPAL | Theory |
|---|---|---|
| $\alpha_s\left(m_\tau^2\right)$ | $0.347 \pm 0.023$ | – |
| $\Delta_V^4$ | $0.0002 \pm 0.0003$ | $0.0008 \pm 0.0003$ |
| $\Delta_A^4$ | $-0.0052 \pm 0.0003$ | $-0.0046 \pm 0.0003$ |
| $\Delta_V^6$ | $0.0256 \pm 0.0034$ | $0.024 \pm 0.013$ |
| $\Delta_A^6$ | $-0.0197 \pm 0.0035$ | $-0.038 \pm 0.020$ |
| $\Delta_V^8$ | $-0.0080 \pm 0.0013$ | $-0.0001$ |
| $\Delta_A^8$ | $0.0041 \pm 0.0020$ | $-0.0001$ |

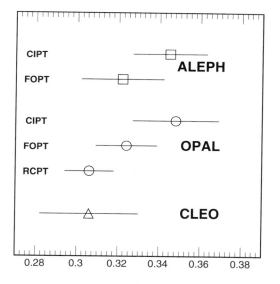

**Fig. 5.38.** Summary of the measurements of $\alpha_s\left(m_\tau^2\right)$: ALEPH [47], OPAL [211], CLEO [130]

errors are smaller than $\pm 0.010$ and are mainly due to the uncertainties in the branching ratios. This error includes the uncertainty in the nonperturbative corrections introduced through the statistical correlation between $R_\tau$ and the spectral moments. Running $\alpha_s$ up to the $Z^0$ pole yields $\alpha_s\left(m_Z\right) = 0.122 \pm 0.002$ [211]. In the course of the running $\alpha_s$ shrinks by about a factor of 3, whereas the error goes down by a factor of 10. This demonstrates the advantage of measuring $\alpha_s$ at a low scale. For comparison Fig. 5.39 shows a compilation of $\alpha_s$ measurements by the Particle Data Group [547].

190   5. Strong Interactions in $\tau$ Decays

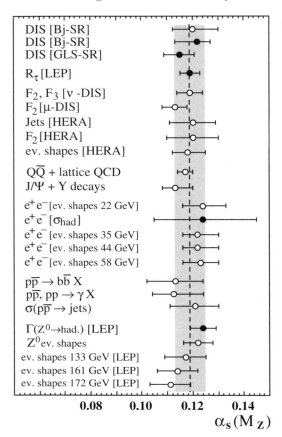

Fig. 5.39. Comparison of measurements of $\alpha_s$ at $m_Z$, from [547]

### 5.4.5 The Running of $\alpha_s$ and Other Tests

This section explains some of the tests carried out to support the validity of the result for $\alpha_s$. Figure 5.38 already presents a cross-check of the uncertainty assigned to the perturbative calculations, due to missing higher orders. Each of the data points in Fig. 5.38 represents a measurement with an error dominated by this perturbative uncertainty and estimated by varying parameters intrinsic to each calculation. The difference between the data points labeled CIPT, FOPT, and RCPT, however is the partial inclusion of different aspects of the higher orders. Therefore one expects the scatter between the different approaches to be of the same size as the theoretical error assigned to each of the points, which is indeed the case.

In order to test the assumption of the global quark–hadron duality, it is possible to use the vector or axial-vector channels separately to measure $\alpha_s$. By doing so, the measurement is restricted to a reduced number of resonances, it becomes less inclusive, and quark–hadron duality is less likely to

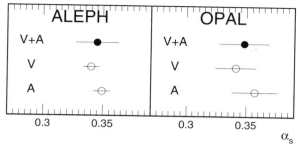

**Fig. 5.40.** Comparison of $\alpha_s$ values from pure vector (V), axial-vector (A) and the sum of both (V+A) from ALEPH [47] and OPAL [211]. (See remarks on the errors in the text)

be justified. If, however, it turns out that the $\alpha_s$ values extracted from pure vector and axial-vector currents are consistent with each other and with the value from the all-inclusive measurement, then this is a good indication that quark–hadron duality is applicable, at least for the all-inclusive measurement [548]. Figure 5.40 shows that comparison of the ALEPH [47] and OPAL [211] measurements. The all-inclusive value (labeled 'V+A') is shown with its full error, whereas the error bars on V and A show only the fraction of the error which is independent between the points (essentially the uncertainty in the hadronic branching ratios). The good agreement between the different values supports quark–hadron duality.

The $\tau$ lepton offers the interesting possibility of studying the running of $\alpha_s$ over a wide range (almost a factor of 2 in $\alpha_s$). To do so, the spectrum is not integrated all the way to $m_\tau^2$, but the integration is stopped at some intermediate value $s_0$ [538, 549]:

$$R_\tau(s_0) = \int_0^{s_0} ds \, \frac{dR_\tau}{ds}. \tag{5.69}$$

Determining $\alpha_s$ from this limited range of $s$ gives $\alpha_s(s_0)$, which can be compared to the running of $\alpha_s$ as predicted by the $\beta$ function. Figure 5.41 shows a comparison based on OPAL data [211] (similar results have been derived by ALEPH [47]). There are four data points, representing $\alpha_s$ at $m_\tau^2$ and at three lower values of $s_0$, for each of the three methods used to calculate $\Delta_{\text{pert}}$. The dominating theoretical errors are omitted from the plot, because they are fully correlated between the points at different $s_0$. There is good agreement with the predicted running down to about $2\,\text{GeV}^2$, and even further in the case of CIPT. This demonstrates the validity of the OPE approach down to values of $s_0$ substantially below $m_\tau^2$.

192    5. Strong Interactions in $\tau$ Decays

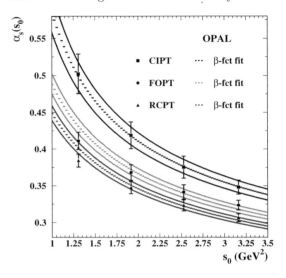

**Fig. 5.41.** The running of $\alpha_s$ from $m_\tau^2$ down to 1.3 GeV$^2$ [211]. *Data points*: measured values of $\alpha_s$ with purely experimental errors. The *dashed curves* indicate the prediction from the four-loop $\beta$ function with $\alpha_s$ adjusted to the three highest points. *Top* to *bottom*: CIPT, FOPT, and RCPT

### 5.4.6 QCD Sum Rules

The difference between the vector and axial-vector spectral functions can be used to test certain QCD sum rules. The sum rules come from dispersion relations between real and absorptive parts of the correlators of the non-strange currents. If the range of integration is extended to large enough $s_0$, i.e. beyond the region of the resonances, then the value given by the sum rule should converge towards a value predicted from chiral symmetry. Four different sum rules are considered here:

$$\frac{1}{4\pi^2}\int_0^{s_0} ds\, \frac{1}{s}\left[v_1(s) - a_1(s)\right] \;\rightarrow\; f_\pi^2 \frac{\langle r_\pi^2 \rangle}{3} - F_A, \quad (5.70)$$

$$\frac{1}{4\pi^2}\int_0^{s_0} ds\, \left[v_1(s) - a_1(s)\right] \;\rightarrow\; f_\pi^2, \quad (5.71)$$

$$\frac{1}{4\pi^2}\int_0^{s_0} ds\, s\left[v_1(s) - a_1(s)\right] \;\rightarrow\; 0, \quad (5.72)$$

$$\frac{1}{4\pi^2}\int_0^{s_0} ds\, s\ln\left(\frac{s}{\lambda^2}\right)\left[v_1(s) - a_1(s)\right] \;\rightarrow\; -\frac{4\pi f_\pi^2}{3\alpha}\left(m_{\pi^\pm}^2 - m_{\pi^0}^2\right). \quad (5.73)$$

The Das–Mathur–Okubo (DMO) sum rule (5.70) [373, 550] is of particular interest for $\tau$ physics. Owing to the suppression of the high-energy end of the spectrum by the $1/s$ weighting factor, its statistical error is reasonably small even for $s_0 = m_\tau^2$. Its prediction is expressed in terms of the pion decay constant $f_\pi = (92.42 \pm 0.26)$ MeV [44], the axial-vector form factor of the pion

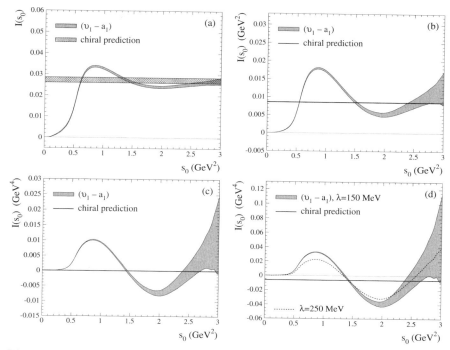

**Fig. 5.42.** Chiral sum rules: **(a)** DMO, **(b)** first Weinberg, **(c)** second Weinberg, and **(d)** electromagnetic splitting of pion masses, from ALEPH data [47]

$F_A = 0.0058 \pm 0.0008$ [44], and its charge radius $\langle r_\pi^2 \rangle = (0.439 \pm 0.008)$ fm$^2$ [551]. It can be seen from Fig. 5.42, where the value given by the sum rule is plotted versus the upper limit of integration, that it is already saturated at $m_\tau^2$. The other sum rules shown in Fig. 5.42 are the first and second Weinberg sum rules ((5.71) and (5.72) [368]) and the electromagnetic mass splitting for the pions ((5.73) [552]). Those three sum rules do not seem to be saturated at $m_\tau^2$ yet.

# 6. $\tau$ Physics at Hadron Colliders

$\tau$ leptons are not only a tool for precision measurements, but also a sensitive probe for new physics. $\tau$ leptons are used in this way by the experiments DØ and CDF at the TEVATRON $p\bar{p}$ collider [553–556] and will be used at future hadron colliders (e.g. the LHC).

## 6.1 Identification of $\tau$ Leptons

### 6.1.1 The Trigger

A major obstacle in $\tau$ physics at a hadron collider is the problem of triggering efficiently on $\tau$ leptons and at the same time keeping the background rate manageable. Different signatures of $\tau$ events can be used:

- Hadronic $\tau$ decays create a narrow jet of a few hadrons. It can be identified at trigger level by a calorimeter cluster above some energy threshold not too large in size, with a small number of tracks pointing to it.
- The energy carried away by the undetected neutrino creates events unbalanced in transverse energy $E_T$. A trigger can look for substantial missing energy $\displaystyle{\not}E_T$.
- At hadron colliders $\tau$ leptons are never the only particles produced in an event. One can thus rely on the other particles of the event to provide a trigger.

During Run Ia[1] the CDF experiment used a trigger based on the identification of hadronic $\tau$ decays, but with the improved luminosity of Run Ib the rate increased so much that this scheme had to be abandoned. A missing-energy $\displaystyle{\not}E_T$ trigger with a 35 GeV threshold was active for the whole of Run I, but it was only fully efficient above 50 GeV (see Fig. 6.1). Also, DØ had a special $\tau$ trigger activated for part of Run I. Despite these obstacles useful data for physics with $\tau$ leptons was collected (see below).

---

[1] The running period of the TEVATRON in 1992–93 was called Run Ia, in which about 20 fb$^{-1}$ of data were collected. Then Run Ib produced approximately 100 fb$^{-1}$ of luminosity in 1994–95.

196    6. τ Physics at Hadron Colliders

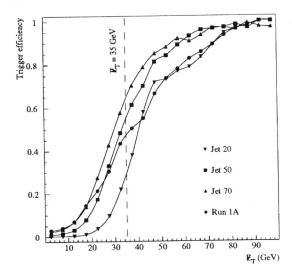

**Fig. 6.1.** Trigger efficiency of the $\not{E}_T > 35$ GeV trigger of the CDF experiment as a function of the missing transverse energy calculated off-line [553]

### 6.1.2 Off-Line Reconstruction

Once τ events have been triggered, they still have to be reconstructed off-line. With no knowledge about the initial energy of the τ leptons it is almost impossible to separate leptonic τ decays from prompt electrons and muons.

The identification of hadronically decaying τ leptons is difficult owing to the much more abundant quark and gluon jets. In general, secondaries from a hadronic τ decay form a narrow, collimated jet with only one or three charged tracks. Hence a few, well-isolated tracks are a good first indication of a τ lepton. Electrons and muons are rejected on the basis of their energy deposition in the calorimeters. The reconstruction of neutral pions from photons detected in the electromagnetic calorimeter supports the selection. Eventually, with the advent of high-resolution vertex detectors, large impact parameters and displayed vertices will enhance the τ identification.

Figure 6.2 shows the charged multiplicity distribution from a sample of monojet events, with large transverse energy of the jet ($15 < E_T < 40$ GeV) and large missing transverse energy in the events ($20 < \not{E}_T < 40$ GeV), from CDF [557]. The sample consists of about equal amounts of $W^\pm$ bosons decaying to $\tau \nu_\tau$ and jets produced by QCD processes. The upper plot shows the sample before the τ identification was applied, the lower plot after. The figure demonstrates the rejection power of the identification cuts.

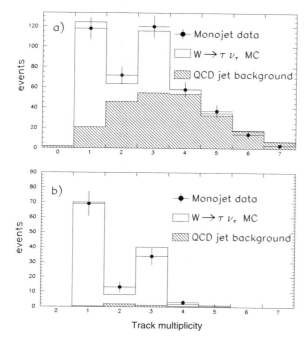

**Fig. 6.2.** Track multiplicity in a sample of monojets (CDF collaboration [557]). **(a)** No $\tau$ identification applied. **(b)** After applying all cuts except track multiplicity (calorimetric selection)

## 6.2 The $\tau$ and the Top Quark

At the TEVATRON collider t quarks are expected to be produced primarily in pairs. In the Standard Model they decay almost 100% into a b quark and a $W^{\pm}$ boson. The events can be classified according to the decay modes of the $W^{\pm}$. The approximate branching ratio into quarks is 6/9 and for each of the three lepton species $e\nu_e$, $\mu\nu_\mu$, and $\tau\nu_\tau$ it is 1/9. The CDF collaboration has searched for t quarks in the dilepton channel where one of the leptons is a $\tau$ [557]. For the trigger they rely on the non-$\tau$ lepton to initiate the readout. The full decay chain is

$$p\bar{p} \to t\bar{t} \to W^+W^-b\bar{b} \to (\tau\nu_\tau)(\ell\nu_\ell)\, b\bar{b}$$
$$\tau \to \text{had}\, \nu_\tau \quad (6.1)$$

where $\ell$ denotes an electron or a muon. Top events in the dilepton channel are expected to have a significant amount of missing transverse energy $\not{E}_T$. They also have a large transverse energy reconstructed from the $\tau$, the other lepton, $\not{E}_T$ for the neutrino, and all the jets. The presence of b quarks supports the association of the event with the production of t quarks. The b quarks are identified through a semileptonic b decay (SLT) or a secondary vertex (SVX).

Altogether CDF has observed two events with an electron and a $\tau$ lepton and another two with a muon and a $\tau$, consistent with their estimate from t quark production plus background. Two of these events have a single b tag

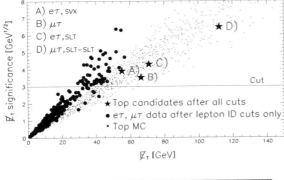

**Fig. 6.3.** The distribution of $\not{E}_T$ significance versus the missing transverse energy $\not{E}_T$ for events with a primary lepton and a $\tau$ candidate (CDF collaboration [557]). Only the four events marked by a star pass the selection cuts for t candidates. Three of them have b-tagged jets

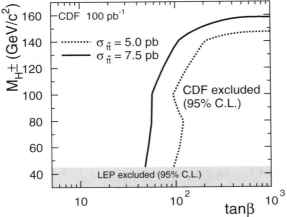

**Fig. 6.4.** Charged-Higgs exclusion region for $m_t = 174$ GeV (CDF collaboration [559])

and one a double tag. Figure 6.3 shows the distribution of the events in the plane of significance of the missing transverse energy $\not{E}_T$ versus $\not{E}_T$ (for details see [557]).

The decay of t quarks into $\tau$ leptons is of particular interest because the existence of charged Higgs bosons $H^\pm$ with $m_{H^\pm} < m_t$ could increase the event rate through the decay chain $t \to H^\pm b \to \tau\nu_\tau b$. For large values of $\tan\beta$ – the ratio of the vacuum expectation values of the two Higgs doublets in SUSY models – the decay of t quarks through $H^\pm$ dominates over the $W^\pm$-mediated decay and the charged Higgs almost exclusively decays to $\tau$ leptons.

The CDF collaboration has optimized the selections for this decay chain and searched their data for such events [558, 559]. The number of events they find is consistent with no charged Higgs boson and they exclude its existence with a mass up to $m_t$ and large values of $\tan\beta$. See Fig. 6.4.

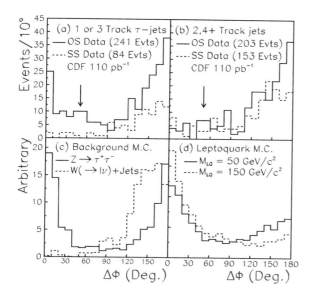

**Fig. 6.5.** Distributions of $\Delta\Phi$, the azimuthal separation between the lepton and the missing transverse energy before imposing a requirement on the hadronic jets (CDF collaboration [571]). **(a)** $\tau$ candidates with one or three tracks, **(b)** with two or four. *Solid line*: both leptons have opposite charge. *Dashed line*: same charge. The *arrow* indicates the cut on $\Delta\Phi$. Plots **(c)** and **(d)** show the Monte Carlo simulations of the background from $Z^0$ and $W^{\pm}$ events and of the signal

## 6.3 Searches

Many extensions of the Standard Model that place quarks and leptons on equal footing predict the existence of leptoquarks (LQ), color-triplet bosons that couple directly to $q\ell$ or $\bar{q}\ell$ pairs [560–568]. The CDF collaboration has searched [569, 570] for third-generation leptoquarks, amongst others, which are pair-produced in $p\bar{p}$ collisions by gluon–gluon fusion or $q\bar{q}$ annihilation. They decay to $b\tau$ [571, 572].

Particularly interesting from the experimental point of view are technicolor models containing a technifamily with technipions which are leptoquarks [567, 573, 574]. Since these are also responsible for the fermion masses in these models, technipions are expected to have Higgs-type couplings to ordinary fermions, i.e. to decay preferentially to third-generation quarks and leptons.

The experimental signature of such events is a pair of oppositely charged $\tau$ leptons accompanied by two jets. Events are selected where one of the two $\tau$ leptons decays hadronically and the other leptonically, satisfying a high-$p_T$ lepton trigger. The hadronic decay of the $\tau$ is identified by the method described above (Sect. 6.1.2) and the leptonic decay by the CDF standard lepton identification. The decays of $\tau$ leptons are a significant source of neutrinos and result in missing transverse energy in leptoquark events. To select $\tau^+\tau^-$ events, CDF uses the variable $\Delta\Phi$, defined as the azimuthal separation between the directions of the missing transverse energy $\not{\!\!E}_T$ and the lepton. The distributions are shown in Fig. 6.5.

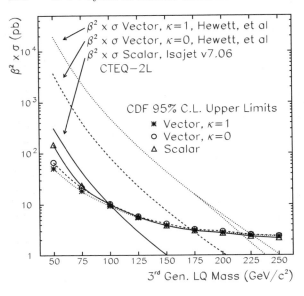

**Fig. 6.6.** Exclusion plot of third generation leptoquarks from CDF [571]. The three *steeper lines* are the theoretical predictions for $\sigma\beta^2$ for different types of leptoquarks; the *symbols* indicate the 95% confidence level upper limits. Leptoquark masses for which the theoretical prediction falls above the line of symbols are excluded. For the vector LQ with $\kappa = 1$ (*dotted line*), the *upper curve* is for $\beta = 1$. The *lower dotted curve* is obtained when the decay mode LQ $\to t\nu_\tau$ is open

The main background comes from two sources: $Z^0 \to \tau^+\tau^-$ decays accompanied by two jets, and QCD processes faking a $\tau$ lepton. The first source is estimated from Monte Carlo simulations; the second can be derived from the data using events where the '$\tau$' and the other lepton have the same charge. After requiring at least two additional jets in the event with $E_T > 10$ GeV, no candidate with oppositely charged leptons survives. The overall efficiency is around 1% and the expected background is 2.4 events.

CDF has set limits on the mass of the LQ using theoretical predictions for $\sigma(p\bar{p} \to LQ\,\overline{LQ})\beta^2$, where $\beta$ is the branching ratio of LQ $\to b\tau$. This is 100% in most of the search range, as the alternative decay LQ $\to t\nu_\tau$ is kinematically forbidden there. The theoretical expectations and the line of exclusion are shown in Fig. 6.6.

The CDF and DØ collaborations are developing $\tau$ lepton triggers for the forthcoming Run II. They want to use the $\tau$ leptons for other interesting searches. There is, for example, the Standard Model Higgs boson, which might be produced at the TEVATRON through gluon–gluon fusion. It has an 8% branching ratio into $\tau^+\tau^-$ in the mass range $80 < m_H < 130$ GeV accessible with Run II. Similarly, non-Standard Model Higgs bosons could be produced and decay via the reaction $p\bar{p} \to b\bar{b}H^0 \to b\bar{b}\,\tau^+\tau^-$. Searches for SUSY particles are another interesting application of $\tau$ leptons. For more details see [555].

## 6.4 W Decays

The largest source of $\tau$ leptons at the TEVATRON is the decay of $W^\pm$ bosons: $W \to \tau \nu_\tau$. Already, the UA1 and UA2 experiments at CERN have determined the branching ratio of this decay [575–577] and recently the DØ collaboration has presented a new, more precise measurement [578] (see also [579]). This provides a crucial test of lepton universality in the Standard Model, when compared to $W \to e\,\nu_e$ or $W \to \mu\,\nu_\mu$.

The DØ collaboration uses the fraction of data collected during Run I when they had a special $\tau$ trigger enabled [578]. They identified $W \to \tau \nu_\tau$ decays by requiring a single $\tau$ jet with transverse energy in the range $25 < E_T < 60$ GeV and a missing $\not{E}_T$ of the event of more than 25 GeV. No other substantial jets were allowed in the event. The $\tau$ jet was identified mainly by exploiting the fine segmentation of the uranium/liquid-argon calorimeter. Several cuts ensured a narrow energy deposition with few tracks pointing to it.

The DØ collaboration selected 1202 events. They expected 106 background events from QCD processes, 81 events from electronic noise in the calorimeter accidentally matching a track, 32 events from $Z^0 \to \tau^+\tau^-$, and 3 misidentified $W \to e\,\nu_e$ decays. This left them with 980 signal events, from which they calculated the product of the $W^\pm$ production cross section and the $W \to \tau \nu_\tau$ branching ratio:

$$\sigma_{W^\pm}\, br(W \to \tau \nu_\tau) = (2.38 \pm 0.19)\text{ nb}. \qquad (6.2)$$

To test lepton universality, the ratio of the $\tau$ and electron charged-current couplings to the $W^\pm$ boson, $g_\tau$ and $g_e$, were determined from

$$\left(\frac{g_\tau}{g_e}\right)^2 = \frac{\sigma_{W^\pm}\, br(W \to \tau \nu_\tau)}{\sigma_{W^\pm}\, br(W \to e\,\nu_e)}. \qquad (6.3)$$

Taking the ratio, the production cross section and the uncertainty in the luminosity cancel, and also in part the acceptance errors. The result is

$$\left(\frac{g_\tau}{g_e}\right)^2 = 1.004 \pm 0.032. \qquad (6.4)$$

Figure 6.7 compares the DØ result to other results obtained at hadron colliders.

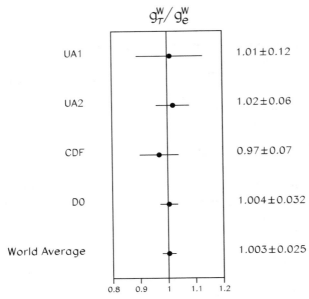

**Fig. 6.7.** Measurements of the ratio of the couplings of the W boson to $\tau$ leptons and electrons at hadron colliders [556, 578]. UA1 [575, 576], UA2 [577], CDF [579], and DØ [578]

# 7. The τ Neutrino

Closely linked to the $\tau$ lepton is the $\tau$ neutrino, its partner in the third generation. The physics of the $\tau$ neutrino is so rich that a complete review goes beyond the scope of this book (see [3, 580–594]). In particular neutrino oscillations, including the very interesting recent measurements from the SUPER-KAMIOKANDE experiment [595–597], will not be discussed here. However, a $\tau$ review would not be complete without at least a brief discussion of those results which are derived from $\tau$ decays.

## 7.1 The Mass

The most important question that can be addressed by studying $\tau$ decays is the search for a neutrino mass in the MeV range. The basic idea is simple: a $\tau$ lepton decays into a $\tau$ neutrino, unobserved in the detector, plus some visible particles $X$. The 4-momentum of the neutrino is bounded by energy conservation and phase space. The energy of the neutrino is given by $E_\tau - E_X$, where $E_\tau$ is the energy of the $\tau$ lepton,[1] and $E_X$ is the visible energy. But, then, the mass of the neutrino cannot be larger than its energy, which gives the first bound:

$$m_{\nu_\tau} \leq E_\tau - E_X. \tag{7.1}$$

On the other hand, the available phase space in a $\tau$ decay restricts the sum of the masses of all particles produced to be less than or equal to the $\tau$ mass. This is the second inequality:

$$m_{\nu_\tau} \leq m_\tau - m_X, \tag{7.2}$$

where $m_X$ is the invariant mass of all visible particles.

In principle, to exclude a neutrino mass of let's say 10 MeV or more, it is sufficient to find a single $\tau$ decay where the visible energy is as close to $E_\tau$ as 10 MeV or the visible mass is that close to $m_\tau$. In practice there are two problems: one has to be sufficiently sure that this one event is not a

---

[1] $E_\tau$ is assumed to be equal to the beam energy in $e^+e^-$ collisions, which is strictly true only in the absence of radiation. The radiative corrections have to be taken into account in an actual analysis.

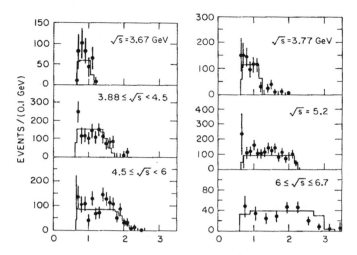

**Fig. 7.1.** Pion energy spectrum (in GeV) measured by MARK II [199]. Background subtraction and efficiency corrections were applied bin-by-bin. The *curves* are the expected spectra for $m_\tau = 1782$ MeV, $m_{\nu_\tau} = 0$, and a branching ratio of $\tau \to \pi\,\nu_\tau$ of 11.7%

background event and, second, $E_X$ and $m_X$ are only measured with a finite resolution. But the basic idea is still true. The limit derived is driven by a few events very close to the endpoint.

Hadronic decays of the $\tau$ lepton with a high multiplicity are preferred. They tend to have the highest invariant masses, and also the energy spectrum is pushed towards the endpoint. However, there is a trade-off as these channels have low branching ratios and more background from q$\bar{\text{q}}$ events, and are harder to reconstruct. With low statistics, typically less than 100 events in a spectrum, it is also a question of luck how close the events are to the endpoint. A problem which has led to considerable discussions between more and less lucky collaborations.

### 7.1.1 Limits from the Endpoint of the Energy Spectrum

The first limits on the mass of the $\tau$ neutrino were derived from the endpoint of the energy spectrum in leptonic decays [149], $\tau \to \pi\nu_\tau$ [199], and $\tau \to 3\pi\nu_\tau$ [97, 598].

Figure 7.1 shows the pion spectrum from $\tau \to \pi\,\nu_\tau$ decays measured by the MARK II collaboration at various center-of-mass energies [199]. Since $\tau \to \pi\,\nu_\tau$ is a two-body decay, the expected pion energy spectrum is flat for monoenergetic, unpolarized $\tau$ leptons produced at a fixed center-of-mass energy. The pions were selected by a rejection technique, i.e. any good charged track was called a pion if it was not identified as an electron by the electromagnetic calorimeter, not identified as muon by the muon chambers, not identified as a kaon or proton by the time-of-flight system, and not identified as a $\tau \to \pi\, n\, \pi^0\, \nu_\tau$ decay by the presence of photons. 2150 events in a 1–1 topology were selected, out of which roughly half were background. Most of the background events populate the lower-momentum regions and therefore do not disturb the neutrino mass measurement. The background has been subtracted from the spectra of Fig. 7.1. Muons with a momentum of less than 700 MeV range out before they reach the muon chambers and become indistinguishable from pions. Hence any track below 700 MeV had to be rejected. The cutoff can be seen in the spectra. The collaboration used the data taken at $\sqrt{s} = 5.2$ GeV, the largest block at a fixed center-of-mass energy, and derived an upper limit on the mass of the $\tau$ neutrino of

$$m_{\nu_\tau} < 250 \text{ MeV} \qquad (95\% \text{ confidence level}). \qquad (7.3)$$

## 7.1.2 Limits from the Endpoint of the Mass Spectrum

It has turned out that the mass spectrum is more sensitive to nonvanishing neutrino masses. Limits have been derived from $\tau \to 4\,\pi\,\nu_\tau$ [192, 194], $\tau \to K^+ K^- \pi^\pm\, \nu_\tau$ [147], and $\tau \to 5\,\pi^\pm\, \nu_\tau$ [60, 91, 164, 191].

The ARGUS collaboration published an improved limit from $\tau \to 5\pi^\pm\, \nu_\tau$ in 1988, which was the best limit for many years (until 1995) [91]. They initially selected 12 $\tau$-pair events in the topology of a one-prong decay versus five prongs. The one-prong decay was restricted to the leptonic channels plus $\tau \to \pi\,\nu_\tau$ and $\tau \to K\,\nu_\tau$.

A critical point in the analysis is the rejection of the $q\bar{q}$ background. The requirement of an isolated track with no photons in one hemisphere greatly reduces this background, but there are some events left. Figure 7.2 shows the distribution of the invariant mass of the five-prong decay versus the momentum of the one-prong. The events in the data in the upper right area of the plot are due to the $q\bar{q}$ background. Everything beyond the line was rejected. After all cuts, the expected background in the final sample was much smaller than one event. To estimate a systematic error ARGUS conservatively assumed that the most sensitive event out of the 12 events surviving all cuts was due to background and removed it from the sample.

Figure 7.3 shows the mass spectrum of the 12 selected events. From these events an upper limit on the mass of the $\tau$ neutrino of 25 MeV was derived, which increases to 35 MeV after inclusion of the systematic errors (95% confidence level). In 1992 the collaboration redid the analysis with a larger data sample and found another 8 events, but none of them near the endpoint. Hence the limit did not improve [84]. At the same time, they had remeasured the $\tau$ mass to be a few MeV smaller than was thought before. As the limit depends on the knowledge of the $\tau$ mass, they recalculated the limit. Their final number is

$$m_{\nu_\tau} < 31 \text{ MeV} \qquad (95\% \text{ confidence level}). \qquad (7.4)$$

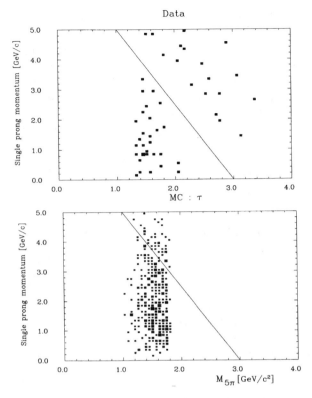

**Fig. 7.2.** Momentum of the one-prong decay versus the mass of the five-prong decay in 1–5 topology events from ARGUS [91]: *upper plot*, data and *lower plot*, $\tau$ Monte Carlo simulation. Events above the *line* were rejected. Most of them are due to $q\bar{q}$ background

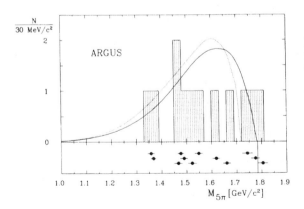

**Fig. 7.3.** $5\pi$ mass spectrum measured by the ARGUS collaboration [91]. The *solid curve* corresponds to the expected shape of a pure phase space decay with $m_{\nu_\tau} = 0$, and the *dashed curve* to the same with $m_{\nu_\tau} = 70$ MeV. Underneath, the reconstructed mass of every event is shown with the corresponding error. The $\tau$ mass is indicated by the *vertical line*

## 7.1.3 Kinematic Reconstruction of the Neutrino

Assume a situation where the $\tau$ lepton decays hadronically in a two-body decay and its 4-momentum $p_\tau$ is known. The 4-momentum of the hadrons $p_{\text{had}}$ is measured in the detector. Then the 4-momentum of the neutrino is given by

$$p_{\nu_\tau} = p_\tau - p_{\text{had}} \tag{7.5}$$

and

$$m_{\nu_\tau}^2 = p_\nu^2. \tag{7.6}$$

This means that the mass of the $\tau$ neutrino could be determined on an event-by-event basis.

Now, in the absence of radiation, the energy and total momentum of the $\tau$ are known from the initial beam energies, but its direction is not known. The method described in Sect. 2.2.4 cannot be applied here, as it assumes a vanishing neutrino mass. A clean method can be developed in 3–3 topologies, i.e. in events where both $\tau$ leptons decay to three charged tracks. The secondary vertices, the points of decay of the two $\tau$ leptons, can be reconstructed from the three prongs. The line connecting the two vertices gives the direction of the $\tau$ leptons. However, owing to the short lifetime of the $\tau$, the lever arm of the measurement is so short that it has not been possible to reconstruct the direction with sufficient resolution with this method.

The OPAL collaboration has published a measurement based on this idea, approximating the $\tau$ direction by the thrust axis [221]. Events are selected where both $\tau$ leptons decay to three charged hadrons. No attempt is made to separate pions from kaons; all hadrons are assumed to be pions (4% of the final sample includes at least one kaon). Events with additional neutral pions are rejected, if there is a significant excess of energy in the electromagnetic calorimeter above that expected from the charged hadrons. The events with an additional $\pi^0$ that survive this cut make up the largest background fraction in the final sample. This fraction is about 15%. Non-$\tau$ backgrounds are very small.

The direction of the $\tau$ leptons is calculated as

$$\hat{p}_\tau = \frac{\boldsymbol{p}_{\text{had}+} - \boldsymbol{p}_{\text{had}-}}{|\boldsymbol{p}_{\text{had}+} - \boldsymbol{p}_{\text{had}-}|} \tag{7.7}$$

and $E_\tau = E_{\text{beam}}$ is assumed. $\boldsymbol{p}_{\text{had}+}$ and $\boldsymbol{p}_{\text{had}-}$ are the sums of the momenta of the three charged hadrons from the decays of the $\tau^+$ and $\tau^-$, respectively. The $\tau$ direction calculated from this approximation differs typically by 7 mrad from the true $\tau$ direction. It gives the dominant contribution to the resolution of $m_{\text{miss}}^2$, where $m_{\text{miss}}^2$ would be $m_{\nu_\tau}^2$ if the proper $\tau$ direction were used. Figure 7.4 shows the measured distribution.

To derive the result, the missing mass is used in combination with information from the missing energy, defined as

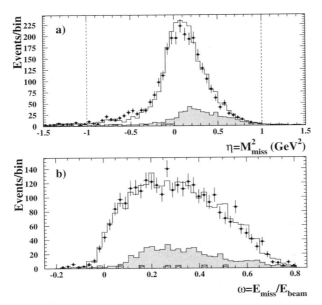

**Fig. 7.4.** The distribution of missing mass squared (**a**) and missing energy (**b**) from $\tau \to 3\,\pi\,\nu_\tau$ versus $\tau \to 3\,\pi\,\nu_\tau$ events (OPAL collaboration [221]). The *points* are data, the *open histogram* is the Monte Carlo expectation for $m_{\nu_\tau} = 10$ MeV, the *gray histogram* is the background from $\tau$ decays, and the *hatched histogram* is the $q\bar{q}$ background. A cut is applied on the missing mass along the *dashed lines*

$$E_{\mathrm{miss}} = E_\tau - E_{\mathrm{had}}. \tag{7.8}$$

The OPAL collaboration finds the data to be consistent with a vanishing neutrino mass and set an upper limit of

$$m_{\nu_\tau} < 35.3\,\mathrm{MeV} \tag{7.9}$$

at the 95% confidence level.

It should be pointed out that there is a subtle difference between the determination of the endpoint of a spectrum and the kinematic reconstruction of the neutrino mass: if there is a nonvanishing neutrino mass in the MeV region this will be tricky to recognize with an endpoint method, despite the fact that these methods currently give better limits. As experiments improve, the limits from the endpoints will no longer improve. It looks as if everybody is being unlucky, not finding events right at the threshold, but it takes a lot more statistics to positively identify a nonvanishing neutrino mass. This is not the case for the kinematic reconstruction. A nonvanishing neutrino mass will shift the peak, and lower limits can be derived just as upper limits are derived now. The missing-mass method is also able to see different components of neutrinos. If, for example, most of the neutrinos are massless, but there is a small admixture of more massive (maybe right-handed) neutrinos, a satellite peak at high $m^2_{\mathrm{miss}}$ will be created.

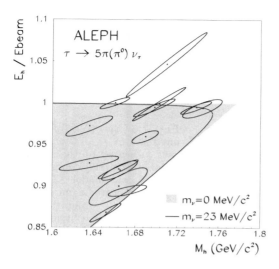

**Fig. 7.5.** Distribution of $m_{\text{had}}$ versus $E_{\text{had}}$ in the decay channel $\tau \to 5\,\pi^{\pm}\,\nu_\tau$ (ALEPH collaboration [49]). The *gray area* is the allowed region for a massless neutrino. The border of the allowed region for a 23 MeV neutrino is also drawn. There is one $\tau \to 5\pi^{\pm}\pi^0\,\nu_\tau$ event in the plot, which is the one with the highest energy. The plot shows only the region of the spectrum close to the endpoint

### 7.1.4 Two-Dimensional Methods

It turns out that the most efficient way of using the data is not to choose between analyzing the energy or mass spectra, but to apply a two-dimensional method and exploit both at the same time. The feasibility of such an approach was first demonstrated in [231] and all recent measurements use it, including the one that currently sets the most stringent limit, by the ALEPH collaboration [49].

The data is plotted as a two-dimensional spectrum of the visible mass versus the visible energy. Figure 7.5 shows the events selected by the ALEPH collaboration in the $\tau \to 5\,\pi^{\pm}\,\nu_\tau$ channel. Each event is plotted as a dot surrounded by the error ellipses (one standard deviation) of the detector resolution. The plot also shows the kinematically allowed range, which is the shaded area in the case of a massless neutrino. For a massive neutrino this range shrinks. For $m_{\nu_\tau} = 23$ MeV the area beyond the line is no longer accessible.

ALEPH derives an upper limit on the mass of the $\tau$ neutrino of 23.1 MeV and a limit of 25.7 MeV by applying the same technique to a sample of $\tau \to 3\,\pi\,\nu_\tau$ events. Combining the two limits, they get

$$m_{\nu_\tau} < 18.2 \text{ MeV} \tag{7.10}$$

at the 95% confidence level [49].

A potential source of systematic errors is the spectral function that is used in the Monte Carlo simulation to generate the expected mass spectra of the events. If in nature there is a resonance right at the phase space limit, it will increase the population near the endpoint and fake a better limit. The problem has been extensively studied [599]. In Fig. 7.6 ALEPH compares its events selected in the $\tau \to 5\,\pi^{\pm}\,\nu_\tau$ channel with two different models.

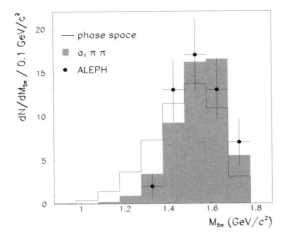

Fig. 7.6. Spectrum of invariant mass in $\tau \to 5\,\pi^{\pm}\,\nu_\tau$ decays measured by the ALEPH collaboration [49] compared to two different models. The *histogram* indicates a pure phase space model, while the *gray area* was obtained by means of an intermediate $a_1\pi\pi$ state

Including the $a_1$ as a subresonance shifts the events towards the edge of phase space. This seems to be slightly preferred by the data. The effect on the limit, however, is negligible. The three pion channel is more sensitive to those effects. If a radial excitation of the $a_1$ meson exists at 1700 MeV as suggested by the DELPHI collaboration [150], then the limit in this channel would increase by 6 MeV.

### 7.1.5 Summary

Many experiments have analyzed $\tau$ decays to search for a sign of a nonvanishing neutrino mass, but to date every result is consistent with a massless neutrino. These direct searches are sensitive to the MeV range. Different techniques have been applied to different channels. An overview of the results from different techniques applied to different channels is given in Fig. 7.7. The most stringent limit of 18.2 MeV (95% confidence level) comes from a two-dimensional method applied to $\tau \to 3\,\pi^{\pm}\,\nu_\tau$ and $\tau \to 5\,\pi^{\pm}\,\nu_\tau$ decays by the ALEPH collaboration [49].

For comparison, the current upper limit on the mass of the electron neutrino is approximately 10 eV and that on the muon neutrino 170 keV. If the KAMIOKANDE effect [597] is due to $\nu_\mu \leftrightarrow \nu_\tau$ oscillations then the mass difference $m_\tau^2 - m_\nu^2$ is on the order of $10^{-3}$ eV and the mass of the $\tau$ neutrino must be below 170 keV. If one assumes further that the muon neutrino is much lighter than the $\tau$ neutrino, then the mass of the $\tau$ neutrino is in the sub-eV range. But this is only speculation.

Cosmological arguments limit the mass of a stable neutrino to values below a few eV [603]. These limits weaken substantially if the neutrinos are allowed to decay [604]. Figure 7.8 shows the exclusions obtained from direct searches and cosmology in the plane of neutrino mass versus neutrino lifetime. The plot is taken from [602] and has been updated in [49].

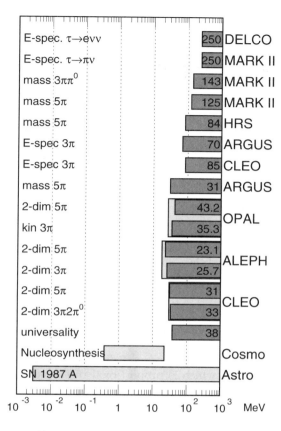

**Fig. 7.7.** Summary of limits on the mass of the $\tau$ neutrino (95% confidence level in units of MeV). The *boxes* mark the excluded regions. The *light boxes* for OPAL, ALEPH, and CLEO give the combined limits from the two measurements, which are 27.6, 18.2, and 30 MeV, respectively. CLEO's individual numbers do not include the systematics; only the combined number does. References (*top to bottom*): DELCO [149], MARK II [192, 194, 199], HRS [164], ARGUS [97], CLEO [598], ARGUS [137], OPAL [213], ALEPH [49], CLEO [113], universality [332, 366], Cosmo [600], Astro [601]

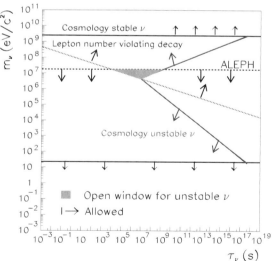

**Fig. 7.8.** Bounds on the mass of $\tau$ neutrinos derived from cosmology (*solid lines*), from the nonobservation of lepton-number-violating decays (*dotted line*), and from the direct searches (*dashed line*) [49, 602]. The *gray area* in the center shows the allowed region for an unstable neutrino

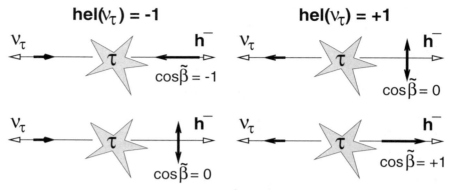

**Fig. 7.9.** Orientation of the spins in the decay $\tau \to 3\pi\nu_\tau$ for neutrinos with positive and negative helicity. The *open arrows* indicate the momenta and the *solid arrows* the spin directions. The spin arrow of the hadron in the *top right* and *bottom left* pictures represents a longitudinally polarized hadron. The angle $\tilde{\beta}$ is the angle between the direction of the momentum and the spin of the hadron

## 7.2 The Helicity

### 7.2.1 Parity Violation in the decay $\tau \to 3\pi\nu_\tau$

It has been pointed out [300, 302, 437, 468] that in some $\tau$ decays it is possible to define a parity-violating asymmetry which is proportional to the helicity of the $\tau$ neutrino. The quantity measured is

$$\gamma_{\mathrm{VA}} = \frac{2 g_\mathrm{v} g_\mathrm{a}}{g_\mathrm{v}^2 + g_\mathrm{a}^2}, \tag{7.11}$$

where $g_\mathrm{v}$ and $g_\mathrm{a}$ are the vector and axial-vector charged-current couplings to the initial leptonic current in $\tau$ decays, i.e.

$$L_\mu = \langle\, \overline{u}(\nu_\tau) \,|\, g_\mathrm{v} \gamma_\mu - g_\mathrm{a} \gamma_\mu \gamma_5 \,|\, u(\tau) \,\rangle. \tag{7.12}$$

For a massless neutrino $\gamma_{\mathrm{VA}}$ is related to the helicity of the neutrino simply by $\gamma_{\mathrm{VA}} = -h_{\nu_\tau}$. In the Standard Model, with purely left-handed neutrinos, one has $g_\mathrm{v} = g_\mathrm{a} = 1$ and hence $\gamma_{\mathrm{VA}} = 1$.

The measurement can be done with any $\tau$ decay into a spin-1 hadronic system with at least three mesons in the final state. In practice it has only been applied to the decay $\tau \to 3\pi\nu_\tau$ and the discussion will be restricted to this channel for simplicity.[2] The basic idea of the measurement is sketched in Fig. 7.9. Owing to conservation of angular momentum, the spin of the neutrino and that of the hadron have to add up to 1/2, the spin of the

---

[2] The $\tau \to 3\pi\nu_\tau$ channel is dominated by the $a_1$ meson, which has spin 1. Upper limits on possible spin-0 contributions to the three-pion final state have been given in [217].

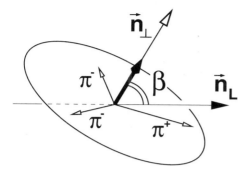

**Fig. 7.10.** Definition of the angle $\beta$ in the decay to three pions in the hadronic rest frame. The vector $\boldsymbol{n}_\perp$ is the normal to the decay plane and $\boldsymbol{n}_\mathrm{L}$ the direction of the boost into the laboratory frame. $\beta$ is the angle between these two vectors

initial $\tau$. The hadron has three possible spin orientations, in the direction of, against, and perpendicular to its momentum, but one orientation leads to a total angular momentum of $3/2$ and is thus forbidden. Figure 7.9 shows the two allowed configurations for both neutrino helicities. There is an asymmetry in the angle $\tilde\beta$ for left-handed neutrinos: negative values are preferred over positive. The asymmetry changes sign for right-handed neutrinos.

The angle $\tilde\beta$ is not directly accessible by experiments; it has to be inferred from the decay of the hadronic system into the three pions, as mentioned in Sect. 4.5.2. The direction of the normal to the decay plane $\boldsymbol{n}_\perp$ serves as a spin analyzer. It is defined in Fig. 7.10. The three pions define a plane in their rest frame owing to momentum conservation. The pions carry no spin, so that the spin of the hadronic system has to be transferred into orbital angular momentum among the pions. This orbital angular momentum is perpendicular to the ordinary momenta of the pions and hence $\boldsymbol{n}_\perp$ indicates the direction of the spin of the hadronic system. In that sense $\beta$ corresponds to $\tilde\beta$.

The angle $\beta$ can be reconstructed by boosting the pions back into the hadronic rest frame. In this frame one has

$$\boldsymbol{n}_\perp = \frac{\boldsymbol{q}_1 \times \boldsymbol{q}_2}{|\boldsymbol{q}_1 \times \boldsymbol{q}_2|}, \qquad (7.13)$$

where $\boldsymbol{q}_1$ and $\boldsymbol{q}_2$ are the momenta of the two like-sign pions. There is, however, an ambiguity in this definition, because the two like-sign pions are indistinguishable (see Fig. 7.11). Interchanging the labels 1 and 2 reverses the direction of $\boldsymbol{n}_\perp$. In practice one avoids the ambiguity by choosing the labels of the pions so that $|q_2| > |q_1|$, which is equivalent to $s_1 > s_2$. In theoretical calculations one has to ensure that the amplitudes are invariant under the exchange of $|q_2|$ and $|q_1|$ (Bose symmetry) and add a factor $\mathrm{sign}(s_1 - s_2)$ when integrating the phase space.

Now, defining the asymmetry in $\cos\beta$ as its first moment, or equivalently the average value of $\cos\beta$, one sees the connection to $\gamma_\mathrm{VA}$ [437]:

$$\langle\cos\beta\rangle = \int \frac{\mathrm{d}\Gamma}{\mathrm{d}\boldsymbol{x}} \cos\beta\, \mathrm{sign}(s_1 - s_2)\, \mathrm{d}\Phi = \gamma_\mathrm{VA} \cos\psi\, W_\mathrm{E}, \qquad (7.14)$$

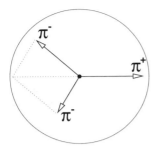

**Fig. 7.11.** The decay plane of the three pions in the hadronic rest frame rotated into the plane of the paper. To define the normal to the decay plane $n_\perp$ one has to specify which of the two negative pions is labeled number 1

where $d\Phi$ is the phase space integral.³ The angle $\psi$ is the angle between $n_L$ and $n_\tau$ in the hadronic rest frame and can be calculated from the energy of the hadrons.⁴ $W_E$ is one of the structure functions of the decay defined in Sects. 5.2.2 and 5.3.2. It is proportional to $\text{Im}(F_1 F_2^*)$, where $F_i$ are the form factors of the hadronic current defined in (5.16). It is an interference term between the two $\rho\pi$ amplitudes, requiring nontrivial phases owing to final-state interactions.

### 7.2.2 Measurements

Using the technique described above, ARGUS was the first experiment to establish parity violation in $\tau$ decays [82, 87]. Similar measurements have been presented in [217, 226, 490]. ARGUS selected 3899 $\tau \to 3\pi\nu_\tau$ decays in the 1–3 topology. The mass spectrum and the projection of the Dalitz plane are shown in Fig. 7.12.

The data was cast into bins in $Q^2$ and $x_{a_1}$, where $x_{a_1}$ is the scaled energy of the hadrons. For each bin an asymmetry was calculated:

$$a(Q^2, x) = \frac{1}{N_i} \sum_i -\cos\beta \, \text{sign}(s_1 - s_2). \qquad (7.15)$$

$N_i$ is the number of events in the bin. ARGUS uses a different sign convention from that in (7.14). Then the numbers were averaged over the $x$ bins:

$$A(Q^2) = \frac{1}{w} \sum_j w_j \frac{a(Q^2, x_j)}{\langle \cos\psi \rangle}, \qquad (7.16)$$

where $\langle \cos\psi \rangle$ is the average value of $\cos\psi$ within the bin and $w_j$ are the statistical weights of $a(Q^2, x_j)$. As expected, the asymmetry was found to have a different sign for $\tau^+$ and $\tau^-$. The asymmetries were corrected for background and then combined, taking into account the sign difference. The result is shown in Fig. 7.13.

---

³ This formula holds for unpolarized $\tau$ leptons. The formula with polarization is given in [437].
⁴ It is different from the angle $\psi$ defined in Sect. 2.2.2 and used in Sect. 3.2.

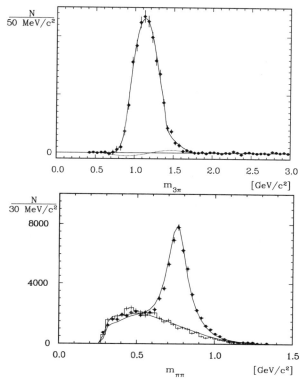

**Fig. 7.12.** Background-corrected three- and two-pion mass spectra (ARGUS collaboration [82]). The two-pion plot (*bottom*) shows the like-sign (*crosses*, one entry per event) and unlike-sign (*dots*, two entries per event) combinations. The *curves* are from a KORALB Monte Carlo simulation. The three-pion spectrum (*top*) is fitted with the IMR model [372]

The same procedure was applied to the Monte Carlo simulation, giving $A^{\mathrm{MC}}(Q^2)$. The asymmetry is proportional to $\gamma_{\mathrm{VA}}$, with $\gamma_{\mathrm{VA}} = 1$ in the Monte Carlo simulation. Hence the result can be extracted from the relation

$$A^{\mathrm{meas}}(Q^2) = \gamma_{\mathrm{VA}}\, A^{\mathrm{MC}}(Q^2). \tag{7.17}$$

The result is

$$\gamma_{\mathrm{VA}} = 1.25 \pm 0.23^{+0.15}_{-0.08}. \tag{7.18}$$

The method relies on the proper prediction of the structure function $W_{\mathrm{E}}$ to extract the size of $\gamma_{\mathrm{VA}}$. It is not so much the shapes of the Breit–Wigners in the models that make a difference, but the proper description of the relative orbital angular momentum between the pions. Figure 7.14 shows a comparison of the predicted asymmetry for a $\rho$ Breit–Wigner with constant width and the Kühn/Santamaria model with and without the inclusion of the $\rho'$

216    7. The τ Neutrino

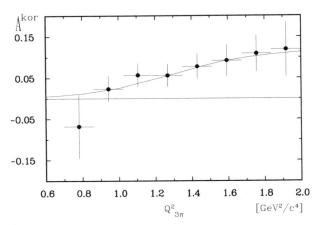

**Fig. 7.13.** The parity-violating asymmetry in $\tau \to 3\,\pi\,\nu_\tau$ measured as a function of the square of the hadronic mass $Q^2$ (ARGUS collaboration [82]). The plot shows the combined asymmetry for $\tau^+$ and $\tau^-$ corrected for background

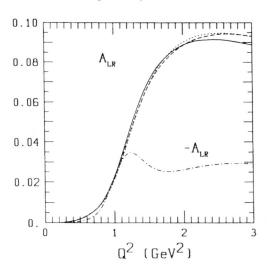

**Fig. 7.14.** The asymmetry function $A(Q^2)$ for different models of the $\rho$ Breit–Wigner, shown as *solid*, *dashed*, and *dotted curves* [300]. The *dashed–dotted* curve shows the prediction for a pure 'D wave' amplitude. It is plotted with reversed sign

(taken from [300]). The predictions line up sufficiently well. If one changes, however, the ratio of the D to the S wave between the $\rho$ and the remaining pion from the ratio predicted in the model to a pure D wave, there is a drastic change. The asymmetry even changes sign. (Some of this model dependence is included in the systematic error.)

The model dependence can be avoided by a simultaneous determination of $\gamma_{\rm VA}$ and the structure functions. But the price in terms of sensitivity lost is high. The OPAL collaboration has performed both analyses. The model-independent result is

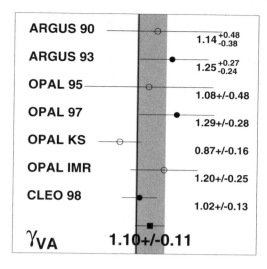

Fig. 7.15. Summary of measurements of the helicity of the $\tau$ neutrino from the parity-violating asymmetry in $\tau \to 3\pi\nu_\tau$. The average from the three measurements, marked by a *solid square* is $1.10 \pm 0.11$. References: ARGUS 90 [87]; ARGUS 93 [82]; OPAL 95 [226]; OPAL 97 model independent calculation, KS, and IMR [217]; CLEO 98 [490]

$$\gamma_{\rm VA} = 1.29 \pm 0.26 \pm 0.11, \tag{7.19}$$

and the results with the Kühn/Santamaria [300] and Isgur/Morningstar/Reader [372] models are

$$\gamma_{\rm VA}^{\rm KS} = 0.87 \pm 0.16 \pm 0.04,$$
$$\gamma_{\rm VA}^{\rm IMR} = 1.20 \pm 0.21 \pm 0.14.$$

Figure 7.15 summarizes the measurements.

In the studies of the Lorentz structure of the charged current in hadronic $\tau$ decays, which will be discussed in the next section, a Michel-type parameter called $\xi_h$ is determined. Under certain assumptions, this parameter is identical to the negative of the neutrino helicity $\xi_h = -h_{\nu_\tau}$. The average is

$$\xi_h = 1.0000 \pm 0.0056. \tag{7.20}$$

The reader is referred to Sect. 8.2.

## 7.3 Electromagnetic Moments

### 7.3.1 The Process $Z^0 \to \nu\bar{\nu}\gamma$

In the framework of the Standard Model the neutrinos have neither an electric charge nor a magnetic or electric dipole moment and hence do not couple to photons. If the Standard Model is extended to allow for massive neutrinos, they get a small magnetic moment of [605]

218    7. The τ Neutrino

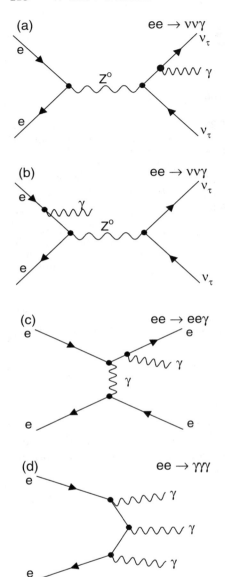

**Fig. 7.16.** Feynman diagrams for the radiation of photons from a magnetic moment of the τ neutrino (**a**) and the main background sources (**b**) – (**d**)

$$\mu_\nu = \frac{3\,e\,G_\mathrm{F}\,m_\nu}{8\,\sqrt{2}\,\pi^2} = 3.20 \times 10^{-19}\,\frac{m_\nu}{[\mathrm{eV}]}\,\mu_\mathrm{B}, \tag{7.21}$$

where $\mu_\mathrm{B}$ is the Bohr magneton. Any magnetic moment beyond this indicates new physics.

One method to search for a magnetic moment of the τ neutrino is to look for photons radiated from the neutrinos in $e^+e^-$ collisions. Figure 7.16 shows

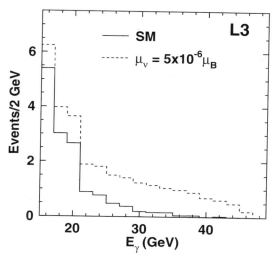

**Fig. 7.17.** Expected energy spectrum of photons from initial-state radiation in $Z^0 \to \nu\bar{\nu}$ and an anomalous magnetic moment of the neutrino in the L3 detector. (The plot is taken from [608]; data and other models of photon sources have been removed from the plot for clarity. The data can be seen in Fig. 7.18)

the Feynman diagram of this process and the major background sources. The largest obstacle is the separation of the signal from neutrino production accompanied by initial-state radiation. The handle used to suppress this background is the energy and angular distribution of the photons. Initial-state radiation tends to be emitted collinear with the initial beams, whereas photons from the signal are radiated along the direction of the neutrinos and thus give a more isotropic distribution [606, 607]. The energy of the initial-state radiation is limited to a value at which the effective energy of the collision after radiation falls below the $Z^0$ peak and neutrino production is largely suppressed. Figure 7.17 shows a simulation of the two processes in the L3 detector [608].

Another source of background is radiative Bhabha scattering (Fig. 7.16c) where both the scattered electron and the positron are lost down the beam pipe. This background can be avoided by restricting the transverse momentum $p_T$ of the photons with respect to the beam. The $p_T$ of the photon has to be balanced. If it is large enough, the electron and/or positron have to be scattered into the detector and the event can be rejected. The same applies to the third background process (Fig. 7.16d).

The L3 collaboration has searched for such additional photons [608]. They selected photons that had at least 15 GeV of energy and a polar angle in the range $20° < \theta_\gamma < 160°$. Otherwise the detector was required to be empty. They selected 14 events, where 14.1 were expected from Standard Model sources. The spectrum and the angular distribution can be seen in Fig. 7.18 in comparison with the Standard Model sources. There is very good agreement with the Standard Model, and additional contributions from a magnetic moment of the $\tau$ neutrino can be limited to

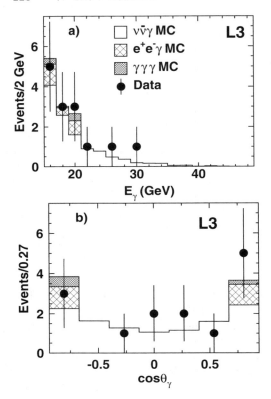

Fig. 7.18. Spectrum and angular distribution of single photons above 15 GeV in the L3 detector [608] in comparison with the Standard Model sources

$$\mu_{\nu_\tau} < 3.3 \times 10^{-6} \, \mu_{\rm B} \tag{7.22}$$

at the 90% confidence level. The collaboration considered only the $\tau$ neutrino, because existing limits on the magnetic moments of the electron neutrino ($\mu_{\nu_e} < 1.8 \times 10^{-10} \, \mu_{\rm B}$) and muon neutrino ($\mu_{\nu_\mu} < 7.4 \times 10^{-10} \, \mu_{\rm B}$) [44] preclude the possibility of observing them at LEP.

Similar (less restrictive) analyses can be found in [609–612].

### 7.3.2 Neutrino Scattering

A different technique has been pursued by the BEBC (WA 66) collaboration [613], utilizing a neutrino beam from a beam dump with a small but calculable prompt $\tau$ neutrino component. The experimental setup is sketched in Fig. 7.19.

400 GeV protons from the CERN SPS were dumped into a large copper block. The block was large enough to contain almost the entire hadron cascade. Few of the long-lived hadrons (pions, kaons), whose decays produce

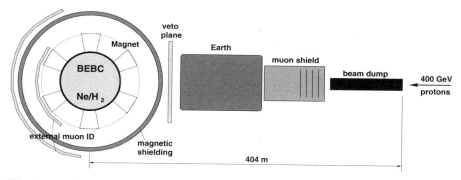

**Fig. 7.19.** Schematic setup of the WA 66 experiment [614]. (Not to scale)

most of the neutrinos in a conventional neutrino beam, had time to decay before being reabsorbed, thus suppressing the 'conventional' electron and muon neutrino flux. The neutrinos produced in such a beam dump stem from the decay of short-lived particles, mainly charmed hadrons. Amongst them are some $D_s$ mesons which produce $\tau$ neutrinos through the decays $D_s \to \tau \nu_\tau$ and $\tau \to \nu_\tau X$.

The Big European Bubble Chamber (BEBC) was located some 400 m downstream behind massive shielding. which absorbed even high-energy muons. The BEBC was filled with a neon–hydrogen mixture of 74 mole percent neon, corresponding to a density of 0.69 g/cm$^3$ and a radiation length of 43 cm. The fiducial volume was 16.6 m$^3$ or 11.5 tons.

A magnetic moment of a neutrino would reveal itself through an increased cross section for elastic scattering on electrons in the chamber. Although, the luminosity and detection efficiency were not high enough to see the interaction of Standard Model $\tau$ neutrinos, the absence of a signal can be used to put a stringent limit on the magnetic moment of the neutrino.

The cross section for elastic scattering of neutrinos on electrons through a magnetic moment is given by [615, 616]

$$\frac{d\sigma}{dE_e} \approx \pi r_e^2 \left(\frac{\mu_\nu}{\mu_B}\right)^2 \left(\frac{1}{E_e} - \frac{1}{E_\nu}\right). \tag{7.23}$$

The integrated cross section for a magnetic moment of $10^{-6}\,\mu_B$ at these energies is several orders of magnitude larger than that for electroweak scattering.

The signature on the film is a backscattered electron with more than 500 MeV of energy, which might initiate a shower in the gas or not, and nothing else. Visually scanning all events twice, a total of three candidates were found. Two of them were positrons, the other an electron. The two positrons are consistent with the background expectation of $2 \pm 1$ events from $\nu_e p \to e^+ n$ where the neutron escaped detection. The kinematics of the electron event match those for elastic scattering and the event is a good candidate for a

$\nu e^- \to \nu e^-$ neutral- or charged-current interaction; $0.5 \pm 0.1$ such events are expected from the Standard Model.

No excess was found and the collaboration derived an upper limit on the magnetic moment of the $\tau$ neutrino of

$$\mu_{\nu_\tau} < 5.4 \times 10^{-7} \, \mu_B \tag{7.24}$$

at the 90% confidence level [613].

### 7.3.3 The Invisible Width of the $Z^0$ Boson

The preceding experiments searched for a magnetic dipole moment, coupling photons to the $\tau$ neutrino. Another question is whether the couplings of the $\tau$ neutrino to the $Z^0$ boson agree with the Standard Model predictions. This situation is different in the sense that there is a tree-level coupling between neutrinos and the $Z^0$ boson in the Standard Model. Even an anomalous weak magnetic dipole moment is introduced through radiative corrections,[5] but it is unmeasurably small.

Deviations from the Standard Model in the couplings of the $\tau$ neutrino to the $Z^0$ boson can be parametrized in terms of a weak magnetic dipole moment or a $\mathcal{CP}$-violating weak electric dipole moment [341, 349]. The coupling is tested by comparing the experimental partial width of the $Z^0$ for neutrinos $\Gamma_{\nu\bar{\nu}}$ with the Standard Model prediction and associating any discrepancy with a dipole moment of the $\tau$ neutrino. Associating the full discrepancy with the $\tau$ neutrino is the most conservative assumption.

Experimentally, $\Gamma_{\nu\bar{\nu}}$ is determined from the following relation:

$$\Gamma_{\nu\bar{\nu}} = \frac{1}{3}\left(\Gamma_{\text{tot}} - \Gamma_{\text{had}} - 3\,\Gamma_{\text{lep}}\right). \tag{7.25}$$

The total width $\Gamma_{\text{tot}}$ is determined from a fit to the $Z^0$ line shape. The hadronic width $\Gamma_{\text{had}}$ is derived from the hadronic cross section at the peak

$$\sigma(e^+e^- \to \text{had}) = \frac{12\pi\, s\, \Gamma_{\text{lep}} \Gamma_{\text{had}}}{(s - m_Z^2)^2 + s^2\,(\Gamma_{\text{tot}}^2/m_Z^2)} \tag{7.26}$$

and from the ratio of the hadronic to the leptonic branching ratio $R_{\text{lep}} = \Gamma_{\text{had}}/\Gamma_{\text{lep}}$, which also gives the leptonic width $\Gamma_{\text{lep}}$.

The authors of [349] derived the following limits, allowing either for a weak magnetic or for a weak electric dipole moment:

$$|\mu_{\nu_\tau}^{Z^0}| < 2.7 \times 10^{-6} \, \mu_B,$$
$$|d_{\nu_\tau}^{Z^0}| < 5.2 \times 10^{-17} \, \text{e cm}, \tag{7.27}$$

at the 95% confidence level.

---

[5] The moments are defined in complete analogy to the moments of the $\tau$ lepton. See Sect. 3.3.

# 8. The Lorentz Structure of the Charged Current

In this chapter the question of the structure of the charged weak current mediating $\tau$ decays will be addressed, i.e. the question of its spin and parity. It will be shown below how it is possible to answer these questions experimentally. A similar analysis of the decay of the muon was a milestone in the verification of our model of weak interactions. It proved to be a pure V–A interaction, i.e. an exchange of a spin-1 boson with maximal parity violation. (For reviews see [617–621].) In the light of the high precision achieved in $\mu$ decays one might ask the question as to whether it is worth redoing the analysis for $\tau$ decays. It is indeed worth doing so, as in the twenty years since the studies of the $\mu$ decays the scope of the question has changed. The $W^\pm$ bosons mediating the interaction are well established, they carry spin 1 and they couple only to left-handed fermions. The question nowadays is: is there anything else beyond the standard $W^\pm$ bosons? There are good reasons to believe that the $\tau$ would be more sensitive to this kind of new physics than the $\mu$.

## 8.1 Generalization of the Weak Current

### 8.1.1 A More General Current

To search for contributions to $\tau$ decays from yet undiscovered interactions, it is necessary to parametrize the new currents under assumptions that are as model-independent as possible. The starting point of the generalization is the description of a leptonic $\tau$ decay through the Standard Model V–A currents.[1] The matrix element with the Fermi constant $G_F$ reads

$$\mathcal{M} = 4\frac{G_F}{\sqrt{2}} \langle \overline{\Psi}_L(\ell^-) | \gamma^\mu | \Psi_R(\overline{\nu}_\ell) \rangle \langle \overline{\Psi}_L(\nu_\tau) | \gamma_\mu | \Psi_L(\tau^-) \rangle. \qquad (8.1)$$

$\Psi_L(\tau^-) = \frac{1}{2}(1 - \gamma_5)\Psi(\tau^-)$ is the wave function of a left-handed $\tau$ lepton and $\Psi_R$ that of a right-handed one. The $\gamma^\mu$ representing a spin-1 current will be replaced by a sum over scalar, vector, and tensor currents, represented by $\Gamma^\kappa = 1, \gamma^\mu$, and $(1/\sqrt{2})\sigma^{\mu\nu} = i/(2\sqrt{2})(\gamma^\mu\gamma^\nu - \gamma^\nu\gamma^\mu)$, respectively. Arbitrary

---

[1] Hadronic decays will be addressed in the next section.

**Table 8.1.** The nonvanishing combinations of chiralities of the fermions in the generalized matrix element of the decay of a negative $\tau$. For positive $\tau$ leptons all chiralities have to be reversed

| Type of current | Chiralities | | | |
|---|---|---|---|---|
| | $\ell^-(\varepsilon)$ | $\bar{\nu}_\ell(\rho)$ | $\tau^-(\lambda)$ | $\nu_\tau(\omega)$ |
| S | L | L | L | R |
|   | L | L | R | L |
|   | R | R | L | R |
|   | R | R | R | L |
| V | L | R | L | L |
|   | L | R | R | R |
|   | R | L | L | L |
|   | R | L | R | R |
| T | L | L | R | L |
|   | R | R | L | R |

chiralities of the participating fermions will be allowed. However, certain restrictions will be applied to conform with general rules of physics (Lorentz invariance, for example) and to keep the complexity of the ansatz manageable. They will be discussed in detail in the next section (Sect. 8.1.2). The matrix element of the generalized current then reads

$$\mathcal{M} = 4 \frac{G_{\tau\ell}}{\sqrt{2}} \sum_{\substack{\kappa=S,V,T \\ \varepsilon,\lambda=R,L}} g^\kappa_{\varepsilon\lambda} \langle \bar{\Psi}_\varepsilon(\ell^-) | \Gamma^\kappa | \Psi_\rho(\bar{\nu}_\ell) \rangle \langle \bar{\Psi}_\omega(\nu_\tau) | \Gamma_\kappa | \Psi_\lambda(\tau^-) \rangle. \quad (8.2)$$

Some of the combinations of chiralities of the four fermions lead to vanishing currents. For example, for a vector current the transition from a left-handed $\tau$ to a right-handed $\nu_\tau$ is not permitted, as

$$\begin{aligned} \langle \bar{\Psi}_R(\nu_\tau) | \gamma_\mu | \Psi_L(\tau^-) \rangle &= \bar{u}_R(\nu_\tau) \gamma^\mu u_L(\tau^-) \\ &= \bar{u}(\nu_\tau) (1-\gamma_5) \gamma^\mu (1-\gamma_5) u(\tau^-) \\ &= \bar{u}(\nu_\tau) \gamma^\mu (1+\gamma_5)(1-\gamma_5) u(\tau^-) \\ &= 0. \end{aligned} \quad (8.3)$$

Here $u$ is a Dirac spinor. It suffices to specify one chirality for each current. We shall use those of the $\tau$ and the charged daughter lepton $\ell$: $\lambda$ and $\varepsilon$. Table 8.1 gives the corresponding neutrino chiralities.

One might have expected two more tensor couplings, $g^T_{LL}$ and $g^T_{RR}$, but these cannot contribute. The simplest way to see that is to use a Fierz transformation [622] to write the currents in the charge retention form

$$\langle \overline{u}_\varepsilon (\ell) | \sigma^{\mu\nu} | v_\rho (\nu_\ell) \rangle \langle \overline{u}_\omega (\nu_\tau) | \sigma_{\mu\nu} | u_\lambda (\tau) \rangle$$
$$= 6 \langle \overline{u}_\omega (\nu_\tau) v_\rho (\nu_\ell) \rangle \langle \overline{u}_\varepsilon (\ell) u_\lambda (\tau) \rangle$$
$$- 2 \langle \overline{u}_\omega (\nu_\tau) \sigma^{\mu\nu} v_\rho (\nu_\ell) \rangle \langle \overline{u}_\varepsilon (\ell) \sigma_{\mu\nu} u_\lambda (\tau) \rangle$$
$$+ 6 \langle \overline{u}_\omega (\nu_\tau) \gamma_5 v_\rho (\nu_\ell) \rangle \langle \overline{u}_\varepsilon (\ell) \gamma_5 u_\lambda (\tau) \rangle.$$

Now, inserting $\varepsilon, \lambda =$ LL or RR causes all three terms to vanish.

There is a total of ten possible contributions to the decay. Each term in the matrix element (8.2) is governed by its own coupling constant $g^\kappa_{\varepsilon\lambda}$, determining its relative strength and phase. The overall strength is given by $G_{\tau\ell}$. The Standard Model amplitude (8.1) is reproduced by setting $g^V_{LL} = 1$, $G_{\tau\ell} = G_F$, and all other coupling constants to zero. There is one overall phase in the matrix element which cannot be observed and this is usually used to make $g^V_{LL}$ positive and real. This leaves ten absolute values and nine phases to be determined by experiment. The advantage of allowing $G_{\tau\ell}$ to be different from $G_F$ is that the $g^\kappa_{\varepsilon\lambda}$ can be normalized now:

$$1 = \frac{1}{4} \left( |g^S_{LL}|^2 + |g^S_{LR}|^2 + |g^S_{RL}|^2 + |g^S_{RR}|^2 \right)$$
$$+ \left( |g^V_{LL}|^2 + |g^V_{LR}|^2 + |g^V_{RL}|^2 + |g^V_{RR}|^2 \right)$$
$$+ 3 \left( |g^T_{LR}|^2 + |g^T_{RL}|^2 \right). \qquad (8.4)$$

This restricts the allowed ranges of the coupling constants to

$$|g^S| \leq 2, \quad |g^V| \leq 1, \quad |g^T| \leq \frac{1}{\sqrt{3}}. \qquad (8.5)$$

### 8.1.2 Restrictions

Although the ansatz of (8.2) is quite general,[2] some restrictions apply. These are:

- In the matrix element of (8.2) the propagator of the boson mediating the interaction is contracted to an effective four-fermion point interaction. This is a valid simplification as long as $Q^2 \ll m^2_{\text{boson}}$, which is certainly a good approximation for the Standard Model[3] and is expected to be an even better approximation for new physics.
- There are exactly four particles participating in the interaction and they are assumed to be spin-1/2 fermions. This is not necessarily the case in any extension of the Standard Model. Figure 8.1 shows an example of a $\tau$ decay in a SUSY model. The $\tau$ decays through a chargino and the two invisible particles in the decay are sneutrinos, which are spin-0 bosons. This decay has a structure which is not included in the matrix element of (8.2).

---

[2] Equation (8.2) is referred to as the most general Lorentz-invariant, local, derivative-free, lepton-number-conserving four-fermion point interaction [619].
[3] The correction to the total decay rate is $2.9 \times 10^{-4}$ [364].

226   8. The Lorentz Structure of the Charged Current

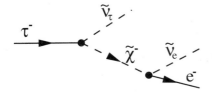

**Fig. 8.1.** A Feynman diagram of the decay of a $\tau$ in a SUSY model. The *dashed lines* represent SUSY particles. The two $\tilde{\nu}$ are sneutrinos and the mediating particle is a chargino

- The lepton quantum numbers are assumed to be conserved at each vertex. The ansatz has been extended to lepton-number-nonconserving currents [623]. The structure of the resulting matrix element is, however, not altered, but the meaning of the couplings $g^{\kappa}_{\epsilon\lambda}$ changes. Each coupling is replaced by the original (given in (8.2)) plus a sum of several new, lepton-number-violating couplings. If such new physics exists, it will still manifest itself in the decays, except for some pathological cases.

- The argument for assuming derivative-free couplings is one of simplicity. Too many different couplings would make it experimentally impossible to disentangle them. The need to include such couplings is further reduced because, in the case of derivatives associated with a vector current, the Dirac equation can be used to approximately rewrite the current as a scalar. This is not possible with derivatives associated with tensor currents. They indeed create new structures which are not included in our ansatz [624–626]. Experiments studying the Lorentz structure of the charged current in the decay of pions and kaons have claimed to see such currents [627, 628]. This could not, however, be verified in $\tau$ decays [629].

- Equation 8.2 represents a tree-level diagram. In general there will be higher-order corrections. But (8.2) is not specific enough to calculate them. To do this, one would have to assume a fully detailed model. This contradicts our attempt to keep the analysis as model-independent as possible. Therefore the experiments use radiative corrections calculated from the Standard Model, as this is the dominating contribution. The corrections are small anyhow[4] and the same can be expected for higher orders in new physics.

- Another assumption will be mentioned here, although it will be used only in the next section. When decay rates are calculated from the matrix element the masses of the participating fermions have to be specified. For the two unobserved particles one simply neglects the mass ($m \ll m_\tau$). This is well justified for the Standard Model neutrinos. But new physics might come with different neutrinos which are not necessarily as light as their Standard Model partners. For example, in left–right symmetric models the right-handed $W^\pm$ comes with a right-handed neutrino which might even be much heavier than the $\tau$. In that case the right-handed $W^\pm$ would not participate in $\tau$ decays.

---

[4] The correction to the total width is $-4.3 \times 10^{-3}$ [358–361].

### 8.1.3 The Michel Parameters

From the ansatz of (8.2) the spectrum of the decay can be calculated. In the rest frame of the decaying $\tau$, integrating over the two neutrinos and the spins of all daughter particles, one gets

$$\frac{\mathrm{d}^2 \Gamma}{\mathrm{d}x^* \mathrm{d}\cos\theta^*} = \frac{G_{\tau\ell}^2 m_\tau^5}{32\,\pi^3} \, x^{*2} \left\{ \left[ (1-x^*) + \rho \left( \frac{8}{9} x^* - \frac{2}{3} \right) + 2\eta \frac{m_\ell}{m_\tau} \frac{(1-x^*)}{x^*} \right] \right.$$
$$\left. - \cos\theta^* \left[ \frac{1}{3}\xi (1-x^*) + \xi\Delta \left( \frac{8}{9} x^* - \frac{2}{3} \right) \right] \right\}.$$
(8.6)

The quantity $x^*$ is the scaled lepton energy,[5] i.e. $x^* = E_\ell^*/E_{\ell,\mathrm{max}}^*$:

$$x_0^* \leq x^* \leq 1,$$
$$E_{\ell,\mathrm{min}}^* \leq E_\ell^* \leq E_{\ell,\mathrm{max}}^*,$$

with

$$E_{\ell,\mathrm{min}}^* = m_\ell, \qquad E_{\ell,\mathrm{max}}^* = \frac{m_\tau^2 + m_\ell^2}{2 m_\tau}. \tag{8.7}$$

Terms of second or higher order in $m_\ell/m_\tau$ and all terms proportional to $m_\nu$ [630] are neglected. The angle $\theta^*$ is measured between the direction of the $\tau$ spin and the momentum of the charged daughter lepton $\ell$. The spectrum depends on the couplings $g_{\epsilon\lambda}^\kappa$ through the four parameters $\eta$, $\rho$, $\xi$, and $\Delta$. They are called the Michel parameters [631, 632]. Their definition is given by the following equations:

$$\alpha^+ = |g_{\mathrm{RL}}^\mathrm{V}|^2 + \frac{1}{16}|g_{\mathrm{RL}}^\mathrm{S} + 6 g_{\mathrm{RL}}^\mathrm{T}|^2,$$
$$\alpha^- = |g_{\mathrm{LR}}^\mathrm{V}|^2 + \frac{1}{16}|g_{\mathrm{LR}}^\mathrm{S} + 6 g_{\mathrm{LR}}^\mathrm{T}|^2,$$
$$\alpha^0 = 8 \operatorname{Re}\left[ g_{\mathrm{RL}}^\mathrm{V}\left( g_{\mathrm{LR}}^{\mathrm{S}\,*} + 6 g_{\mathrm{LR}}^{\mathrm{T}\,*}\right) + g_{\mathrm{LR}}^\mathrm{V}\left( g_{\mathrm{RL}}^{\mathrm{S}\,*} + 6 g_{\mathrm{RL}}^{\mathrm{T}\,*}\right) \right],$$
$$\beta^+ = |g_{\mathrm{RR}}^\mathrm{V}|^2 + \frac{1}{4}|g_{\mathrm{RR}}^\mathrm{S}|^2,$$
$$\beta^- = |g_{\mathrm{LL}}^\mathrm{V}|^2 + \frac{1}{4}|g_{\mathrm{LL}}^\mathrm{S}|^2,$$
$$\beta^0 = -4 \operatorname{Re}\left( g_{\mathrm{RR}}^\mathrm{V} g_{\mathrm{LL}}^{\mathrm{S}\,*} + g_{\mathrm{LL}}^\mathrm{V} g_{\mathrm{RR}}^{\mathrm{S}\,*} \right),$$
$$\gamma^+ = \frac{3}{16}|g_{\mathrm{RL}}^\mathrm{S} - 2 g_{\mathrm{RL}}^\mathrm{T}|^2,$$
$$\gamma^- = \frac{3}{16}|g_{\mathrm{LR}}^\mathrm{S} - 2 g_{\mathrm{LR}}^\mathrm{T}|^2, \tag{8.8}$$

and

---

[5] All quantities with a $^*$ refer to the rest frame of the $\tau$.

228    8. The Lorentz Structure of the Charged Current

$$\eta = \alpha^0 - 2\beta^0,$$
$$\rho = \frac{3}{4}\left(\beta^- + \beta^+\right) + \left(\gamma^+ + \gamma^-\right),$$
$$\xi = 3\left(\alpha^- - \alpha^+\right) + \left(\beta^- - \beta^+\right) + \frac{7}{3}\left(\gamma^+ - \gamma^-\right),$$
$$\xi\Delta = \frac{3}{4}\left(\beta^- - \beta^+\right) + \left(\gamma^+ - \gamma^-\right). \tag{8.9}$$

In the Standard Model the values of $\eta$, $\rho$, $\xi$, and $\Delta$ are 0, 3/4, 1, and 3/4, respectively. Some combinations of nonstandard currents are given in Table 8.2 with their corresponding couplings and Michel parameters. The transformation from couplings to Michel parameters is unique, but the inverse is not, i.e. different combinations of couplings can give identical Michel parameters.

For illustration Fig. 8.2 shows the Standard Model spectrum in the rest frame of the $\tau$ in two dimensions ($p_L$ and $p_T$ with respect to the $\tau$ spin). For emission of the lepton against the direction of the spin the matrix element is independent of $x^*$, whereas in the direction of the spin it rapidly falls with $x^*$. Figure 8.3 shows the spin-independent part, including a few extensions of the Standard Model.

The parameter $\eta$ has its biggest impact at low energies, whereas the other parameters affect more the high end of the spectrum. This becomes more pronounced as the spectrum is boosted into the laboratory frame. Therefore $\eta$ is often called the low-energy parameter.

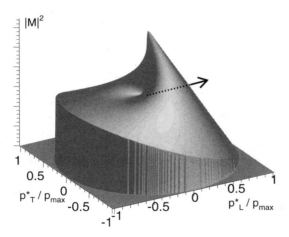

**Fig. 8.2.** Polar diagram of the Standard Model spectrum of leptonic $\tau$ decays ((8.6), $m_\ell = 0$). The two *horizontal axes* show the components of the momentum of the charged daughter lepton with respect to the $\tau$ spin. The *arrow* indicates the direction of the $\tau$ spin ($\theta^* = 0$). The distance from the origin gives $x^*$; the angle to the arrow in the horizontal plane gives $\theta^*$. The height of the surface represents the magnitude of the matrix element. The phase space element $x^{*2}dx^*d\cos\theta^*d\phi^*$ is not included in the plot

## 8.1 Generalization of the Weak Current

**Table 8.2.** Some examples of coupling constants and corresponding Michel parameters. The left column gives the structure of the two leptonic currents, the center the coupling constants, and the right column the resulting Michel parameters. Different combinations can lead to the same Michel parameters, e.g. rows 1, 7, and 12

| Type of interaction $\ell\, \nu_\ell$ vertex $\otimes$ $\tau\, \nu_\tau$ vertex | Coupling constants | $\rho$ | $\xi$ | $\xi\Delta$ | $\eta$ |
|---|---|---|---|---|---|
| V−A $\otimes$ V−A | $g^V_{LL} = 1$ | $\frac{3}{4}$ | 1 | $\frac{3}{4}$ | 0 |
| V+A $\otimes$ V+A | $g^V_{RR} = 1$ | $\frac{3}{4}$ | −1 | $-\frac{3}{4}$ | 0 |
| V $\otimes$ V | $g^V_{LL} = g^V_{RL} = g^V_{LR} = g^V_{RR} = \frac{1}{2}$ | $\frac{3}{8}$ | 0 | 0 | 0 |
| A $\otimes$ A | $g^V_{LL} = -g^V_{RL} = -g^V_{LR} = g^V_{RR} = \frac{1}{2}$ | $\frac{3}{8}$ | 0 | 0 | 0 |
| V−A $\otimes$ V+A | $g^V_{LR} = 1$ | 0 | 3 | 0 | 0 |
| V+A $\otimes$ V−A | $g^V_{RL} = 1$ | 0 | −3 | 0 | 0 |
| S+P $\otimes$ S−P | $g^S_{LL} = 2$ | $\frac{3}{4}$ | 1 | $\frac{3}{4}$ | 0 |
| S−P $\otimes$ S+P | $g^S_{RR} = 2$ | $\frac{3}{4}$ | −1 | $-\frac{3}{4}$ | 0 |
| S $\otimes$ S | $g^S_{LL} = g^S_{RL} = g^S_{LR} = g^S_{RR} = 1$ | $\frac{3}{4}$ | 0 | 0 | 0 |
| P $\otimes$ P | $-g^S_{LL} = g^S_{RL} = g^S_{LR} = -g^S_{RR} = 1$ | $\frac{3}{4}$ | 0 | 0 | 0 |
| S+P $\otimes$ S+P | $g^S_{LR} = 2$ | $\frac{3}{4}$ | −1 | $-\frac{3}{4}$ | 0 |
| S−P $\otimes$ S−P | $g^S_{RL} = 2$ | $\frac{3}{4}$ | 1 | $\frac{3}{4}$ | 0 |
| T $\otimes$ T | $g^T_{LR} = g^T_{RL} = \sqrt{\frac{1}{6}}$ | $\frac{1}{4}$ | 0 | 0 | 0 |
| 50% V−A $\otimes$ V−A <br> 50% S $\otimes$ S | $g^V_{LL} = g^S = \sqrt{\frac{1}{2}}$ | $\frac{3}{4}$ | $\frac{1}{2}$ | $\frac{3}{8}$ | $\frac{1}{4}$ |
| 50% V−A $\otimes$ V−A <br> 50% S−P $\otimes$ S+P | $g^V_{LL} = \sqrt{\frac{1}{2}},\ g^S_{RR} = \sqrt{2}$ | $\frac{3}{4}$ | 1 | 0 | $\frac{1}{2}$ |
| 50% V−A $\otimes$ V−A <br> 50% V+A $\otimes$ V−A | $g^V_{LL} = g^V_{RL} = \sqrt{\frac{1}{2}}$ | $\frac{3}{8}$ | −1 | $\frac{3}{8}$ | 0 |
| 50% V+A $\otimes$ V+A <br> 50% V−A $\otimes$ V+A | $g^V_{RR} = g^V_{LR} = \sqrt{\frac{1}{2}}$ | $\frac{3}{8}$ | 1 | $-\frac{3}{8}$ | 0 |
| 50% V−A $\otimes$ V+A <br> 50% V+A $\otimes$ V−A | $g^V_{LR} = g^V_{RL} = \sqrt{\frac{1}{2}}$ | 0 | 0 | 0 | 0 |
| 67% V−A $\otimes$ V+A <br> 33% V+A $\otimes$ V−A | $g^V_{LR} = \sqrt{\frac{2}{3}},\ g^V_{RL} = \sqrt{\frac{1}{3}}$ | 0 | 1 | 0 | 0 |
| 75% S±P $\otimes$ S±P <br> 25% T $\otimes$ T | $g^S_{LR} = g^S_{RL} = \sqrt{\frac{3}{2}},$ <br> $g^T = -\sqrt{\frac{1}{24}}$ | 1 | 0 | 0 | 0 |
| 12.5% S±P $\otimes$ S±P <br> 50% V∓A $\otimes$ V±A <br> 37.5% T $\otimes$ T | $g^S_{LR} = g^S_{RL} = g^V_{LR} = g^V_{RL} = \frac{1}{2},$ <br> $g^T = \frac{1}{4}$ | 0 | 0 | 0 | 1 |

230     8. The Lorentz Structure of the Charged Current

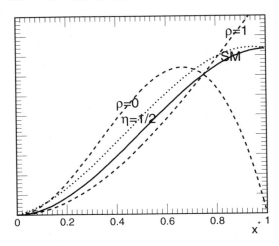

Fig. 8.3. The spectrum of the decay $\tau \to \mu\,\nu_\mu\nu_\tau$ in the rest frame of the $\tau$ averaged over its spin. The *solid line* represents the Standard Model, the *dashed lines* take $\rho = 0$ and 1 and the *dotted line* $\eta = 1/2$

Integrating the spectrum gives the total decay width, which depends only on one of the four Michel parameters: $\eta$:

$$\Gamma_{\text{tot}} = \frac{G_{\tau\ell}^2\, m_\tau^5}{192\,\pi^3}\left(1 + 4\eta\,\frac{m_\ell}{m_\tau}\right). \tag{8.10}$$

The Michel parameters include contributions of the form $|g^\kappa_{\epsilon\lambda}|^2$, which are due to a single diagram, as well as interference terms. For example $\gamma^+$ has a term $g^{\text{S}}_{\text{RL}}{}^*g^{\text{T}}_{\text{RL}}$ representing the interference between the two currents. It is interesting to realize where they come from. Two amplitudes will only interfere if the helicities of all final-state particles are identical. For the massless neutrinos this is equivalent to identical chiralities, which can be looked up in Table 8.1. There are four groups of couplings with identical neutrino chiralities: $(g^{\text{S}}_{\text{LL}}, g^{\text{V}}_{\text{RR}})$, $(g^{\text{S}}_{\text{RR}}, g^{\text{V}}_{\text{LL}})$, $(g^{\text{S}}_{\text{LR}}, g^{\text{V}}_{\text{RL}}, g^{\text{T}}_{\text{LR}})$, and $(g^{\text{S}}_{\text{RL}}, g^{\text{V}}_{\text{LR}}, g^{\text{T}}_{\text{RL}})$. Selecting those where the chiralities of the daughter leptons $\ell$ match in addition gives the unsuppressed interferences. There are only two such pairs of couplings: $(g^{\text{S}}_{\text{LR}}, g^{\text{T}}_{\text{LR}})$ and $(g^{\text{S}}_{\text{RL}}, g^{\text{T}}_{\text{RL}})$. For the other combinations the charged daughter lepton has to be produced with the 'wrong' helicity, i.e. with its spin pointing in the direction of flight for a left-handed lepton. This introduces a suppression factor of $(1/2)(1-\beta) \propto (m_\ell/m_\tau)$. This is the origin of the parameter $\eta$. The suppression factor is explicit in (8.6).

Of particular interest are interference terms between the Standard Model coupling and a new interaction. They are linear in the new coupling and therefore more sensitive to it than the direct terms. Unfortunately there is only one such term and it is helicity-suppressed: the $g^{\text{V}}_{\text{LL}}g^{\text{S}}_{\text{RR}}{}^*$ term in $\eta$.

There is an interesting geometrical interpretation of the normalization condition (8.4) [633]. If one ignores $\eta$ for the moment, the remaining three parameters $\rho$, $\xi$, and $\Delta$ span a three-dimensional space. The normalization condition in that space is represented by the tetrahedron shown in Fig. 8.4. Any combination of Michel parameters that falls inside the tetrahedron is

8.1 Generalization of the Weak Current    231

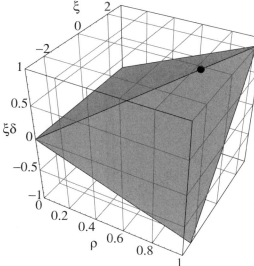

**Fig. 8.4.** The physically allowed ranges of the Michel parameters $\rho$, $\xi$, and $\eta$ in a geometrical representation. A tetrahedron is formed

allowed, any combination outside forbidden. This geometrical representation will prove useful in the next section.

As the rest frame of the $\tau$ cannot be reconstructed in leptonic decays, the spectrum has to be boosted into the laboratory and integrated over the unobservable angle $\theta^*$ (see Sect. 4.5). With $P_\tau$ being the polarization of the $\tau$ leptons, one gets

$$\frac{\mathrm{d}\Gamma}{\mathrm{d}x} = \frac{G_{\tau\ell}^2 m_\tau^5}{192\,\pi^3} \left\{ f_0(x) + \rho f_1(x) + \eta \frac{m_\ell}{m_\tau} f_2(x) - P_\tau \left[ \xi g_1(x) + \xi \Delta g_2(x) \right] \right\},$$
(8.11)

with the following polynomials:

$$f_0(x) = 2 - 6\,x^2 + 4\,x^3,$$
$$f_1(x) = -\frac{4}{9} + 4\,x^2 - \frac{32}{9}\,x^3,$$
$$f_2(x) = 12\,(1-x)^2,$$
$$g_1(x) = -\frac{2}{3} + 4\,x - 6\,x^2 + \frac{8}{3}\,x^3,$$
$$g_2(x) = \frac{4}{9} - \frac{16}{3}\,x + 12\,x^2 - \frac{64}{9}\,x^3.$$
(8.12)

Figure 8.5 illustrates the Standard Model spectrum and its variations with the Michel parameters.

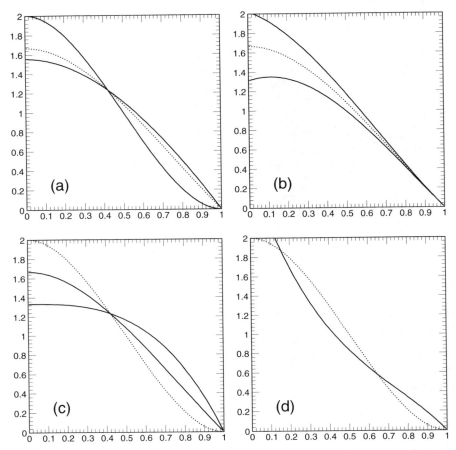

**Fig. 8.5.** The Michel spectrum boosted into the laboratory frame as a function of $x$. *Dotted line*: Standard Model, $(\rho, \eta, \xi, \Delta) = (3/4, 0, 1, 3/4)$; **(a)** variation of $\rho$, (0 and 1, 0, 1, 3/4); **(b)** variation of $\eta$ using $m_\mu$ as $m_\ell$, $(3/4, \pm 1/2, 1, 3/4)$, **(c)** variation of $\xi$, $(3/4, 0, 0$ and $-1, 3/4)$, **(d)** variation of $\Delta$, $(3/8, 0, 1, -3/8)$. Figures **(a)** and **(b)** have no polarization, **(c)** and **(d)** maximum negative polarization $(P_\tau = -1)$

### 8.1.4 The Complete Determination of Leptonic Decays

First a minimal set of measurements necessary to prove that the current is indeed (within experimental limits) a pure 'V–A' current will be described [619]. Then a few more measurements which will improve the precision in the intermediate steps will be discussed.

The first step is to verify the overall normalization, i.e. to check whether the Fermi constant in $\tau$ decays is the same as in $\mu$ decays:

$$\frac{\mathcal{B}_\ell}{\tau_\tau}\frac{192\,\pi^3}{m_\tau^5} = G_{\tau\ell}^2\left(1-4\eta\,\frac{m_\ell}{m_\tau}\right) \stackrel{?}{=} G_F^2. \tag{8.13}$$

$\mathcal{B}_\ell$ is the leptonic branching ratio and $\tau_\tau$ the lifetime of the $\tau$. In the case of $\tau \to e\,\nu_e\nu_\tau$ decays, $m_e/m_\tau$ is so small that the $\eta$ term can safely be neglected. For $\tau \to \mu\,\nu_\mu\nu_\tau$, however, a good limit on $\eta$ is necessary or it will spoil the test.

The next step is to measure the Michel parameters from the spectrum. How that is done will be explained in more details in the following sections. From the Michel parameters one can calculate the following quantity:

$$\begin{aligned}Q_R^\tau &= \frac{1}{2}\left(1+\frac{1}{9}(3\xi-16\xi\Delta)\right) \\ &= \frac{1}{4}|g_{RR}^S|^2 + \frac{1}{4}|g_{LR}^S|^2 + |g_{RR}^V|^2 + |g_{LR}^V|^2 + 3|g_{LR}^T|^2.\end{aligned} \tag{8.14}$$

From the second line one sees that this is the probability that a right-handed $\tau$ participates in the interaction. $Q_R^\tau$ is zero in the Standard Model, and if one finds experimentally a value consistent with zero, one can put upper limits on all five couplings $g_{\epsilon R}^\kappa$ contributing to $Q_R^\tau$. This measurement also yields $\eta$, needed in (8.13).

The simplest way to proceed from a theoretical point of view is to measure the polarization of the daughter lepton from the decays. This gives another Michel-type parameter $\xi'$, which is related to $Q_R^\ell$ – the probability that a right-handed daughter lepton is produced in the interaction – by

$$\begin{aligned}Q_R^\ell &= \frac{1}{2}(1-\xi') \\ &= \frac{1}{4}|g_{RR}^S|^2 + \frac{1}{4}|g_{RL}^S|^2 + |g_{RR}^V|^2 + |g_{RL}^V|^2 + 3|g_{RL}^T|^2.\end{aligned} \tag{8.15}$$

This gives upper limits on three more couplings ($g_{R\lambda}^\kappa$), if one finds a value for $\xi'$ consistent with 1, the Standard Model expectation. There are two ideas on how to do the measurement, either by using a muon polarimeter at a b or $\tau$ charm factory or by studying the photon energy and angular distributions from radiative decays. Both are challenging projects from the experimental point of view. For more details see [619, 634].

By now all couplings involving right-handed charged fermions are excluded, and those left over are $g_{LL}^S$ and the Standard Model coupling $g_{LL}^V$. These two can only be distinguished by a measurement on at least one of the neutrinos [618, 635], which is, however, out of present experimental reach.

The quantities $Q_R^\tau$ and $Q_R^\ell$ are not the only quantities that can be used to set limits on couplings. Any combination of Michel parameters which is a positive definite function of the couplings can do that. The geometrical representation of the parameters in Fig. 8.4 now becomes useful for finding such combinations [636, 637]. Distances in space are always positive and therefore any geometrical distance within the tetrahedron is such a quantity. For example, the distances from the faces on the top and the front are

$$\rho - \xi\Delta = \frac{3}{2}|g_{RR}^V|^2 + \frac{3}{8}|g_{LR}^S - 2g_{LR}^T|^2 + \frac{3}{8}|g_{RR}^S|^2,$$

$$1 - \rho + \frac{1}{3}\xi - \frac{7}{9}\xi\Delta = 2|g_{LR}^V|^2 + \frac{1}{2}|g_{RR}^V|^2 + \frac{1}{8}|g_{RR}^S|^2 + \frac{1}{8}|g_{LR}^S + 6\,g_{LR}^T|^2. \quad (8.16)$$

The Standard Model value is on the intersection of the two faces and therefore the expectation for both quantities is zero. The distance to the other two faces is positive too, but the expectation does not vanish and therefore the limits will be less stringent.

### 8.1.5 Universality

To this point $\tau \to \ell\nu_\ell\nu_\tau$ has been a symbol for either $\tau \to e\nu_e\nu_\tau$ or $\tau \to \mu\nu_\mu\nu_\tau$, but no relation was assumed between the two and, in general, there is no such relation. One can easily think of new physics which affects only one channel or affects the two in different ways. For example, any Higgs-type boson couples proportionally to the mass and therefore would only be visible in the $\tau \to \mu\,\nu_\mu\nu_\tau$ channel. But there are also many ideas about extensions to the Standard Model where the couplings are universal. Therefore the data is usually analyzed to give two results, one assuming universality between the two channels and one treating them as independent. If one goes one step further and assumes universality also with $\mu$ decays, then there is no need to measure the Michel parameters again. The current precision in $\tau$ decays is less than that in $\mu$ decays. It would hardly contribute to the average.

Experimenters use the word 'universality' in connection with Michel parameters with a slightly modified meaning. In general we speak of universality if the tree-level couplings of the fermions to the bosons are universal, i.e. independent of the fermion generation. But here we speak of universality if the couplings $g_{\epsilon\lambda}^\kappa$ and therefore the Michel parameters are universal. These are, however, effective couplings which include the effects of higher-order corrections, mixing angles, phase space corrections, and others. A model with universal tree-level couplings might still give nonuniversal Michel parameters. Figure 8.6 shows such an example (admittedly an exotic one): a $\tau$ lepton decays through a lepton-flavor-changing neutral current. Even with universal tree-level couplings there could be a Cabibbo-type mixing, which probably prefers the $\tau \to \mu$ transition over $\tau \to e$.

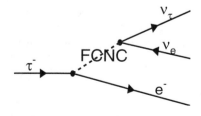

**Fig. 8.6.** A Feynman diagram of the decay of a $\tau$ through a flavor-changing neutral current (FCNC) violating lepton number

## 8.2 Hadronic Decays

### 8.2.1 The $\tau^- \to \pi^-/K^-\nu_\tau$ Decay

The hadronic decays of the $\tau$ lepton provide another opportunity to study the Lorentz structure of the charged current. The decay $\tau^- \to \pi^- \nu_\tau$ is the simplest case. In the spirit of Sect. 8.1.1 the matrix element can be generalized to

$$\mathcal{M} = \frac{G_{\tau\pi}}{\sqrt{2}} \sum_{\substack{\kappa=S,V \\ \lambda=R,L}} g_\lambda^\kappa \, \langle \overline{\Psi}_\omega(\nu_\tau) | \, \Gamma^\kappa \, | \Psi_\lambda(\tau^-) \rangle \, J_\kappa^\pi. \tag{8.17}$$

The boson mediating the decay can either be a scalar or a vector, but it is not possible to couple the pion to a tensor current. The subscript of the coupling indicates the chirality of the decaying $\tau$. The currents are

$$\begin{aligned} J_S^\pi &= \cos\theta_C f_\pi m_\pi, \\ J_V^\pi &= \cos\theta_C f_\pi p_\mu^\pi, \end{aligned} \tag{8.18}$$

with the Cabibbo angle being given by $\cos\theta_C$ and the four-momentum of the pion being $p^\pi$. A factor $f_\pi$ – the pion decay constant – has been removed from all couplings and made explicit in the currents, so that in the Standard Model $g_L^V = 1$ and all other couplings vanish. The factor $m_\pi$ in the scalar current is introduced for dimensional reasons. The spectrum calculated from these equations in the rest frame of the $\tau$ is

$$\begin{aligned} \frac{d\Gamma}{d\cos\theta^*} &= \frac{G_{\tau\pi}^2 f_\pi^2 \cos^2\theta_C}{32\pi} m_\tau^3 \left(1 - \frac{m_\pi^2}{m_\tau^2}\right)^2 (1 + \xi_\pi \cos\theta^*), \\ \Gamma &= \frac{G_{\tau\pi}^2 f_\pi^2 \cos^2\theta_C}{16\pi} m_\tau^3 \left(1 - \frac{m_\pi^2}{m_\tau^2}\right)^2. \end{aligned} \tag{8.19}$$

Boosting into the laboratory frame simply replaces $\cos\theta^*$ by $P_\tau(2x - 1)$ in (8.19).

The spectrum depends on only one Michel-type parameter,

$$\xi_\pi = |g_L^V|^2 - |g_R^V|^2 + \frac{m_\pi^2}{m_\tau^2}\left(|g_L^S|^2 - |g_R^S|^2\right) + 2\frac{m_\pi}{m_\tau}\text{Re}\left(g_L^V g_R^{S*} - g_R^V g_L^{S*}\right), \tag{8.20}$$

which is the negative of the neutrino helicity $\xi_\pi = -h_\nu$ in the case of massless neutrinos ($\xi_\pi = 1$ in the Standard Model). The normalization condition for the couplings reads

$$1 = |g_L^V|^2 + |g_R^V|^2 + \frac{m_\pi^2}{m_\tau^2}\left(|g_L^S|^2 + |g_R^S|^2\right) + 2\frac{m_\pi}{m_\tau}\text{Re}\left(g_L^V g_R^{S*} + g_R^V g_L^{S*}\right). \tag{8.21}$$

In general it appears impossible to disentangle the scalar from the vector current using the shape of the spectrum.[6] Therefore the experiments usually

---

[6] In principle it is possible from the total width, but it would require a theoretical prediction of $f_\pi$, which is currently not possible, or an experimental measurement which is independent of scalar currents.

ignore extensions of the Standard Model with charged scalar bosons. The same formalism can be used to describe the decay $\tau \to K \nu_\tau$.

### 8.2.2 The Two-Pion Channel

Just as in the $\tau \to \pi \nu_\tau$ channel, experiments usually make the simplification that the pions are produced by a pure vector current. Then the spectrum in the rest frame of the $\tau$ reads [437]

$$d\Gamma = \frac{G_{\tau\rho}^2 m_\tau \cos^2 \theta_C}{128\,\pi^3} \left(1 - \frac{Q^2}{m_\tau}\right)^2 |\boldsymbol{p}_\pi|\, L_B W_B \frac{dQ^2}{|Q|} \frac{d\cos\theta^*}{2} \frac{d\alpha}{2\pi} \frac{d\cos\beta}{2}. \tag{8.22}$$

The kinematic quantities $\theta^*$, $\alpha$, and $\beta$ are defined in [437]. $Q^2$ is the invariant mass of the two pions and $\boldsymbol{p}_\pi$ the three-momentum of one of the pions. A boost into the laboratory frame replaces $\cos\theta^*$ by $P_\tau(2x-1)$. $L_B$ is a function of the kinematic quantities and a single Michel-type parameter $\xi_\rho$.[7] As in the one-pion channel, this parameter represents the negative of the neutrino chirality $\xi_\rho = -h_\nu$ and is 1 in the Standard Model.

$W_B$ is a hadronic structure function which describes the shape of the $\rho$ resonance or any other structure one might find in the mass spectrum. It introduces a new complication: it can currently not be predicted from theory. Either one has to use a model or one has to measure $W_B$ simultaneously with $\xi_\rho$. Both approaches are feasible. The structure of the decay is reasonably well understood, so that a model approach does not introduce sizeable systematic uncertainties. On the other hand, a simultaneous measurement is not too complicated, as $W_B$ depends on one kinematic quantity only: $Q^2$. The spectrum is binned in $Q^2$ and $\xi_\rho$ is measured separately for each bin. The population of the bins gives $W_B$.

In the laboratory frame $\xi_\rho$ appears only as a product with $P_\tau$. Rearranging the terms of the spectrum, one can write

$$\frac{d\Gamma}{d\boldsymbol{x}} = f(\boldsymbol{x}) + \xi_\rho\, P_\tau\, g(\boldsymbol{x}), \tag{8.23}$$

where $\boldsymbol{x} = (x, Q^2, \alpha, \beta)$ are all the kinematic variables,[8] and $d\boldsymbol{x}$ is defined accordingly. $\xi_\rho$ can be extracted by a fit to this spectrum. The decays $\tau \to K\pi\nu_\tau$ can be treated in exactly the same fashion.

The restriction to vector currents only is well justifiable in the one-pion channel. Tensor currents cannot occur and scalar currents do not introduce new structures associated with additional Michel-type parameters into the

---

[7] $\rho$ is used here as a label, since the channel is known to be dominated by the $\rho$ resonance. It is, however, not required that the two pions indeed form a $\rho$.

[8] In practice $\boldsymbol{x}$ is replaced by $\omega$, the optimal observable from the polarization measurement (see Sect. 4.5).

spectrum. They only affect the interpretation of the final result. The situation is different in the two-pion channel: scalar currents are experimentally distinguishable from the vector currents. They do modify the shape of the spectrum and are associated with new Michel-type parameters. Also, tensor currents can be present in this channel. Ignoring these additional currents restricts the generality of the ansatz and is hard to justify. This is especially true in a simultaneous measurement with the leptonic decays, looking for scalar and tensor currents there. A study which includes at least the scalar currents has been presented in [638].

### 8.2.3 The Three-Pion Channel

The situation is very similar to the two-pion channel. There are two charge combinations which can be looked at: $\tau^- \to \pi^-\pi^-\pi^+\nu_\tau$ and $\tau^- \to \pi^-\pi^0\pi^0\nu_\tau$. In the laboratory frame the spectrum can be written as

$$\frac{\mathrm{d}\Gamma}{\mathrm{d}x} = f(x) + \xi_{a_1} P_\tau g(x), \tag{8.24}$$

where $x$ is now a vector of seven kinematical quantities and $\xi_{a_1}$ is again the negative of the chirality of the neutrino emitted in the decay. The comment about scalar and tensor currents in the two-pion channel applies here, too.

### 8.2.4 Universality

In general the parameters $\xi_\pi$, $\xi_\rho$, and $\xi_{a_1}$ should be treated as being independent. It is not obvious that new physics has, for example, the same preference of for vector states over pseudoscalars as the Standard Model, and therefore it might affect one of the channels more or less than the others. But experimentally this means one has to fit for three hadronic parameters instead of one (labeled $\xi_\mathrm{had}$), which reduces the precision.

One would also like to derive an independent result for the Cabibbo-suppressed channels, as Higgs-type couplings would essentially affect only these. A separation of pions and kaons is experimentally possible, but has to be paid for by a loss of statistics and precision.

## 8.3 Spin-Dependent Terms

### 8.3.1 Measurements from Unpolarized $\tau$ Leptons

In the simplest experimental situation one looks at the energy spectrum of the charged leptons from the decay of unpolarized $\tau$ leptons. From (8.11) one sees that the dependence on the Michel parameters $\xi$ and $\Delta$ vanishes and one measures $\rho$ and $\eta$ alone. With the same approach, the hadronic decays do

238   8. The Lorentz Structure of the Charged Current

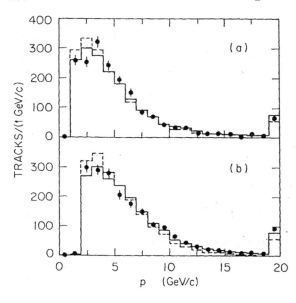

Fig. 8.7. Momentum spectrum for (a) $\tau \to e\, \nu_e \nu_\tau$ and (b) $\tau \to \mu\, \nu_\mu \nu_\tau$ as measured by the MAC collaboration [639]. The *solid* and *dashed* curves are the Monte Carlo predictions for the momentum spectra when $\rho = 0.75$ and $\rho = 0$, respectively

not depend on any Michel-type parameter at all ((8.19), (8.22), and (8.24)). This is the situation for experiments running at $e^+e^-$ machines away from the $Z^0$ resonance and analyzing single $\tau$ leptons, not $\tau$ pairs.

A measurement by the MAC collaboration of leptonic decays might illustrate the procedure [639]. The MAC experiment ran at the PEP storage ring at SLAC at a center-of-mass energy of 29 GeV. Figure 8.7 shows the spectrum of electrons and muons from $\tau$ decays. The spectra are compared to the theoretical predictions for values of $\rho$ of 3/4 (solid histogram) and 0 (dashed). The electron spectrum does not depend on $\eta$, owing to the helicity suppression, and the muon spectrum does not have much sensitivity either, because they had to require a minimum momentum of $2\,\text{GeV}/c$ for the muon identification ($x \approx 0.14$). So they were left with one parameter to extract, which they did by simply calculating the average energy of the spectra. The averages depend linearly on $\rho$, and they obtained

$$\rho_e = 0.62 \pm 0.17 \pm 0.14,$$
$$\rho_\mu = 0.89 \pm 0.14 \pm 0.08, \qquad (8.25)$$

in agreement with the Standard Model within errors ($\rho_{\text{SM}} = 0.75$). This was one of the first measurements of Michel parameters (see also [88, 101, 149, 640]).

The sensitivity to the Michel parameter $\eta$ (and $\rho$) could be increased by boosting back into the $\tau$ rest frame. But, owing to the neutrinos escaping detection, this can not be achieved experimentally. Instead, a technique of boosting back into a pseudorest frame was first applied by the ARGUS collaboration [79]. This method regains some of the sensitivity.

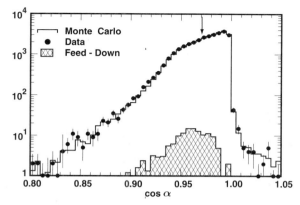

**Fig. 8.8.** The distribution of the angle between the estimated and the true $\tau$ direction from a measurement by the CLEO collaboration [121]: data (*dots*), Monte Carlo simulation (*solid histogram*), and background contamination from multi-$\pi^0$ decay modes (*hatched region*). The *arrow* indicates the minimum requirement. Events with $\cos\alpha > 1$ result from measurement errors and are discarded

$\tau$ pairs are tagged by a hadronic decay of one of the $\tau$ leptons. The direction of the hadron is used as an estimate of the $\tau$ direction. The error of that estimate can be quantified. The opening angle between the direction of the $\tau$ and the direction of the hadron $\alpha$ can be calculated from

$$\begin{aligned}p_\nu^2 &= (p_\tau - p_{\text{had}})^2 \\ &= p_\tau^2 + p_{\text{had}}^2 - 2p_\tau p_{\text{had}} \\ &= m_\tau^2 + m_{\text{had}}^2 - 2E_\tau E_{\text{had}} + 2|\boldsymbol{p}_\tau||\boldsymbol{p}_{\text{had}}|\cos\alpha = m_\nu^2 = 0.\end{aligned} \quad (8.26)$$

Figure 8.8 shows the distribution of events in the angle $\alpha$ from a measurement by the CLEO collaboration [121]. They used events with $\tau \to \pi\pi\nu_\tau$ as the hadronic tag and a leptonic decay of the other $\tau$. Tags with $0.97 \leq \cos\alpha \leq 1.00$ gave an acceptable estimate of the direction of the $\tau$ and were used for this analysis. The other $\tau$ was expected to fly into the opposite direction with a Lorentz $\gamma$ factor calculated from $E_\tau = E_{\text{beam}}$. The electrons and muons were then boosted back in the estimated direction. Figure 8.9 shows the resulting muon spectrum. To further enhance the sensitivity to $\eta$, CLEO has developed an additional muon identification for low-energy muons ($x_\mu^* < 0.6$). This creates the step in the spectrum. The result is [121]

$$\begin{aligned}\rho_\mu &= 0.747 \pm 0.048 \pm 0.044, \\ \eta_\mu &= 0.010 \pm 0.149 \pm 0.171.\end{aligned} \quad (8.27)$$

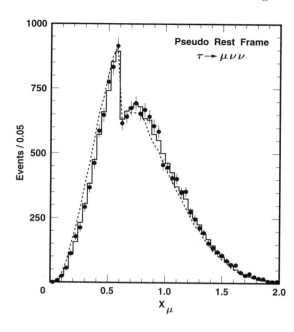

**Fig. 8.9.** The scaled pseudorest frame energy spectrum with the data (*dots*) and fit function (*solid histogram*). The *dashed line* represents the $\eta = 1$ Monte Carlo spectrum. Events with $x_\mu > 1$ result from the imperfect reconstruction of the $\tau$ direction. The addition of the low-momentum muons results in the discontinuity observed at $x_\mu = 0.6$. CLEO collaboration [121]

### 8.3.2 Natural Polarization

For experiments running at the $Z^0$ pole the $\tau$ leptons are polarized owing to parity violation associated with the $Z^0$ boson. This gives some sensitivity to $\Delta$ and the $\xi$ parameter, though the sensitivity is small. Therefore the LEP collaborations use the technique described in the next section instead. At the SLC, with a longitudinally polarized electron beam, the polarization achieved is much higher (see Fig. 8.10). It depends on the production angle $\cos\theta_\tau$ of the $\tau$ lepton. Hence the data is analyzed in two dimensions: $x$ versus $\cos\theta_\tau$. Figure 8.11 shows the spectra of the $\tau \to \ell\nu_\ell\nu_\tau$ and $\tau \to \pi\nu_\tau$ events measured by SLD [246]. For illustration the spectra have been summed over the hemispheres with positive and negative $\tau$ polarization. The results are derived from a fit taking into account the full $\theta_\tau$ dependence of the polarization (see Table 8.3).

### 8.3.3 Spin correlations

At higher energies the helicities of the two $\tau$ leptons in an event are strongly correlated. The spins point in the same direction. These spin correlations can be employed to increase the sensitivity to the spin-dependent part of the spectrum [442, 641]. For the decay of a $\tau^\pm$ with helicity $h_{\tau^\pm}$ the decay distribution can be converted into the form

$$H^-(x_-) = f(x_-) - h_{\tau^-}\, g(x_-),$$

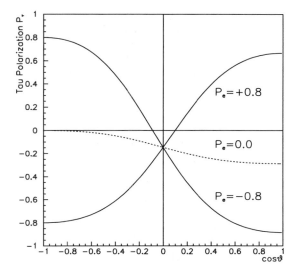

**Fig. 8.10.** The polarization of $\tau$ leptons produced at the Stanford Linear Collider (SLC) versus the production angle with and without beam polarization [246]

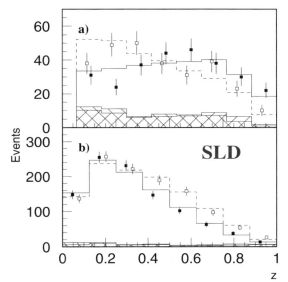

**Fig. 8.11.** (a) $\tau \to \pi \nu_\tau$ and (b) $\tau \to \ell \nu_\ell \nu_\tau$ decay spectra. The *solid squares* and *line* represent the sum of the measured and Monte Carlo spectra, respectively, for $\tau$ decays in the forward direction with $P_e < 0$ and in the backward direction with $P_e > 0$. The *open squares* and *dashed line* are the sum of the measured and Monte Carlo spectra, respectively, for $\tau$ decays in the backward direction with $P_e < 0$ and in the forward direction with $P_e > 0$. The *hatched regions* represent the estimated backgrounds in the two combinations. The Monte Carlo spectra were generated with the Standard Model values for the $\tau$ decay parameters [246]

**Table 8.3.** Michel parameters given by decay channels and as combined results, compared with the Standard Model (SM) prediction (SLD collaboration [246])

|  | SM | $\tau \to \pi \nu_\tau$ | $\tau \to \rho \nu_\tau$ | Hadrons combined |
|---|---|---|---|---|
| $\xi_{had}$ | 1 | $0.81 \pm 0.17 \pm 0.02$ | $0.99 \pm 0.12 \pm 0.04$ | $0.93 \pm 0.10 \pm 0.04$ |
|  |  | $\tau \to e \nu_e \nu_\tau$ | $\tau \to \mu \nu_\mu \nu_\tau$ | $\tau \to \ell \nu_\ell \nu_\tau$ combined |
| $\rho$ | $\frac{3}{4}$ | $0.71 \pm 0.14 \pm 0.05$ | $0.54 \pm 0.28 \pm 0.14$ | $0.72 \pm 0.09 \pm 0.03$ |
| $\eta$ | 0 |  | $-0.59 \pm 0.82 \pm 0.45$ |  |
| $\xi$ | 1 | $1.16 \pm 0.52 \pm 0.06$ | $0.75 \pm 0.50 \pm 0.14$ | $1.05 \pm 0.35 \pm 0.04$ |
| $\xi\Delta$ | $\frac{3}{4}$ | $0.85 \pm 0.43 \pm 0.08$ | $0.82 \pm 0.32 \pm 0.07$ | $0.88 \pm 0.27 \pm 0.04$ |

$$H^+(x_+) = f(x_+) + h_{\tau^+} \, g(x_+) \,. \tag{8.28}$$

The spin-dependent parts $g(x)$ depend on $\xi$ and $\Delta$ in leptonic decays and on $\xi_{had}$ in hadronic decays. Now a $\tau^-$ with negative helicity is accompanied by a $\tau^+$ with positive helicity and vice versa. The polarization[9] gives the fractions of events in the two classes: $(1 - P_\tau)/2$ of the events have a $\tau^-$ with negative helicity, and in $(1 + P_\tau)/2$ cases its helicity is positive. Therefore the combined spectrum of both $\tau$ leptons reads

$$\begin{aligned}I(x_+, x_-) &= \frac{1+P_\tau}{2} [f(x_+) - g(x_+)][f(x_-) - g(x_-)] \\ &+ \frac{1-P_\tau}{2} [f(x_+) + g(x_+)][f(x_-) + g(x_-)] \\ &= f(x_+)f(x_-) + g(x_+)g(x_-) \\ &- P_\tau [f(x_+)g(x_-) + g(x_+)f(x_-)] \,. \end{aligned} \tag{8.29}$$

Figure 8.12 illustrates the correlation for $\tau^- \to e^- \bar\nu_e \nu_\tau$ versus $\tau^+ \to \pi^+ \nu_\tau$ decays. The left histogram is the Standard Model expectation, with $\xi_e = \xi_\pi = 1$ and $P_\tau = -0.14$. High-energy electrons are preferentially emitted in combination with high-energy pions. For the right histogram $\xi_e$ and $\xi_\pi$ are both set to zero; then the $g(x)$ functions vanish and the correlation is gone.

A measurement by the L3 collaboration might serve as an example [170]. The L3 collaboration analyzed the decay channels $\tau \to e \nu_e \nu_\tau$, $\tau \to \mu \nu_\mu \nu_\tau$, $\tau \to \pi \nu_\tau$, and $\tau \to \pi \pi \nu_\tau$, taking advantage of the correlations whenever both $\tau$ leptons were identified in one of the channels. If only one of the $\tau$ leptons was identified the event was still used in a single-arm fit as described in the last section. From a sample of roughly $150\,000$ $\tau$ pairs they obtained the numbers of utilizable events quoted in Table 8.4. For the leptonic decays and $\tau \to \pi \nu_\tau$ they used the spectrum in the scaled energy $x$ to analyze the data, and for $\tau \to \pi \pi \nu_\tau$ they used the optimal observable $\omega_\rho$ from the polarization measurements. Figure 8.13 shows the correlation between $x_\pi$ and $\omega_\rho$. The events were cut into five bins in $x_\pi$ and for each bin the spectrum in $\omega_\rho$

---

[9] The polarization $P_\tau$ is defined as the average helicity of the negative $\tau$.

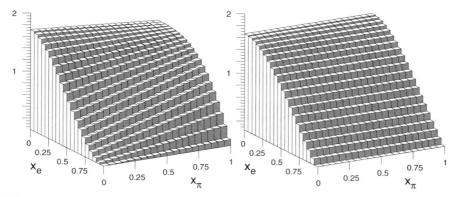

**Fig. 8.12.** Lepton–hadron correlations with Standard Model Michel parameters (*left histogram*) and $\xi_e = \xi_\pi$ set to zero (*right histogram*)

**Table 8.4.** Numbers of selected events from a sample of roughly 150 000 $\tau$ pairs in a measurement by the L3 collaboration [170]. $X$ represents an unidentified $\tau$ decay

|         | $e\nu\nu$ | $\mu\nu\nu$ | $\pi\nu$ | $\pi\pi\nu$ | $X$  |
|---------|-----------|-------------|----------|-------------|------|
| $e\nu\nu$   | 558 | 1574 | 1489 | 2817 | 6176 |
| $\mu\nu\nu$ |     | 437  | 1002 | 1892 | 3897 |
| $\pi\nu$    |     |      | 456  | 1921 | 4088 |
| $\pi\pi\nu$ |     |      |      | 1816 | 7500 |

was plotted. The correlation between the two kinematic variables is clearly visible. The result obtained is (Standard Model expectations in brackets):

$$\begin{aligned}
\rho &= 0.724 \pm 0.043 \pm 0.021 \quad (0.75), \\
\eta &= 0.26 \pm 0.19 \pm 0.08 \quad (0), \\
\xi &= 0.51 \pm 0.19 \pm 0.09 \quad (1), \\
\xi\Delta &= 0.62 \pm 0.14 \pm 0.06 \quad (0.75), \\
\xi_{\text{had}} &= 1.065 \pm 0.033 \pm 0.016 \quad (1).
\end{aligned} \quad (8.30)$$

### 8.3.4 Fitting Michel Parameters

The Michel parameters are extracted from the measured spectra by a fit of the theoretical expectations ((8.11), (8.19) boosted into the laboratory, (8.23), and (8.24)) to the data. The data are cast into histograms, which are one-dimensional for the measurements from a single $\tau$ (Sects. 8.3.1 and 8.3.2) and two-dimensional for the correlation measurements (Sect. 8.3.3). Usually a binned maximum-likelihood fit with Poissonian errors is applied to allow for a fine binning. For the sake of simplicity, only the $\tau \to \pi\pi\nu_\tau$ decay will be considered in the following, where the spectrum (8.23) is

244    8. The Lorentz Structure of the Charged Current

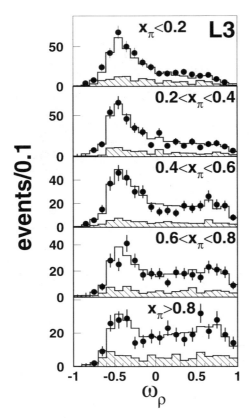

Fig. 8.13. Correlation spectra between $\tau \to \pi \nu_\tau$ and $\tau \to \pi \pi \nu_\tau$ events, measured by L3 [170]. Distributions of $\omega_\rho$ for different slices in the pion energy. The *points* are data, the *open histogram* is the fit result. The *shaded histogram* shows the background.

$$\frac{\mathrm{d}\Gamma}{\mathrm{d}\omega} = f(\omega) + \xi_\rho\, P_\tau\, g(\omega)\,. \tag{8.31}$$

In order to avoid a regeneration of the full spectrum for every variation of the Michel parameters during the fit, events are generated according to $f(\omega)$ and $g(\omega)$ separately, once. These parts of the spectrum are then added in the fit, weighted with the current value of the Michel parameter and the polarization.

The theoretical formulas cannot be fitted directly to the data. They have to be corrected for radiation and experimental effects, such as the finite resolution, efficiency, and background. The radiative corrections and the resolution in $\omega$ are rather tricky to handle. Different approaches have been applied, three of which will be mentioned.

**Convolution.** The radiative corrections can be approximated by a radiator function, which gives the probability that the initial electrons, the $\tau$, or its daughters lose a certain fraction of energy owing to radiation (see e.g. [642]). Bremsstrahlung in the material of the detector can also be handled by such a radiator function.

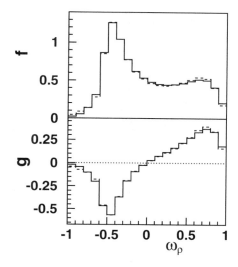

Fig. 8.14. Kinematical functions used to fit the $\tau \to \pi\pi\nu_\tau$ spectrum by the L3 collaboration [170]. *Dashed histogram*: generator spectrum including radiative corrections. *Solid histogram*: spectrum convoluted with the experimental resolution

The experimental resolution can be parametrized by a smearing function which gives the probability that a decay with a true value of $\omega$ is reconstructed at $\omega + \Delta\omega$.

The spectrum is then convoluted with the radiator and smearing functions to give the prediction for the data. This approach, although correct in principle, simplifies the corrections to a degree which becomes problematic at very high precision. Variations of such an approach can be found in [78, 175]. Figure 8.14 shows the two functions $f, g$ before and after convolution with the smearing function (L3 collaboration [175]).

**Reweighting.** The second approach uses a reweighting technique on fully simulated events. To generate a spectrum with $\xi_\rho$ modified from the value used in the simulation ($\xi_\rho^0$), the decay probability of the event has to be calculated from (8.31) for the new and old parameters. Each event gets a weight given by

$$w = \frac{f(\omega) + \xi_\rho\, P_\tau\, g(\omega)}{f(\omega) + \xi_\rho^0\, P_\tau\, g(\omega)} \tag{8.32}$$

and is rehistogrammed to give the new prediction.

Care has to be taken to properly define the energies in case of radiation. The weights have to be calculated with the energy of the $\tau$ remaining after radiation during the $\tau$ production and the energy of its daughters before radiation in the detector material. This preserves the radiative corrections of the $\tau$ production and the corrections owing to radiation in the detector material. The smaller corrections to the $\tau$ decay are more problematic as they cannot be associated clearly with the initial $\tau$ or its final daughter(s).

This approach has the advantage of treating the experimental effects of efficiency, resolution, and background by means of the full details of the detector simulation in use. A disadvantage of the method might be that in certain corners of the phase space, where the original spectrum (usually the Standard Model spectrum) predicts only few events, the weights can get quite large and introduce large statistical fluctuations. An analysis using this procedure is given in, for example, [643].

**Reduction to Base Spectra.** In this approach each spectrum is represented as a superposition of several basic spectra. In our example one could choose the two spectra with $\xi_\rho = +1$ and $-1$ as a basis. Let us call them $d\Gamma_{+1}$ and $d\Gamma_{-1}$. A spectrum with an arbitrary value of $\xi_\rho$ is then generated by a superposition

$$d\Gamma_{\xi_\rho} = a_+ d\Gamma_{+1} + a_- d\Gamma_{-1} \tag{8.33}$$

with

$$\begin{aligned} a_+ + a_- &= 1, \\ a_+ - a_- &= \xi_\rho. \end{aligned} \tag{8.34}$$

Then in the fit the coefficients $a_+$ and $a_-$ are varied (respecting the normalization $a_+ + a_- = 1$) to maximize the likelihood and, finally $\xi_\rho$ is calculated from (8.34).

There are two basic spectra necessary to represent a hadronic decay, five for a leptonic decay and 25 for a lepton–lepton correlation. These spectra are simulated with the full detector simulation and all radiative corrections implemented.

In order to reduce the effort of simulation, events are reused whenever possible. All basic spectra of a given decay channel can be generated simultaneously. An event is kept if it fits into any of the basic spectra and flags are stored according to which it belongs to (these are often several). The spectra are created in a second process from all events with the corresponding flag set. One never has to simulate more than twice the events of the Standard Model spectrum.

The advantage of this approach is that all corrections are treated according to the best capabilities of the simulation; the disadvantage is that one has to simulate special events for the analysis. This method was first applied in [210].

## 8.4 The Current Experimental Situation

### 8.4.1 Model-Independent Results

In addition to some early measurements which mainly determined the Michel parameter $\rho$ from leptonic decays [79, 81, 88, 101, 149, 440, 640], several collaborations presented full analyses of the Lorentz structure of the charged weak current: ALEPH [643], ARGUS [76], CLEO [118], DELPHI [629], L3 [175], SLD [246], and OPAL [210]. The SLD measurement [246] benefits from the

## 8.4 The Current Experimental Situation 247

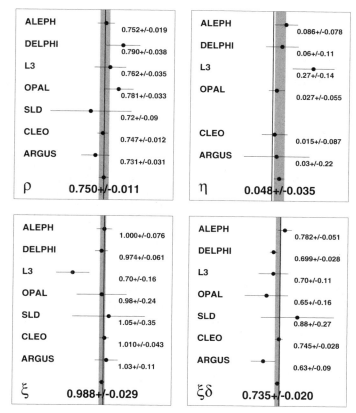

**Fig. 8.15.** World averages of the Michel parameters in leptonic $\tau$ decays under the assumption of universality between $\tau \to e\,\nu_e\nu_\tau$ and $\tau \to \mu\,\nu_\mu\nu_\tau$. The average is indicated by the number at the bottom and the *shaded band*. The *solid line* is the Standard Model expectation $(3/4, 0, 1, 3/4)$

polarization of the initial beams at the SLC; all others use correlations to improve their sensitivity.

The production of the $\tau$ pairs through a pure vector current is assumed in all measurements. The polarization of the $\tau$ leptons is either inferred from other neutral-current data or determined in the measurement together with the Michel parameters. It is possible to study the Lorentz structure of the weak neutral current in $\tau$ production, too (see [644]), but not simultaneously with the charged current. The number of parameters would be too large.

The current averages of the parameters are shown in Figs. 8.15 and 8.16, obtained from fits which assume universality between the decays. More details and the averages without the assumption of universality can be found in [357]. The measurements show large correlations between the different parameters,

248   8. The Lorentz Structure of the Charged Current

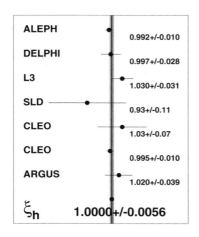

Fig. 8.16. World average of the parameter $\xi_{\text{had}}$ under the assumption of universality. The average is indicated by the number at the bottom and the *shaded band*. The *solid line* is the Standard Model expectation (1)

in particular between $\rho$ and $\eta$ and also between $\xi$ and $\Delta$. These have to be taken into account in interpreting the results.

With the procedures described in Sect. 8.1.4 one can now derive limits on the couplings of the leptonic decays. Figure 8.17 shows these limits under the assumption of universality between the two leptonic decays. The limits without this assumption, and more details can be found in [357].

### 8.4.2 Interpretation in Terms of Models

In the following a few results derived from the measurements of the Michel parameters in the framework of specific models will be quoted. This list is by no means considered to be complete.

**Pure Vector Currents.** Hebbeker and Lohmann have studied the chirality structure of $\tau$ decays assuming it is a pure vector current [645]. They showed that under this assumption it is indeed a pure left-handed interaction, obtaining

$$L_\tau = 0.999 \pm 0.003,$$
$$R_\tau = 0.00 \pm 0.08, \qquad (8.35)$$

where $L_\tau$ is the strength of the left- and $R_\tau$ of the right-handed coupling relative to the Standard Model.

**No Tensor Currents.** Pich and Silva have analyzed the measurements under various simplifying assumptions, all excluding tensor currents from the interpretation [646]. For example, if they assume that there is only a single charged $W^\pm$ boson, they can show that its coupling strength is at least 87% of the Standard Model value for $\tau \to e\,\nu_e\nu_\tau$ and 83% for $\tau \to \mu\,\nu_\mu\nu_\tau$ (90% confidence level).

8.4 The Current Experimental Situation    249

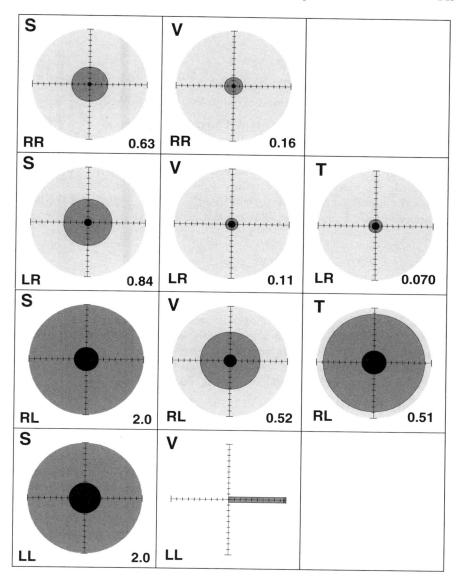

**Fig. 8.17.** Limits on the coupling constants $g^\kappa_{\epsilon\lambda}$ for leptonic decays under the assumption of universality. The *upper letter* in each box indicates the type of coupling (scalar/vector/tensor), the *lower two letters* the chirality of the $\tau$ (*right letter*) and the daughter lepton (*left letter*). The *light gray circles* define the physically allowed range of the couplings and the *darker area* is the region still consistent with the measurements of the Michel parameters (90% confidence level). These limits on the couplings are also printed in the *lower right corner* of each box. The *black circles* in the center indicate for comparison the corresponding limits from $\mu$ decays

250   8. The Lorentz Structure of the Charged Current

If they allow for an additional scalar boson, its couplings to right-handed $\tau$ leptons are constrained by

$$|g_{\rm LR}^{\rm S}|^2 + |g_{\rm RR}^{\rm S}|^2 < 0.79 \quad \text{for} \quad \tau \to e\,\nu_e\nu_\tau,$$
$$|g_{\rm LR}^{\rm S}|^2 + |g_{\rm RR}^{\rm S}|^2 < 0.56 \quad \text{for} \quad \tau \to \mu\,\nu_\mu\nu_\tau, \tag{8.36}$$

at the 90% confidence level (the maximum value allowed is 2).

**Lepton-Flavor-Violating Neutral Currents.** The decay of a $\tau$ lepton is not necessarily a charged-current interaction. Figure 8.6 shows the decay through a neutral current, which then has to violate lepton number conservation. Pich and Silva studied such models in the same paper as mentioned above [646]. For example, if they assume that there is such a neutral current besides the Standard Model $W^\pm$, then the $g_{\rm RR}^{\rm S}$ and $g_{\rm LL}^{\rm S}$ vanish and the other two couplings are bounded by (90% confidence level):

$$\begin{array}{ll} \tau \to e\,\nu_e\nu_\tau & \tau \to \mu\,\nu_\mu\nu_\tau \\ |g_{\rm LR}^{\rm S}| < 0.19 & |g_{\rm LR}^{\rm S}| < 0.22 \\ |g_{\rm RL}^{\rm S}| < 0.19 & |g_{\rm RL}^{\rm S}| < 0.16. \end{array} \tag{8.37}$$

**Left–Right Symmetry.** In left–right symmetric models there is a second charged $W^\pm$ boson coupling exclusively to the right-handed fermions. To account for parity violation at low energies, it has to be heavier than the Standard Model $W^\pm$ boson. It affects the Lorentz structure of both leptonic and hadronic $\tau$ decays. A lower bound on such a right-handed boson can be set at 236 GeV at the 90% confidence level. For details see [357].

**Charged Higgs Bosons.** Dova et al. studied $\tau$ decays in the framework of type II Higgs models [333]. In these models there are three neutral Higgs bosons and a pair of charged Higgs bosons $H^\pm$. The charged Higgs bosons can mediate $\tau$ decays. They mainly modify the $\eta$ parameter in $\tau \to \mu\,\nu_\mu\nu_\tau$ [647, 648]:

$$\eta_\mu = -\frac{m_\mu m_\tau}{2}\left(\frac{\tan\beta}{m_{\rm H^\pm}}\right)^2. \tag{8.38}$$

A lower limit of 2.5 GeV can be derived for $m_{\rm H^\pm}/\tan\beta$ at the 90% confidence level [357]. Figure 8.18 summarizes the excluded regions in the parameter space of $m_{\rm H^\pm}$ versus $\tan\beta$ from various experiments. The branching ratio of the decay $B \to \tau\nu$ gives a similar limit to the one from $\tau$ decays [649, 650].

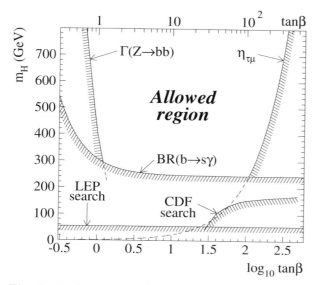

**Fig. 8.18.** Excluded regions for the mass of a charged Higgs $m_H$ and $\tan\beta$ in type II Higgs models. Each line shows the bound achieved from the measurement indicated. From $\tau$ decays, the region at high $\tan\beta$ is excluded [333]

# 9. Searching for $\mathcal{CP}$ Violation

Understanding the origin of $\mathcal{CP}$ violation is one of the most important outstanding questions in particle physics. Historically, it came as a surprise when in 1964 Christenson and his collaborators found a small fraction of $K_L^0$ mesons to decay to two pions, thereby violating the $\mathcal{CP}$ symmetry [651]. Although $\mathcal{CP}$ violation has been incorporated into the Standard Model [448] and is now able to describe the effect observed by Christenson, there is no deeper understanding of its origin. Also, from the experimental point of view, very little is known to date. The system of the $K^0$ mesons is still the only place where $\mathcal{CP}$ violation has been observed (for reviews see [652–661]).

Many authors have proposed alternative models to the Standard Model (see [662–664] and references therein). Additional motivation to go beyond the Standard Model comes from the observed baryon asymmetry in the universe. It has been suggested that the amount of $\mathcal{CP}$ violation in the Standard Model is not sufficient to create such a big excess of matter, but that is still under discussion [665–667]. The present bounds on the electric dipole moments of the electron and the neutron imply quite significant constraints on these alternative models, and for some of them the limits on the electric dipole moments of the $\tau$ lepton, to be described below, are even more restrictive.

## 9.1 $\mathcal{CP}$ Violation in $\tau$ Production

### 9.1.1 Models of $\mathcal{CP}$ Violation

A stable particle $X$ cannot have an electric dipole moment unless both $\mathcal{T}$ and $\mathcal{P}$ symmetry are broken. Writing the electrical dipole moment $\boldsymbol{d}_X$ as

$$\boldsymbol{d}_X = \int \boldsymbol{x}\, \rho(\boldsymbol{x})\, \mathrm{d}^3 x, \tag{9.1}$$

one sees that it is even under $\mathcal{T}$ and odd under $\mathcal{P}$ transformations. Now, for a particle at rest, the orientation of $\boldsymbol{d}_X$ has to be related to the orientation of the spin, i.e. $\boldsymbol{d}_X$ has to be proportional to $\boldsymbol{s}$. But $\boldsymbol{s}$ is odd under $\mathcal{T}$ and even under $\mathcal{P}$ transformations. Therefore, if $\boldsymbol{d}_X$ does not vanish, $\mathcal{T}$ and $\mathcal{P}$ have to be violated, which means that $\mathcal{CP}$ is violated if $\mathcal{CPT}$ holds.

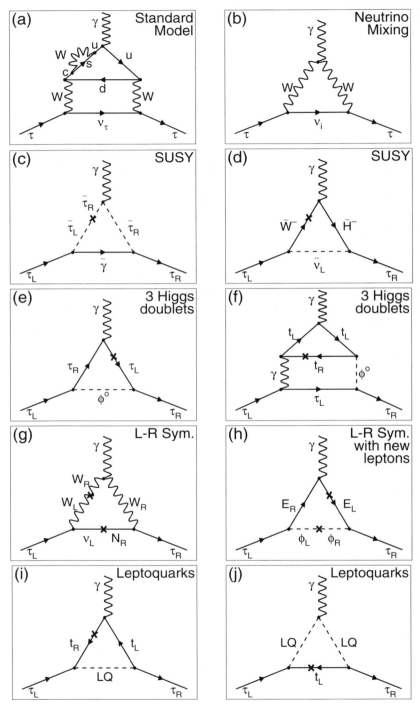

**Fig. 9.1.** $\mathcal{CP}$ violating vertex corrections. A *cross* marks a chirality flip. *Wavy lines* represent spin-1 bosons; *dashed lines* represent spin-0 bosons. See text for details

Detectable $\mathcal{CP}$ violation in $\tau$ production requires new $\mathcal{CP}$-violating interactions involving leptons. These interactions would induce electric dipole form factors into the production amplitude through radiative corrections (see Sects. 3.3 and 9.1.2 for details).

The typical order of magnitude of these form factors can be guessed. As radiative corrections, one expects the amplitudes to be of order $\alpha/\pi$ with respect to the $\mathcal{CP}$-conserving tree-level amplitude. Furthermore, the dipole form factors describe chirality-flipping interactions and are therefore proportional to some fermion mass with respect to a typical energy scale of the reaction: $m_\mathrm{f}/\sqrt{s}$. The fermion mass is not necessarily $m_\tau$, it can be the mass of a fermion in a loop.

These $\mathcal{CP}$-violating dipole form factors lead to an additional contribution to the production cross section proportional to $|d_\tau|^2$, with an angular distribution $\mathrm{d}\sigma/\mathrm{d}\cos\theta_\tau \propto \sin^2\theta_\tau$ (Sect. 3.3). Because of the quadratic dependence this is not very sensitive to small dipole moments. On top of that, it is not considered a test of $\mathcal{CP}$. A magnetic dipole moment, for example, gives rise to a cross section proportional to $\sin^2\theta_\tau$, too. The truly $\mathcal{CP}$-violating contribution originates from the interference of the $\mathcal{CP}$-violating amplitude with the Standard Model amplitude. This contribution to the cross section is linear in $d_\tau$.

The Standard Model and some of its extensions will be discussed below.

**Standard Model.** In the Standard Model $\mathcal{CP}$ violation arises from the complex couplings in the Cabibbo–Kobayashi–Maskawa (CKM) matrix [448]. It is able to describe all phenomena observed so far. There is no $\mathcal{CP}$ violation in leptonic couplings. Nevertheless, $\mathcal{CP}$ violation in the quark sector induces a small electric dipole moment of the leptons. One has to go at least to three-loop order to generate a nonvanishing contribution [668] (see Fig. 9.1a). A crude estimate gives

$$|d_\tau(\mathrm{SM})| \lesssim 10^{-34} e\,\mathrm{cm}, \tag{9.2}$$

which is undetectably small by many orders of magnitude.

**Massive Neutrinos.** If at least two of the three neutrinos are massive (and their masses are different), $\mathcal{CP}$ violation can occur in the lepton sector. There could be mixing between the neutrinos with a CKM-type mixing matrix. This could introduce $\mathcal{CP}$ violation into the lepton sector in complete analogy to the $\mathcal{CP}$ violation in the quark sector.

Figure 9.1b shows a diagram involving neutrino couplings. Such a one-loop diagram does not, however, generate $\mathcal{CP}$ violation. The effect of this diagram is proportional to $(V_\mathrm{lept})_{3i}(V^*_\mathrm{lept})_{3i}$, which is independent of possible $\mathcal{CP}$-violating phases. Here $V_\mathrm{lept}$ is the leptonic mixing matrix and $i$ the flavor of the intermediate neutrino. Even at two-loop order there is no $\mathcal{CP}$-violating effect [669]. The three-loop effect can safely be neglected, so that neutrino masses as a source of $\mathcal{CP}$ violation are undetectable through a dipole moment of the $\tau$ lepton [670].

**Supersymmetry (SUSY).** In supersymmetric models [671–680] there are several potential sources of $\mathcal{CP}$ violation besides the phase of the CKM matrix and possibly massive neutrinos. $\mathcal{CP}$-violating phases might arise from complex parameters in the superpotential and in the supersymmetry-breaking terms which introduce phases into the lepton–slepton–gaugino and lepton–slepton–Higgsino couplings [681, 682]. These phases can generate an electric dipole moment of the $\tau$ lepton at the one-loop level [683–690].

Two examples are shown in Fig. 9.1c, d. There are too many possible scenarios for supersymmetry and too many unknown parameters in these models to give a firm prediction. But diagrams with a photino exchange, like the one in Fig. 9.1c, are expected to give the largest contributions, if the photino is indeed the lightest stable supersymmetric particle. The contribution from photino exchange is proportional to

$$d_\tau \propto \frac{m_\tau |A_{\tilde{\tau}}|}{m_{\tilde{\gamma}}^3} \sin(\phi_{\tilde{\tau}} - \phi_{\tilde{\gamma}}), \tag{9.3}$$

where $|A_{\tilde{\tau}}|$, $\phi_{\tilde{\tau}}$ and $m_{\tilde{\gamma}}$, $\phi_{\tilde{\gamma}}$ are the magnitudes and phases of the stau and photino mass matrices (see [664] for details). The contribution is linear in the lepton mass. The constraint from the electric dipole moment of the electron ($0.18 \pm 0.16 \times 10^{-26}\,e$ cm [44]) is more stringent than that from the $\tau$ lepton ($|d_\tau| < 3.1 \times 10^{-16}$ ecm [44]), assuming similar values for the SUSY parameters of the three families. The constraint form the neutron is even more sensitive. A bold guess gives $10^{-25}\,e\,\text{cm} \times \sin(\phi_e - \phi_{\tilde{\gamma}})$ for $d_e$, which might indicate that the phase difference is not maximal.

**$\mathcal{CP}$ Violation in the Higgs Sector.** Higgs models of $\mathcal{CP}$ violation are motivated by the idea of linking the origin of $\mathcal{CP}$ violation to the origin of the masses, i.e. the mechanism of spontaneous symmetry breaking [691–697]. In these models the Lagrangian before symmetry breaking is $\mathcal{CP}$ invariant. There have to be at least two Higgs multiplets. The vacuum expectation values which they acquire through spontaneous symmetry breaking are complex and might have a phase relative to each other. The complex vacuum expectation values lead, after diagonalization of the fermion mass matrices, to $\mathcal{CP}$-violating Yukawa couplings of Higgs particles to fermions. Care has to be taken in these models to avoid large flavor-changing neutral currents.

Probably the most interesting model which allows for natural flavor conservation is the one due to Weinberg [696]. It has three Higgs doublets, from which four charged and five neutral Higgs bosons arise. The neutral Higgs bosons can generate $\mathcal{CP}$-violating interactions as they are mixtures of $\mathcal{CP}$ even and $\mathcal{CP}$ odd states. A one-loop diagram is shown in Fig. 9.1e ($\phi^0$ is one of the Higgs bosons). The contribution is proportional to $m_\ell^3/m_\phi^2$ and might be as large as $10^{-23}\,e$ cm for the $\tau$ lepton. Two powers in $m_\ell$ arise from the two couplings of the $\tau$ lepton to the Higgs boson and the third from the helicity flip. The contribution to the electric dipole moment of the electron is much smaller and the $\tau$ lepton gives a more stringent constraint.

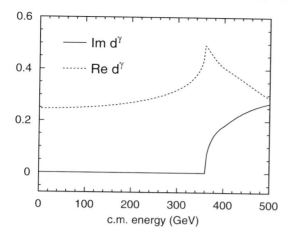

Fig. 9.2. The real and imaginary parts of the dipole form factor of the $\tau$ lepton in a leptoquark model [662] in units of $10^{-18}\,e$ cm

For the $\tau$ lepton the one-loop contribution is expected to dominate. But two-loop corrections with top-quark loops might be more important for the electron (see Fig. 9.1f). They avoid one of the weak electron Higgs couplings and the helicity flip is provided more easily by the top quark. Their contribution is proportional to $m_\ell m_t^2/m_\phi^2$ [698].

**Left–Right Symmetric Models.** Models based on the gauge group $SU(2)_L \times SU(2)_R \times U(1)$ are symmetric between left- and right-handed fermions [699, 700]. The symmetry is broken when the weak bosons $Z_R^0$, $W_R^\pm$ coupling to the right-handed fermions acquire a mass substantially larger than $Z_L^0$, $W_L^\pm$, which couple to the left-handed fermions.

In the simplest models $\mathcal{CP}$ violation might manifest itself through phases in the complex mass matrices of $W_R - W_L$ and/or the neutrinos. In order to generate a sizable dipole moment a Dirac mass term connecting the left-handed $\nu_L$ with the heavy right-handed $N_R$ is needed. The see-saw mechanism [701, 702] is used to suppress the mass of the light neutrino mass eigenstate.

Diagrams like the one shown in Fig. 9.1g generate electric dipole moments. The contribution is linear in the mass of the lepton and depends on the parameters of the neutrino mass matrix and the mixing angle between the left- and right-handed $W^\pm$ bosons. It is probably not larger than $10^{-23}\,e$ cm for the $\tau$ lepton and is therefore too small to be currently observed.

In more complicated models there are more sources of $\mathcal{CP}$ violation. For example, Fig. 9.1h shows a diagram where a new charged lepton E and Higgs bosons are involved [703].

**Leptoquarks.** The largest dipole moments of $\tau$ leptons are expected from leptoquark models [662, 704–711]. Leptoquarks are spin-0 or spin-1 bosons that turn leptons into quarks and vice versa. Figure 9.2 shows the real and imaginary parts of the electric dipole form factor generated by a spin-0 leptoquark. Diagrams are shown in Figs. 9.1i, j. The dipole form factor is of order $2 \times 10^{-19}\,e$ cm. Above the $t\bar{t}$ threshold it develops an imaginary part.

One might expect the couplings to be of a Higgs type, i.e. proportional to the fermion masses. The chirality flip is provided by the top quark with its large mass. Then the dipole moment scales like

$$d_e : d_\mu : d_\tau = m_u^2 m_e : m_c^2 m_\mu : m_t^2 m_\tau \tag{9.4}$$

and the $\tau$ lepton is clearly the most sensitive fermion.

There is no general connection between the electromagnetic and weak electric dipole moments $d_\tau^\gamma$ and $d_\tau^{Z^0}$. The precise relation depends on the model in question and its parameters, but one expects them to be of the same order of magnitude. In that sense the previous estimates are valid for both $d_\tau^\gamma$ and $d_\tau^{Z^0}$.

### 9.1.2 A Phenomenological Ansatz

The weak and electric dipole moments of the $\tau$ lepton and the dipole form factors have been introduced in Sect. 3.3. The effects of $\mathcal{CP}$ violation are parametrized either by an effective Lagrangian [339, 340, 712, 713]

$$\mathcal{L}_{\text{eff}} = -\frac{i}{2}\,\tilde{d}_\tau^{Z^0}\,\overline{\Psi}_\tau \sigma_{\mu\nu}\gamma_5 \Psi_\tau Z^{\mu\nu} - \frac{i}{2}\,\tilde{d}_\tau^\gamma\,\overline{\Psi}_\tau \sigma_{\mu\nu}\gamma_5 \Psi_\tau F^{\mu\nu}, \tag{9.5}$$

where $Z^{\mu\nu}$ and $F^{\mu\nu}$ are the weak and electric field tensors and $\tilde{d}_\tau^{Z^0}$ and $\tilde{d}_\tau^\gamma$ are the weak and electric dipole moments, or by a form factor to the currents

$$J_\mu^{\mathcal{CP}}(q) = e\, d_\tau(q^2)\, \langle\, \overline{u}_\tau(p')\, |\, \sigma_{\mu\nu} \gamma_5 q^\nu\, |\, u_\tau(p)\, \rangle, \tag{9.6}$$

where $d_\tau(q^2)$ is the weak or electric dipole form factor, depending on whether the current is coupled to a photon or a $Z^0$ boson ($q = p' - p$).

It is important to realize that this is simply a parametrization of $\mathcal{CP}$-violating effects in the coupling of $\tau$ leptons to a photon or a $Z^0$ boson. It is the extension of the Standard Model interaction by the $\mathcal{CP}$-violating interaction with the smallest mass dimension possible. This is of dimension 6 and hence the interaction is proportional to $1/\Lambda^2$, where $\Lambda$ is the scale of the new interaction. The parametrization does not make use of any model of $\mathcal{CP}$ violation by any means.

Any cross section will get an additional contribution from the square of the matrix elements calculated from these currents. A weak dipole moment[1] will, for example, change the partial width of the decay of the $Z^0$ boson into $\tau$ leptons. The precise measurements of $\Gamma(Z^0 \to \tau\tau)$ can be used to limit $d_\tau^{Z^0}$. The Standard Model prediction for the partial width is [408]

$$\Gamma(Z^0 \to \tau^+\tau^-) = (83.64 \pm 0.13)\text{ MeV}; \tag{9.7}$$

the measured width is [408]

---

[1] It has become common to call the weak electric dipole form factor at $q^2 = m_Z^2$ the weak dipole moment.

$$\Gamma(Z^0 \to \tau^+\tau^-) = (84.10 \pm 0.27) \text{ MeV}, \tag{9.8}$$

which is roughly in agreement with the prediction. A weak dipole moment would give an additional contribution to $\Gamma(Z^0 \to \tau\tau)$ of

$$\Delta\Gamma(Z^0 \to \tau^+\tau^-) = \frac{m_Z^3}{24\pi} |d_\tau^{Z^0}(m_Z^2)|^2, \tag{9.9}$$

which is limited by the above measurement to about 1 MeV (95% confidence level). From this, a limit of

$$|d_\tau^{Z^0}(m_Z^2)| < 2 \times 10^{-17} e \text{ cm} \tag{9.10}$$

can be derived (95% confidence level).

This contribution is quadratic in $d_\tau$. It is $\mathcal{CP}$ symmetric and the above limit is not a test of $\mathcal{CP}$ violation, as many other kinds of new physics could cause a deviation in $\Gamma(Z^0 \to \tau\tau)$ from the Standard Model prediction. It is the interference term between the $\mathcal{CP}$-violating amplitude and the Standard Model amplitude which introduces $\mathcal{CP}$ violation into the reaction. This interference term is linear in $d_\tau$.

### 9.1.3 Observables

To truly test $\mathcal{CP}$ invariance in any kind of process, a $\mathcal{CP}$-odd observable is needed, i.e. an observable $\mathcal{O}$ that changes sign under $\mathcal{CP}$ transformation:

$$\mathcal{O} \xrightarrow{\mathcal{CP}} -\mathcal{O}. \tag{9.11}$$

Then, if $\mathcal{CP}$ parity is conserved, the expectation value of such an observable has to vanish. Or – turning the argument around – if an experiment measures an expectation value different from zero, $\mathcal{CP}$ violation has been observed.

The question of how big the deviation from zero will be and whether it is observable at all is a question of sensitivity. Parametrizing the $\mathcal{CP}$ violation through the dipole moments, the expectation value of any $\mathcal{CP}$-odd observable will be proportional to the dipole moment:[2]

$$\langle \mathcal{O} \rangle = c_{\text{sens}}\, d_\tau. \tag{9.12}$$

The constant $c_{\text{sens}}$ is called the sensitivity. The larger it is, the more sensitive the experiment will be. It is usually determined by applying the analysis to Monte Carlo samples with known amounts of $\mathcal{CP}$ violation.

In a series of papers [339, 340, 712, 713], Bernreuther and Nachtmann proposed the use of certain tensor observables to search for $\mathcal{CP}$ violation in the

---

[2] The linearity in $d_\tau$ is an approximation which is only valid as long as the $\mathcal{CP}$-violating amplitude is small compared to the Standard Model amplitude, i.e. for small $d_\tau$.

production of $\tau$ pairs and estimated the achievable sensitivity (see also [714–716]). The $\mathcal{CP}$-odd observable used was

$$T_{ij} = (\boldsymbol{p}_+ - \boldsymbol{p}_-)_i \, (\boldsymbol{p}_+ \times \boldsymbol{p}_-)_j + (\boldsymbol{p}_+ - \boldsymbol{p}_-)_j \, (\boldsymbol{p}_+ \times \boldsymbol{p}_-)_i, \tag{9.13}$$

where $\boldsymbol{p}_+$ and $\boldsymbol{p}_-$ are the momenta of the visible decay products of the positive and negative $\tau$ lepton, respectively, and $i, j$ are Cartesian coordinates. A similar observable $\hat{T}_{ij}$ can be constructed from the unit momenta.

The expectation values of the off-diagonal elements of the tensor have to vanish and the three diagonal elements have to sum to zero. These are useful relations for investigating systematic errors. The sensitivity of $T_{33}$ to $\mathcal{CP}$ violation is twice as large as and opposite in sign to $T_{11}$ and $T_{22}$. Hence, $T_{33}$ (or $\hat{T}_{33}$) is the preferred observable. Several experiments have used these to investigate $\mathcal{CP}$ violation in $\tau$ production [62, 70, 237, 717].

Although reasonably sensitive, these are not the best observables for $\mathcal{CP}$ violation. Optimal observables can be defined in the sense that they incorporate all available information [438, 718, 719]. The optimal observables are

$$\mathcal{O}^{\text{Re}} = \frac{\text{Re}\,(A_{\mathcal{CP}})}{A_{\text{SM}}}, \tag{9.14}$$

$$\mathcal{O}^{\text{Im}} = \frac{\text{Im}\,(A_{\mathcal{CP}})}{A_{\text{SM}}}. \tag{9.15}$$

There are two observables: the first one is sensitive to the real part of the dipole form factor, the second to the imaginary part. They are formed from the ratio of the $\mathcal{CP}$-violating amplitude to the Standard Model amplitude. The amplitudes can be written down in terms of the momentum and spin vectors of the $\tau$ leptons (see [228] for the full formulas):

$$\text{Re}\,(A_{\mathcal{CP}}) = (\hat{p}_\tau \hat{p}_e)\,[\hat{p}_\tau \times (\boldsymbol{s}_+ - \boldsymbol{s}_-)]\,\hat{p}_e,$$
$$\text{Im}\,(A_{\mathcal{CP}}) = (\hat{p}_\tau \hat{p}_e)\,[(\hat{p}_\tau \boldsymbol{s}_+)(\hat{p}_e \boldsymbol{s}_-) - (\hat{p}_\tau \boldsymbol{s}_-)(\hat{p}_e \boldsymbol{s}_+)]. \tag{9.16}$$

The vectors $\boldsymbol{s}_+$ and $\boldsymbol{s}_-$ are the spin vectors of the $\tau^+$ and $\tau^-$, respectively, $\hat{p}_\tau$ is a unit vector in the direction of the momentum of the positive $\tau$ lepton and $\hat{p}_e$ the direction of the initial positron beam. The negative $\tau$ lepton is assumed to be back-to-back with the positive $\tau$ lepton and the initial beams unpolarized.

The direction of the $\tau$ leptons is reconstructed using the technique described in Sect. 2.2.4. The spin vectors are derived from the momenta of the visible decay products of each $\tau$ lepton. The detailed procedure depends on the decay channel in question (see [228]). Briefly, the spin vector is chosen to be the one which maximizes the probability of generating the observed configuration of decay products. Optimal observables have been applied to search for $\mathcal{CP}$ violation in $\tau$ production in [218, 228, 720].

**Table 9.1.** Identification probabilities of the maximum-likelihood selection of the one-prong channels in the OPAL analysis [218]. There are four channels identified: $\tau^- \to e^- \bar{\nu}_e \nu_\tau$, $\tau^- \to \mu^- \bar{\nu}_\mu \nu_\tau$, $\tau \to \pi \nu_\tau$, and $\tau \to \pi \pi \nu_\tau$. The vertical list of true channels gives only the major decay modes. Therefore the columns do not sum to unity

| Monte Carlo $\tau$ decay channel | Identification probabilities (%) | | | | | | | |
|---|---|---|---|---|---|---|---|---|
| | Barrel | | | | End caps | | | |
| 1-prong | e | $\mu$ | $\pi$ | $\rho$ | e | $\mu$ | $\pi$ | $\rho$ |
| e | 95.3 | 0.0 | 0.4 | 1.2 | 89.2 | 0.0 | 1.3 | 2.5 |
| $\mu$ | 0.5 | 92.3 | 1.0 | 0.6 | 0.1 | 89.8 | 3.3 | 0.7 |
| $\pi$ | 1.6 | 2.3 | 78.0 | 9.6 | 2.8 | 6.0 | 68.9 | 11.9 |
| $\rho$ | 0.3 | 0.2 | 7.8 | 61.4 | 0.7 | 0.2 | 10.7 | 48.6 |
| $a_1 \to \pi \, 2\pi^0$ | 0.0 | 0.0 | 0.7 | 21.9 | 0.2 | 0.0 | 1.8 | 19.1 |

### 9.1.4 Experimental Results

There have been several attempts to identify $\mathcal{CP}$ violation in $\tau$ production [62, 70, 172, 218, 228, 237, 717, 720, 721] (for summaries see [722, 723]). No effect has been seen. The currently most stringent limit has been derived by the OPAL collaboration [218].

From the OPAL collaboration's full data set from the LEP-I period $\tau$ pairs were selected and then the decays of the two $\tau$ leptons were classified independently by a likelihood method. Table 9.1 gives the efficiencies of the four one-prong channels that were used, together with the three-prong decay $\tau \to 3 \pi \, \nu_\tau$.

The $\tau$ direction was reconstructed using the method described in Sect. 2.2.4. The ambiguity was resolved where ever possible. (See Fig. 2.10 in Sect. 2.2.4. It is taken from this analysis.)

Then the spins were reconstructed and from them the values of the optimal observables of (9.15) were calculated. Figure 9.3 shows the distribution for the $\tau^+\tau^- \to \pi^+\pi^- \nu_\tau \bar{\nu}_\tau$ channel from Monte Carlo simulations with a weak dipole moment of 0 and $5 \times 10^{-17}$ e cm. The distribution becomes asymmetric for a nonvanishing dipole moment and the mean value (expectation value) is shifted from zero.

The sensitivity is different for each combination of decay channels of the two $\tau$ leptons owing to different analyzing powers for the $\tau$ spins, different accuracies in the reconstruction of the $\tau$ direction and different detector resolutions and backgrounds. Therefore, each combination of decay channels was treated independently up to the determination of $d_\tau$. Only then were the individual values for $d_\tau$ averaged. Figure 9.4 shows the sensitivities to the real part of the weak dipole moment for events in the central section of the detector. The values are clearly higher for the semihadronic channels, owing to their better spin-analyzing power. For $\pi$ and $a_1$ there are two entries. The first one is the early OPAL data, where a three-dimensional vertex detector was not yet implemented and the ambiguity in the $\tau$ direction could not be resolved. The sensitivity is reduced in this case.

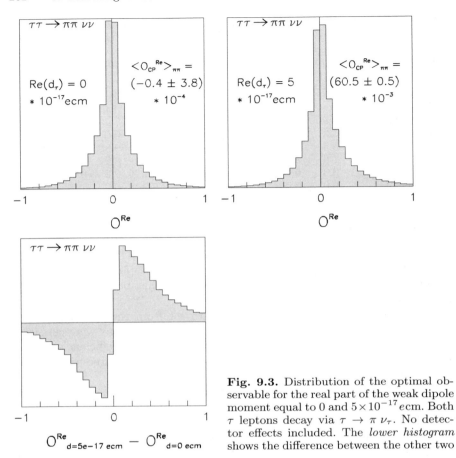

**Fig. 9.3.** Distribution of the optimal observable for the real part of the weak dipole moment equal to 0 and $5\times 10^{-17}$ ecm. Both $\tau$ leptons decay via $\tau \to \pi \nu_\tau$. No detector effects included. The *lower histogram* shows the difference between the other two

Figure 9.5 shows the corresponding results for the weak dipole moments in each channel. They scatter statistically around zero. No $CP$ violation was observed.

It is important to ensure that the detector itself does not fake a $CP$ violation or, even worse, obscures the observation of true $CP$ violation by some detector effect. To test the $CP$ symmetry of the detector the analysis was applied to events which cannot show a true $CP$ violation. These could be $e^+e^- \to e^+e^-$ or $e^+e^- \to \mu^+\mu^-$ events, which have no spin-analyzing power, or $\tau$ pairs constructed from two $\tau$ leptons with matching characteristics (for example a back-to-back configuration) from two different events. By combining $\tau$ leptons from different events the spin correlations are artificially destroyed and, as a consequence, a true $CP$ violation can no longer be observed, only detector effects.

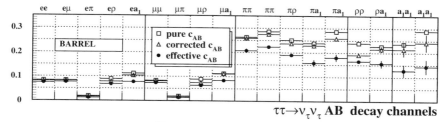

**Fig. 9.4.** Sensitivities of $\langle \mathcal{O}^{\mathrm{Re}} \rangle$ to the real part of the weak dipole moment (OPAL collaboration [218]). Three values are given for each channel: tree level without detector effects (*squares*), including detector resolution and radiative corrections (*triangles*), and including the background (*dots*)

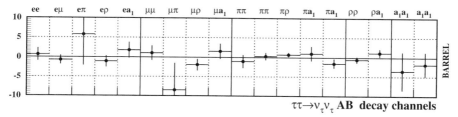

**Fig. 9.5.** Real part of the weak dipole moment measured channel by channel in the central part of the detector in units of $10^{-17}\, e\,\mathrm{cm}$ (OPAL collaboration [218]). These values were averaged with the other two detector regions, not shown here, to obtain the final result

The final result is

$$\mathrm{Re}\,(d_\tau^{Z^0}) = (0.72 \pm 2.46 \pm 0.24) \times 10^{-18}\, e\,\mathrm{cm},$$
$$\mathrm{Im}\,(d_\tau^{Z^0}) = (3.5 \pm 5.7 \pm 0.8) \times 10^{-18}\, e\,\mathrm{cm}.$$

The first error is statistical, the second systematics. The values are consistent with zero.

It has been pointed out [724, 725] that an initial beam polarization would enhance the sensitivity substantially, especially for the imaginary part of the weak dipole form factor. The SLD collaboration has a polarization of the electron beam from the SLC of 73% available, on average. With only about 10% of the statistics compared to OPAL, their preliminary result is [721]

$$\mathrm{Re}\,(d_\tau^{Z^0}) = (18.3 \pm 7.4 \pm 2.6) \times 10^{-18}\, e\,\mathrm{cm},$$
$$\mathrm{Im}\,(d_\tau^{Z^0}) = (-6.6 \pm 4.0 \pm 0.2) \times 10^{-18}\, e\,\mathrm{cm},$$

roughly consistent with zero.

Figures 9.6 and 9.7 summarize the measurements. The limits are

$$|\mathrm{Re}\,(d_\tau^{Z^0})| < 3.2 \times 10^{-18}\, e\,\mathrm{cm},$$
$$|\mathrm{Im}\,(d_\tau^{Z^0})| < 9.4 \times 10^{-18}\, e\,\mathrm{cm}, \tag{9.17}$$

at the 95% confidence level.

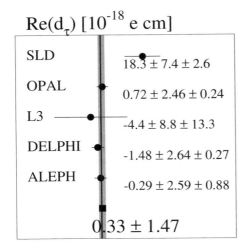

Fig. 9.6. Summary of measurements of the real part of the weak dipole moment at $s = m_Z^2$

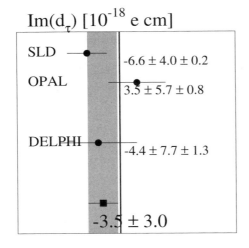

Fig. 9.7. Summary of measurements of the imaginary part of the weak dipole moment at $s = m_Z^2$

One can define a parameter $\epsilon'_\tau$ for the decay $Z^0 \to \tau^+\tau^-$ as the ratio of the $\mathcal{CP}$-violating decay width to the Standard Model decay width, similar to the parameter $\epsilon'$ of direct $\mathcal{CP}$ violation in the decay of neutral kaons. Then the limit on the real part of $d_\tau^{Z^0}$ transforms into a limit on $\epsilon'_\tau$ of $2.9 \times 10^{-4}$.

## 9.2 $\mathcal{CP}$ Violation in $\tau$ Decays

### 9.2.1 Theoretical Background

The first section of this chapter has dealt with $\mathcal{CP}$ violation in the neutral-current couplings of the $\tau$ lepton, which would manifest itself in the production of the $\tau$ pairs. But $\mathcal{CP}$ violation can also occur in the decay of $\tau$ leptons through the charged current [726, 727]. The situation is somewhat similar to direct $\mathcal{CP}$ violation in the decay of neutral kaons or neutral or charged B mesons. However, we are dealing with a leptonic reaction, so that the phase of the Cabibbo–Kobayashi–Maskawa matrix only comes into play through higher-order corrections and creates only a negligible effect.

Taking the $\mathcal{CP}$ violation to be a phase difference between an amplitude and its $\mathcal{CP}$ conjugate,[3] there are certain assumptions that have to be fulfilled to turn the phase difference into a difference in decay width between $\tau^+$ and $\tau^-$. There must be at least two different amplitudes contributing to the decay that interfere with each other. Each amplitude can be decomposed into a magnitude and two phases, one which changes sign under $\mathcal{CP}$ transformation ($\phi_i$) and a second phase which is $\mathcal{CP}$ invariant ($\Delta_i$):

$$A_i = |A_i|\, e^{i\phi_i}\, e^{i\Delta_i}. \tag{9.18}$$

The second $\mathcal{CP}$-invariant phase is created by final-state interactions between the outgoing particles in hadronic $\tau$ decays.[4] There have to be at least two different $\mathcal{CP}$ phases $\phi_i$ and two different strong phases $\Delta_i$ involved to create an observable rate difference between $\tau^+$ and $\tau^-$ [729, 730].

Several decay channels have been discussed [731–733]. The most promising seems to be the Cabibbo-suppressed $\tau \to K \pi \nu_\tau$ decay [730, 733]. There is a Standard Model vector amplitude dominated by the $K^*(892)$ resonance. There could be in addition a $\mathcal{CP}$-violating amplitude through a charged Higgs boson or a scalar leptoquark, which would probably be dominated by the $K_0^*(1430)$ meson. Then the $\mathcal{CP}$-violating phases would be different between the two amplitudes and the strong phases would be different owing to the different resonances, and a rate difference could be observed. Figure 9.8 shows the Feynman diagrams of the two interfering amplitudes.

A rate difference between the decays of $\tau^+$ and $\tau^-$ is the simplest, but not the only possibility for observing $\mathcal{CP}$ violation in $\tau$ decays. $\mathcal{CP}$-violating observables can also be constructed from the spin of the decaying $\tau$ lepton and its visible decay products [728, 729, 734, 735]. The spin of the decay products can be derived from the measured momenta of these particles. The spin of the initial $\tau$ lepton can be inferred either from spin correlations with the other $\tau$ lepton in the event or from the beam polarization, if the initial beams are polarized. This method not only is sensitive to the occurrence of $\mathcal{CP}$ violation, but can also shed light on its mechanisms, once $\mathcal{CP}$ violation is observed.

### 9.2.2 The First Measurement

The CLEO collaboration has published an initial investigation of $\mathcal{CP}$ violation in $\tau \to K \pi \nu_\tau$ [112] in the decay chain $\tau^- \to K_S^0 \pi^- \nu_\tau$, $K_S^0 \to \pi^+ \pi^-$

---
[3] Any other form would violate $\mathcal{CPT}$.
[4] Leptonic $\tau$ decays will not be discussed here (see [728]).

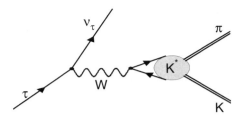

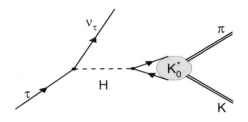

**Fig. 9.8.** Feynman diagrams of the decay $\tau \to K\pi\nu_\tau$. The *upper diagram* shows the Standard Model amplitude, the *lower* is an example of a possible new $\mathcal{CP}$-violating amplitude mediated by a charged Higgs boson

and its charge-conjugate decay. The events were selected in a 1–3 topology. Stringent cuts were applied to reduce the non-$\tau$ background – mainly two-photon events and continuum quark–antiquark production – to an acceptable level. Then the $K_S^0$ was identified by two tracks forming a secondary vertex well separated from the primary in the plane perpendicular to the beam. Events with additional showers above some energy thresholds were eliminated in order to reject events with additional $\pi^0$ or $K_L^0$ mesons. No attempt was made to reject $\tau \to K K_S^0 \nu_\tau$ events. Figure 9.9 shows the mass distribution of the $K_S^0$ candidates. The signal region shows the prominent $K_S^0$ mass peak. Out of these events 52.5% are $\tau \to \pi K_S^0 \nu_\tau$ and 12.4% $\tau \to K K_S^0 \nu_\tau$. The main $\tau$ background comes from $\tau \to 3\pi\nu_\tau$ with a fake secondary vertex. The three-pion channel also dominates the sidebands. See Table 9.2.

Instead of simply looking for a rate difference between $\tau^+ \to \pi^+ K_S^0 \bar{\nu}_\tau$ and $\tau^- \to \pi^- K_S^0 \nu_\tau$ CLEO has chosen to analyze an asymmetry in the decay angles proposed in [730]:

$$A_{\mathcal{CP}} = \frac{N^+(\cos\beta\cos\psi) - N^-(\cos\beta\cos\psi)}{N^+(\cos\beta\cos\psi) + N^-(\cos\beta\cos\psi)}. \tag{9.19}$$

The angle $\beta$ has been defined in Sect. 4.5.2 (4.29), and $\psi$ in Sect. 7.2 (7.14). $\psi$ is closely related to the angle $\theta^*$ from (4.28) (see [730] for more details). $N^+$ indicates the number of decays observed from the positive $\tau$ with $\cos\beta\cos\psi$ in a certain interval, and $N^-$ indicates the corresponding number from the negative $\tau$. The interval chosen was simply the region where $\cos\beta\cos\psi$ was either positive or negative.

Table 9.3 gives the resulting asymmetries. They are somewhat different from zero for both intervals, and the sidebands show a similar asymmetry.

## 9.2 $\mathcal{CP}$ Violation in $\tau$ Decays

**Table 9.2.** Signal and sideband mode composition: $f_{\text{mode}}^{\text{signal,sideband}}$ is the fraction of the total signal or sideband sample for a particular mode; $\alpha$ is the approximate magnitude of asymmetry expected relative to the $\tau \to \pi\, K_S^0\, \nu_\tau$ mode. The last column gives the dilution factor expected in the asymmetry when the measured asymmetry in the sideband control sample is subtracted from the measured asymmetry in the signal sample. CLEO collaboration, see [112] for details

| Tau mode | $\alpha$ | $f_{\text{mode}}^{\text{signal}}$ | $f_{\text{mode}}^{\text{sideband}}$ | $(f_{\text{mode}}^{\text{signal}} - f_{\text{mode}}^{\text{sideband}})\alpha$ |
|---|---|---|---|---|
| $K_S^0(\pi^+\pi^-)\pi^-\nu_\tau$ | 1 | $0.525 \pm 0.057$ | $0.043 \pm 0.005$ | $0.4820 \pm 0.0570$ |
| $K_S^0 K^- \nu_\tau$ | $1/20$ | $0.124 \pm 0.036$ | $0.009 \pm 0.003$ | $0.0060 \pm 0.0020$ |
| $a_1^- \nu_\tau$ | $1/80$ | $0.106 \pm 0.003$ | $0.620 \pm 0.013$ | $-0.0064 \pm 0.0002$ |
| $K_S^0 \pi^- \pi^0 \nu_\tau$ | $1/4$ | $0.066 \pm 0.016$ | $0.006 \pm 0.002$ | $0.0150 \pm 0.0040$ |
| $K_S^0 K_L^0 \pi^- \nu_\tau$ | $1/80$ | $0.055 \pm 0.018$ | $0.003 \pm 0.001$ | $0.0007 \pm 0.0002$ |
| $K_S^0 K^- \pi^0 \nu_\tau$ | $1/20$ | $0.030 \pm 0.008$ | $0.003 \pm 0.001$ | $0.0014 \pm 0.0004$ |
| $\pi^+\pi^-\pi^-\pi^0\nu_\tau$ | $1/20$ | $0.028 \pm 0.002$ | $0.167 \pm 0.007$ | $-0.0070 \pm 0.0004$ |
| $K^-\pi^+\pi^-\nu_\tau$ | $1/4$ | $0.008 \pm 0.003$ | $0.043 \pm 0.007$ | $-0.0090 \pm 0.0020$ |
| Others | 0 | $0.012 \pm 0.002$ | $0.071 \pm 0.017$ | 0 |
| $q\bar{q}$ | 0 | $0.044 \pm 0.003$ | $0.037 \pm 0.003$ | 0 |
| Total | – | $1.00 \pm 0.07$ | $1.00 \pm 0.00$ | $0.48 \pm 0.06$ |

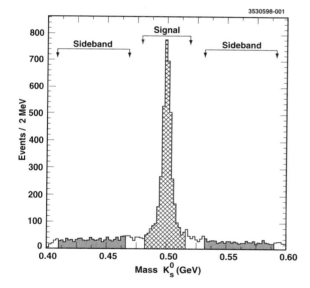

**Fig. 9.9.** Distribution of invariant mass of $K_S^0 \to \pi^+ \pi^-$ candidates (CLEO collaboration [112])

This indicates a detector effect. It probably originates from low-momentum pions which interact differently with the detector depending on whether they are positive or negative. Several cross-checks were made to support this interpretation. The asymmetry determined from the sidebands was finally subtracted from the signal region to correct for the detector effect.

**Table 9.3.** Observed asymmetries in signal and sideband regions from the CLEO analysis [112]

| $A_{\mathcal{CP}}$ | $\cos\beta\cos\psi < 0$ | $\cos\beta\cos\psi > 0$ |
|---|---|---|
| Signal | $0.058 \pm 0.023$ | $0.024 \pm 0.021$ |
| Sideband | $0.049 \pm 0.030$ | $0.034 \pm 0.033$ |

The sensitivity of the measurement was determined from a Monte Carlo simulation [283, 297], which was modified to include a $\mathcal{CP}$-violating scalar coupling. Two models were used for the corresponding hadronic current, either a full dominance of the $K_0^*(1430)$ meson or a pure phase space distribution. It turns out that the sensitivity is lower with the pure phase space distribution, so that it is conservative to use this model. The pure sensitivity is diluted by about 50% owing to the background and the necessity to subtract the asymmetry of the sidebands (see Table 9.2).

No $\mathcal{CP}$ violation was observed. The final result was expressed as an upper limit on the strength of a scalar boson exchange with $\mathcal{CP}$-violating couplings with respect to the strength of the Standard Model couplings. This limit is

$$g_{\mathcal{CP}} \sin\theta_{\mathcal{CP}} < 1.7 \quad (95\% \text{ confidence level}), \qquad (9.20)$$

where $\theta_{\mathcal{CP}}$ is the $\mathcal{CP}$-violating phase relative to the Standard Model.

# 10. Rare and Forbidden Decays

This chapter deals with the experimental searches for certain rare or forbidden decays of the $\tau$ lepton. The first section describes the search for second-class currents, which are expected from the Standard Model at a very low level and have not been observed so far. The second section summarizes the limits on decay modes violating the conservation laws of lepton or baryon numbers. The last section collects experimental results associated with leptons of a new, fourth generation and a composite structure of the $\tau$ lepton.

## 10.1 Second-Class Currents

The hadronic currents in $\tau$ decays can be separated into first- and second-class currents according to their $G$ parity (see Sect. 5.1.9, [449]). In the Standard Model the first class ($J^{PG} = 0^{++}, 0^{--}, 1^{+-}, 1^{-+}$) is expected to dominate $\tau$ decays, with a tiny second class ($J^{PG} = 0^{+-}, 0^{-+}, 1^{++}, 1^{--}$) which vanishes in the limit of perfect isospin symmetry.[1]

There have been many searches for second-class currents in nuclear $\beta$ decays and muon capture by nuclei (see e.g. [736–739]), but the expected effects are small and difficult to extract. No clear signal has been seen yet. Conversely, it was pointed out soon after the discovery of the $\tau$ lepton that the $\tau$ lepton provides a clean way to look for second-class currents [740, 741]. There are two interesting decay modes,

$$\tau \to a_0\,(980)\,\nu_\tau \to \eta\,\pi\,\nu_\tau,$$
$$\tau \to b_1\,(1235)\,\nu_\tau \to \omega\,\pi\,\nu_\tau,$$

with quantum numbers $J^{PG} = 0^{+-}$ and $1^{++}$, respectively. The $\eta\pi$ final state is of particular interest as it can only be produced through a second-class current, no matter what the intermediate resonance or the angular momentum between the $\eta$ and the $\pi$ is. This is different for the $\omega\pi$ final state, where there is a first-class contribution associated with the $\omega\pi$ p wave. The second-class current proceeds through an s or a d wave.

In the Standard Model the decay constants associated with the second-class currents are proportional to the mass difference between the up and

---
[1] These currents can be found in Cabibbo-suppressed decay modes.

270  10. Rare and Forbidden Decays

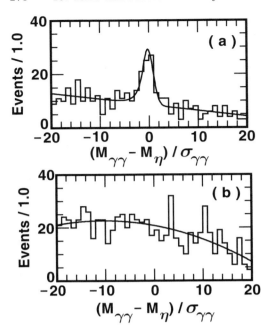

Fig. 10.1. The invariant-mass spectrum of $\eta$ candidates in $\tau \to$ had $\eta\, \nu_\tau$ (CLEO collaboration [126]). Each $\eta$ candidate is accompanied by a charged kaon in (a) and by a pion in (b). For details see [126]

down quarks and therefore small. Precise estimates of the branching ratios are difficult, but their order of magnitude can be given. They are expected to be suppressed by a factor of $10^{-4}$ with respect to $\tau \to \pi\, \nu_\tau$ [742–751].

In 1987 the HRS collaboration published a branching ratio of $(5.1 \pm 1.5)\,\%$ for the decay $\tau \to \eta\, \pi\, \nu_\tau$, clearly in excess of the Standard Model prediction [452], but it could not be explained by non-Standard Model contributions either [446, 745, 748]. The result was contradicted by more precise measurements shortly after [89, 202, 752]

The CLEO collaboration has produced the most stringent limit on the decay $\tau \to \eta\, \pi\, \nu_\tau$ to date [126]. They have searched for the $\eta$ meson in its decay to two photons, which amounts to about 40% of its total branching ratio. They selected events with a charged hadron and two photons in one hemisphere and a lepton (electron or muon) on the opposite side. The event was excluded if there was an additional photon with more than 100 MeV reconstructed. The restriction to the lepton tags is necessary to reject background from $e^+e^- \to q\bar{q}$. The $dE/dx$ and time-of-flight information was used to separate the remaining events into $\eta K$ and $\eta\pi$. In order to suppress the background further, each photon was required to have at least 150 MeV of energy; to have a lateral shower profile consistent with a photon, unlikely to be a fragment of a nearby shower; and not to combine to a $\pi^0$ with any other photon in the event. Figure 10.1 shows the spectrum of the invariant mass of the photon pairs of the selected events. It is plotted as a deviation

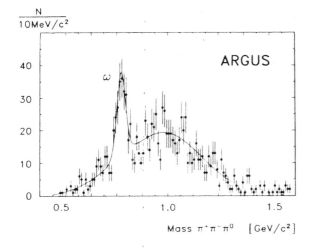

Fig. 10.2. Distribution of the invariant mass of the $\pi^-\pi^+\pi^0$ subsystem in $\tau \to 4\pi\nu_\tau$ events from ARGUS [94]. There are two entries per event

from the mass of the $\eta$ in standard deviations, where the error is calculated from the measured angle and energy resolution of each photon. The upper plot shows the photon pairs accompanied by a charged kaon. An $\eta$ signal is clearly visible. The CLEO collaboration extracted a branching ratio of $(2.6 \pm 0.5 \pm 0.4) \times 10^{-4}$, with the first error statistical and the second systematic (mainly from $q\bar{q}$ background). This is the first observation of this decay mode. The lower plot shows the photon pairs associated with a charged pion. There is no signal visible and CLEO derives a limit (95% confidence level) of

$$br\left(\tau \to \eta\,\pi\,\nu_\tau\right) < 1.4 \times 10^{-4}. \tag{10.1}$$

The ARGUS collaboration was the first to identify the $\tau \to \omega\,\pi\,\nu_\tau$ decay mode and study the angular distribution for second-class currents [94]. They selected $\tau^- \to \pi^-\,\pi^-\,\pi^+\,\pi^0\,\nu_\tau$ events and identified the $\omega$ in its dominating decay $\omega \to \pi^+\,\pi^-\,\pi^0$. The mass spectrum of the two possible neutral combinations of three pions in these events is shown in Fig. 10.2. They found 146 events in the $\omega$ peak and calculated a branching ratio of $(1.5 \pm 0.3 \pm 0.3)\,\%$. To separate any contributions from a second-class current the angular momentum between the $\omega$ and the pion has to be determined. For $l = 1$ the total quantum numbers are $J^{PG} = 1^{-+}$, i.e. an ordinary-vector current, whereas for $l = 0, 2$ one has a second-class axial-vector current $1^{++}$. The angular distribution of the pion in the $\omega$ rest frame reveals the difference. Figure 10.3 shows the distribution of the cosine of the angle of the pion to the normal of the decay plane of the $\omega$. The events are consistent with a pure vector current and no evidence for a second-class current is found. A similar, more precise analysis of this decay can be found in [55].

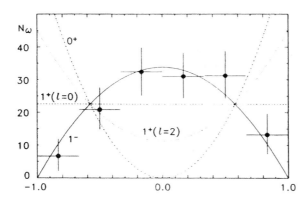

**Fig. 10.3.** The cosine of the angle between the direction of the pion and the normal of the decay plane of the $\omega$ in the $\omega$ rest frame in $\tau \to 4\pi\nu_\tau$ measured by the ARGUS collaboration [94]. Also shown is the expected behavior for the states $J^P = 0^+, 1^-$ (first class) and $1^+$ (second class)

## 10.2 Forbidden $\tau$ Decays

In the Standard Model the lepton number, i.e. the difference between the number of leptons and the number of antileptons, is conserved for each generation separately. However, there is no fundamental motivation for this lepton flavor conservation in this theory. There is no symmetry associated with the lepton family number. Many extensions of the Standard Model naturally introduce lepton flavor violations. Among them are models that involve heavy neutral leptons [753–759], left–right symmetries [760–762], supersymmetry [763–770], or superstrings [771–774]. The expected branching ratios depend on the parameters of these models and are typically very small, no larger than $10^{-5}$. For a review on the topic see [775–779].

Figure 10.4 shows three examples of Feynman diagrams for the decay $\tau \to \mu\gamma$. In Fig. 10.4a the $\tau$ neutrino emitted at the first vertex oscillates into a $\mu$ neutrino and is reabsorbed at the second vertex, leading to a $\tau \to \mu$ transition [753]. A photon is radiated to bring the virtual muon onto the mass shell. Replacing the photon by a $Z^0$ boson generates a number of other lepton-flavor-violating decays such as $\tau \to \mu$ e e or $\tau \to \mu\pi^0$, with the $Z^0$ boson decaying to an $e^+e^-$ pair or a $\pi^0$.

In the light of the results of the Super-KAMIOKANDE experiment, claiming evidence for oscillations of atmospheric $\mu$ neutrinos [597], if these are indeed $\nu_\mu \leftrightarrow \nu_\tau$ oscillations,[2] lepton-flavor-violating $\tau$ decays are no longer speculations about new physics. They have to happen at some level of branching ratio [780–782]. At what level depends on the model: their observation could be around the corner, but if there are no new particles and it is merely the case that the Standard Model neutrinos gain mass, then there could be a

---

[2] Currently it is unclear whether the deficit seen by KAMIOKANDE is due to $\nu_\mu \to \nu_\tau$ transitions or $\nu_\mu \to \nu_s$, where $\nu_s$ is a new, sterile neutrino. There seem to be experimental weak hints from neutral-current reactions supporting the first interpretation.

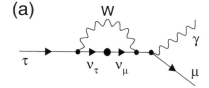
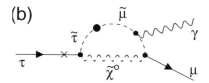
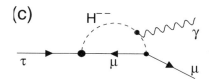

Fig. 10.4. Feynman diagrams of the lepton-flavor-violating decay $\tau \to \mu\,\gamma$ introduced by (**a**) a heavy neutrino [753], (**b**) mixing in the SUSY slepton sector [756], and (**c**) a doubly charged Higgs boson in a left–right symmetric model [760]

GIM-type (Glashow–Iliopoulos–Miani) mechanism which makes these rates as low as $10^{-40}$ [783–786].

The second diagram of Fig. 10.4 shows a lepton flavor violation introduced through a loop correction from flavor-nondiagonal couplings in the slepton sector of a SUSY model [756]. The decay of the third diagram comes from a flavor-blind, doubly charged Higgs boson in a left–right symmetric model [760].

Constraints on lepton number violation come from various reactions, in particular from neutrinoless muon decays: $br\,(\mu \to e\,\gamma) < 4.9 \times 10^{-11}$, $br\,(\mu \to e\,2\,\gamma) < 7.2 \times 10^{-11}$, and $br\,(\mu \to e\,e\,e) < 1.0 \times 10^{-12}$. But the rates might depend on the mass and the generation number of the decaying lepton, thus enhancing $\tau$ lepton decay rates. Also, the larger mass of the $\tau$ allows for new decay channels which are kinematically forbidden for the muon.

Lepton number violations through neutrinoless $\tau$ decays are the decay modes which are experimentally accessible, because they leave a very distinct signal in the detector: now that there are no more unobserved particles in the decay, the invariant mass of the decay products equals the $\tau$ mass and their energy adds up to the beam energy. The experimental sensitivity is driven by the number of $\tau$ leptons available. Therefore the CLEO experiment at the CESR storage ring is the best place to study these decays and the CLEO collaboration has indeed studied a large number of different neutrinoless decay modes [114, 119, 122] (see Table 10.1).

As an example, their search for the decay $\tau \to e\,\gamma$ will be discussed (for details see [122]). CLEO selected a sample of isolated electrons accompanied by a single photon recoiling against a one-prong $\tau$ decay. The main back-

274    10. Rare and Forbidden Decays

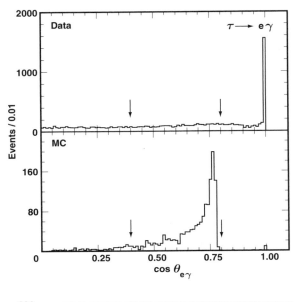

**Fig. 10.5.** Distribution of the cosine of the angle between the electron and the photon in $\tau \to e\gamma$. CLEO collaboration [122]. *Upper plot*, data, *lower plot* simulation of the signal. The *arrows* indicate the cuts

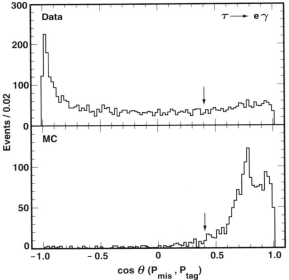

**Fig. 10.6.** The cosine of the angle between the missing momentum and the tagging track in the data (*upper plot*) and in a Monte Carlo simulation of the signal (*lower plot*). CLEO collaboration [122]. The region below 0.4 was rejected

ground comes from radiative Bhabha scattering where the photon and the electron accidentally make a $\tau$ mass. To reduce this background they rejected events where both tracks were identified as electrons. They made use of the special kinematics of the events searched for to suppress background further. Figure 10.5 shows the cosine of the angle between the photon and the elec-

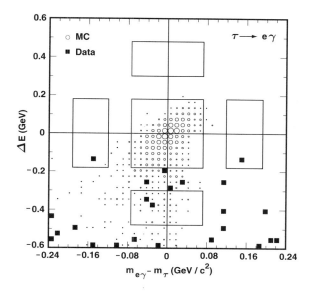

**Fig. 10.7.** The distribution of $\Delta E = E_{e\gamma} - E_{\text{beam}}$ versus $\Delta m = m_{e\gamma} - m_\tau$ for the final events in the search for $\tau \to e\gamma$ from CLEO [122]. The *square* in the *center* of the plot indicates the signal region. The four *rectangles* were used to estimate the background (sidebands). The *solid squares* represent the data; the *open circles* show a Monte Carlo simulation of the signal

tron for the data and a Monte Carlo simulation of the signal. Owing to the two-body character of the decay, the photon is separated from the electron by at least 0.2 in $\cos\theta_{e\gamma}$ in the signal. The arrows indicate the window accepted for further analysis. Even with no neutrino in $\tau \to e\gamma$ there is still a neutrino produced by the other $\tau$, which creates a missing momentum roughly in the direction of the recoiling track. This distinguishes the signal from Bhabha scattering and two-photon processes. Figure 10.6 shows the distribution for the data and a simulated signal, together with the cut.

After a few more cuts, the invariant mass and energy of the electron–photon pairs were plotted against each other (see Fig. 10.7). A signal would manifest itself as an enhancement of events at the beam energy and the $\tau$ mass. The resolution in these quantities was determined from Monte Carlo simulation and the number of events within 2.5 standard deviations of the signal was counted. The CLEO collaboration found no events and derived an upper limit on the branching ratio of

$$br\,(\tau \to e\,\gamma) < 2.7 \times 10^{-6} \quad \text{at the 90\% confidence level.} \tag{10.2}$$

Apart from the neutrinoless decay modes mentioned in Table 10.1, the ARGUS collaboration has searched for decays which violate baryon number conservation in addition [85]. They found the following limits (90% confidence level):

$$\begin{aligned}
br\,(\tau^- \to \overline{p}\,\gamma) &< 2.9 \times 10^{-4}, \\
br\,(\tau^- \to \overline{p}\,\pi^0) &< 6.6 \times 10^{-4}, \\
br\,(\tau^- \to \overline{p}\,\eta) &< 1.3 \times 10^{-3}.
\end{aligned}$$

## 10. Rare and Forbidden Decays

**Table 10.1.** Limits on lepton-number-violating, neutrinoless decay modes (90% confidence level) [44]

| Mode | Limit | |
|---|---|---|
| $e^- \gamma$ | $< 2.7 \times 10^{-6}$ | [122] |
| $\mu^- \gamma$ | $< 3.0 \times 10^{-6}$ | [122] |
| $e^- \pi^0$ | $< 3.7 \times 10^{-6}$ | [119] |
| $\mu^- \pi^0$ | $< 4.0 \times 10^{-6}$ | [119] |
| $e^- K^0$ | $< 1.3 \times 10^{-3}$ | [197] |
| $\mu^- K^0$ | $< 1.0 \times 10^{-3}$ | [197] |
| $e^- \eta$ | $< 8.2 \times 10^{-6}$ | [119] |
| $\mu^- \eta$ | $< 9.6 \times 10^{-6}$ | [119] |
| $e^- \rho^0$ | $< 2.0 \times 10^{-6}$ | [114] |
| $\mu^- \rho^0$ | $< 6.3 \times 10^{-6}$ | [114] |
| $e^- K^*$ | $< 5.1 \times 10^{-6}$ | [114] |
| $\mu^- K^*$ | $< 7.5 \times 10^{-6}$ | [114] |
| $e^- \overline{K}^*$ | $< 7.4 \times 10^{-6}$ | [114] |
| $\mu^- \overline{K}^*$ | $< 7.5 \times 10^{-6}$ | [114] |
| $e^- \phi$ | $< 6.9 \times 10^{-6}$ | [114] |
| $\mu^- \phi$ | $< 7.0 \times 10^{-6}$ | [114] |
| $e^- e^+ e^-$ | $< 2.9 \times 10^{-6}$ | [114] |
| $e^- \mu^+ \mu^-$ | $< 1.8 \times 10^{-6}$ | [114] |
| $e^+ \mu^- \mu^-$ | $< 1.5 \times 10^{-6}$ | [114] |
| $\mu^- e^+ e^-$ | $< 1.7 \times 10^{-6}$ | [114] |
| $\mu^+ e^- e^-$ | $< 1.5 \times 10^{-6}$ | [114] |
| $\mu^- \mu^+ \mu^-$ | $< 1.9 \times 10^{-6}$ | [114] |
| $e^- \pi^+ \pi^-$ | $< 2.2 \times 10^{-6}$ | [114] |
| $e^+ \pi^- \pi^-$ | $< 1.9 \times 10^{-6}$ | [114] |
| $\mu^- \pi^+ \pi^-$ | $< 8.2 \times 10^{-6}$ | [114] |
| $\mu^+ \pi^- \pi^-$ | $< 3.4 \times 10^{-6}$ | [114] |
| $e^- \pi^+ K^-$ | $< 6.4 \times 10^{-6}$ | [114] |
| $e^- \pi^- K^+$ | $< 3.8 \times 10^{-6}$ | [114] |
| $e^+ \pi^- K^-$ | $< 2.1 \times 10^{-6}$ | [114] |
| $e^- K^+ K^-$ | $< 6.0 \times 10^{-6}$ | [114] |
| $e^+ K^- K^-$ | $< 3.8 \times 10^{-6}$ | [114] |
| $\mu^- \pi^+ K^-$ | $< 7.5 \times 10^{-6}$ | [114] |
| $\mu^- \pi^- K^+$ | $< 7.4 \times 10^{-6}$ | [114] |
| $\mu^+ \pi^- K^-$ | $< 7.0 \times 10^{-6}$ | [114] |
| $\mu^- K^+ K^-$ | $< 1.5 \times 10^{-5}$ | [114] |
| $\mu^+ K^- K^-$ | $< 6.0 \times 10^{-6}$ | [114] |
| $e^- \pi^0 \pi^0$ | $< 6.5 \times 10^{-6}$ | [119] |
| $\mu^- \pi^0 \pi^0$ | $< 1.4 \times 10^{-5}$ | [119] |
| $e^- \eta \eta$ | $< 3.5 \times 10^{-5}$ | [119] |
| $\mu^- \eta \eta$ | $< 6.0 \times 10^{-5}$ | [119] |
| $e^- \pi^0 \eta$ | $< 2.4 \times 10^{-5}$ | [119] |
| $\mu^- \pi^0 \eta$ | $< 2.2 \times 10^{-5}$ | [119] |

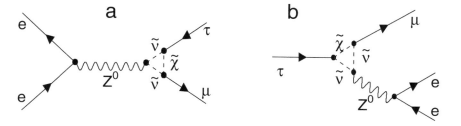

**Fig. 10.8.** Feynman diagram of a lepton-flavor-violating decay of a $Z^0$ boson through a sneutrino loop in a SUSY model (**a**) and the same vertex in a lepton-flavor-violating $\tau$ decay (**b**).

## 10.3 Flavor-Changing Neutral Currents

Lepton-flavor-violating decays of the $Z^0$ boson are intimately linked to the neutrinoless $\tau$ decays discussed in the preceding section. Figure 10.8 is intended to illustrate the relation. In diagram a it shows the decay of a $Z^0$ into a pair of sneutrinos in a SUSY model, which are converted into two regular leptons through the exchange of a chargino. The sneutrinos are allowed to have flavor-nondiagonal couplings and therefore the two leptons in the final state can be of different generations [787]. Diagram b is exactly the same process but with the in- and out-going leptons rearranged to form a lepton-flavor-violating $\tau$ decay. (For reviews see [775, 788].)

Lepton-flavor-violating $Z^0$ decays can be generated by tree-level couplings to new bosons [789, 790] or be introduced through flavor-nondiagonal effects in loop corrections [787, 791–797]. The branching ratios for $Z^0 \to \tau\mu$, $Z^0 \to \tau\,\mathrm{e}$, and $Z^0 \to \mu\,\mathrm{e}$ expected from such models might be as large as $10^{-12}$ to $10^{-6}$.

From the experimental point of view the search for $Z^0 \to \mu\,\mathrm{e}$ is very simple. One expects an electron with the full beam energy recoiling against a muon with the same energy. Such events can be selected with high efficiency and almost no background and the limit is driven by the number of $Z^0$ bosons available. The search for $Z^0 \to \tau\mu$ and $Z^0 \to \tau\mathrm{e}$ is more difficult, because there is a Standard Model background. To simplify the notation, only $Z^0 \to \tau\,\mathrm{e}$ will be discussed here. The decay $Z^0 \to \tau\mu$ is similar. Again one expects a $\tau$ lepton with missing momentum due to undetected neutrinos in the decay, recoiling against an electron with the full beam energy. But in the Standard Model it can happen that in a $\tau$ pair one of the two $\tau$ leptons decays through $\tau \to \mathrm{e}\,\nu_e\nu_\tau$ and the neutrinos take only a negligible amount of energy. Then one sees an electron with the full beam energy recoiling against a typical $\tau$. Such decays at the endpoint of the spectrum of $\tau \to \mathrm{e}\,\nu_e\nu_\tau$ are kinematically suppressed, but not forbidden. Therefore, if one looks at the spectrum of electrons recoiling against a $\tau$, one expects to see the Standard Model shape with a small peak right at $E_\mathrm{e} = E_\mathrm{beam}$, if there are lepton-flavor-violating

278    10. Rare and Forbidden Decays

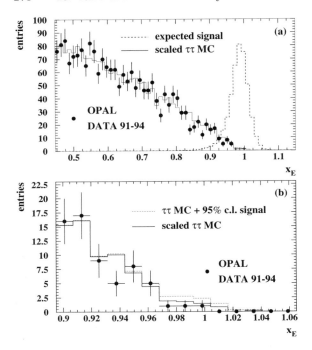

Fig. 10.9. Distribution of energy normalized to $E_{\text{beam}}$ of electron candidates in the search for $Z^0 \to \tau\,e$ from the OPAL collaboration [798]. The *points with error bars* are data; the *solid histogram* is the Standard Model Monte Carlo simulation; the *dashed histogram* is the expected shape of a signal (derived from $e^+e^- \to e^+e^-$ events in the data)

$Z^0$ decays (see Fig. 10.9 for an illustration). The limit will depend crucially on the momentum resolution of the detector. The smaller one can make the window of the signal around $E_{\text{beam}}$, the less background from the Standard Model one gets. Another important aspect is the rejection of background from $Z^0 \to e^+e^-$. If one of the electrons is misidentified as a pion this looks like a perfect signal event.

Figure 10.9 shows the spectrum of electrons recoiling against a $\tau$ from data obtained by the OPAL collaboration [798]. In Fig. 10.9a one sees the spectrum with the Standard Model background from $\tau \to e\,\nu_e\nu_\tau$ as a solid histogram, including $4.1 \pm 1.7$ events expected from $Z^0 \to e^+e^-$. The dashed histogram indicates the shape of a signal, with arbitrary normalization. The peak is shifted slightly below 1 owing to radiation. Figure 10.9b shows the region of the signal on an enlarged scale. No signal is found; the shape of the signal is now normalized to the number of events representing the 95% confidence-level upper limit.

Searches for lepton-flavor-violating $Z^0$ decays have been performed by all four LEP collaborations [798–801], UA1 [575], and, at lower energies, by MARK II [802]. The limits are summarized in Table 10.2.

**Table 10.2.** Limits on lepton-flavor-violating decays of the $Z^0$ boson in units of $10^{-6}$ at the 95% confidence level

|         |       | $\mu$ e | $\tau$ e | $\tau \mu$ |
|---------|-------|---------|----------|------------|
| ALEPH   | [799] | < 26    | < 120    | < 100      |
| DELPHI  | [800] | < 2.5   | < 22     | < 12       |
| L3      | [801] | < 6.0   | < 13     | < 19       |
| OPAL    | [798] | < 1.7   | < 9.8    | < 17       |

## 10.4 Excited Leptons

One approach to the question of the origin of the mass of the fundamental particles is to postulate the existence of a further layer of substructure in nature such that the quarks and leptons, and possibly the gauge bosons, are composite objects. The fundamental particles at this deeper level are referred to as preons. But at present no satisfactory preon model that explains the observed spectrum of quarks, leptons, and gauge bosons exists.

The most striking evidence of compositeness would be the discovery of excited states of quarks or leptons. In this section the search for excited leptons ($\ell^*$) will be briefly described. For reviews and phenomenological models see [803–809].

As the spectrum of the excited fermions depends crucially on the dynamics of the model in question, the lowest-lying state can have various quantum numbers. The experimental searches usually assume spin 1/2 and isospin 0 or 1/2. There are theoretical arguments [805] for assuming that the excited fermions have vector-type couplings. The alternative is to assume a chirality structure identical to that of the ground-state fermions. This alternative gives a lower production cross section and therefore more conservative limits. Experimentalists usually follow this path.

Excited fermions couple to the gauge bosons. The strength is assumed to be that of the ground-state fermions. They also couple to the ground states through an effective magnetic-dipole coupling. The relevant piece of the Lagrangian is

$$\mathcal{L}_{\text{eff}} = \sum_{V=\gamma, Z^0, W^\pm} \frac{e}{2\Lambda} \bar{\Psi}_{\ell^*} \sigma^{\mu\nu} \left( c_V - d_V \, \gamma_5 \right) \Psi_\ell \partial_\mu V_\nu. \quad (10.3)$$

Here $c_V$ and $d_V$ are arbitrary couplings and $\Lambda$ is the composite mass scale. The precise measurements of $g-2$ of the muon impose $|c_V| = |d_V|$ and the absence of the electric dipole moments of the fermions suggests that both $c_V$ and $d_V$ are real. Assuming further that the excited charged lepton forms a weak isospin doublet with the excited neutrino, we have $c_V = d_V$ and

$$c_\gamma = t_3 \, f + \frac{Y}{2} \, f',$$

$$c_{Z^0} = t_3 \, f \, \cot \theta_W - \frac{Y}{2} \, f' \tan \theta_W,$$

280   10. Rare and Forbidden Decays

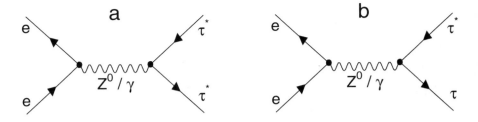

**Fig. 10.10.** Feynman diagrams for the production of (**a**) a pair of and (**b**) a single excited $\tau$ lepton

**Table 10.3.** Branching ratios of the decay of excited charged leptons in % for two masses of the excited state: $m_{\ell^*} = 80$ and $160$ GeV

| Decay channel | 80 GeV | | 160 GeV | |
|---|---|---|---|---|
| | $f=f'$ | $f=-f'$ | $f=f'$ | $f=-f'$ |
| $\ell^* \to \ell\,\gamma$ | 100.0 | 0.0 | 38.6 | 0.0 |
| $\ell^* \to \ell\, Z^0$ | 0.0 | 0.0 | 8.4 | 35.3 |
| $\ell^* \to \nu_\ell\, W^\pm$ | 0.0 | 100.0 | 53.0 | 64.7 |

$$c_{W^\pm} = \frac{f}{\sqrt{2}\,\sin\theta_W}, \qquad (10.4)$$

where $t_3$ is the third component of the weak isospin, $Y$ the hypercharge, and $\theta_W$ the Weinberg angle; $f$ and $f'$ are unknown couplings.

At $e^+e^-$ colliders excited leptons can be produced either in pairs or as a single state through the Feynman diagrams[3] of Fig. 10.10. For pair production, neglecting the anomalous magnetic moments of the excited states, the cross section does not depend on the composite scale. In the absence of a signal, limits can be set directly on the masses of the excited states. The pair production is kinematically restricted to masses below half the center-of-mass energy. For single production the kinematic reach extends to almost the full center-of-mass energy. The cross section depends on the couplings $f$ and $f'$. A negative search result implies limits on these couplings as a function of the mass of the excited state.

Excited leptons decay instantaneously by radiating either a photon, a $Z^0$ boson, or a $W^\pm$. The branching ratios are functions of the couplings $f$ and $f'$ and therefore depend on the model. For masses of the excited state below $m_{W^\pm}$ the decay $\ell^* \to \ell\gamma$ should be dominant, except for $f = -f'$, where it is forbidden. Table 10.3 gives the branching ratios for $m_{\ell^*} = 80$ and $160$ GeV. The experimental signature of the production of excited leptons is a pair of standard leptons accompanied by one or two isolated bosons. The

---

[3] For the first generation there is also a t-channel contribution.

spectrum of the invariant mass between the daughter lepton and the boson should exhibit a sharp peak at the mass of the excited state.

All four LEP collaborations [799, 810–816] and also TOPAZ and AMY at TRISTAN [817, 818] have searched for excited leptons. The most sensitive limits come from the runs at the highest center-of-mass energies [814–816].

The OPAL collaboration has searched for single and pair production of excited leptons in the data taken during the 1996 run at center-of-mass energies of 161, 170, and 172 GeV [816]. They considered the decay modes $\ell^* \to \ell\,\gamma$ and $\ell^* \to \nu_\ell\,W^\pm$ with the $W^\pm$ decaying either leptonically or into quarks.

For the photonic decays, two identified leptons of the same flavor and one or two photons with at least 1 GeV of energy were required in the event. The main background comes from lepton pair production with final-state radiation. Therefore the photon was required to be isolated from both leptons by $\cos\theta_{\ell\gamma} < 0.95$ (0.90 for the $e^*$ search). The momenta of the leptons and photons were recalculated from their scattering angles while applying the beam constraint, conserving energy and momentum. In the case of the $\tau^*$ search this compensates for the missing neutrinos in the $\tau$ decays.

Then the invariant masses of the lepton–photon pairs were calculated. The distributions from the search for the production of a single excited muon or $\tau$ lepton are shown in Fig. 10.11. The dashed histogram indicates the peak expected from a signal for $m_{\ell^*} = 140$ GeV. There are two entries per event. The proper combination makes the peak at 140 GeV; the other combination causes the kinematic reflection at 60 to 80 GeV. The resolution in the $\tau$ channel is reduced owing to the missing neutrinos.

No evidence for excited leptons was found. Figure 10.12 shows the exclusion plot for the couplings $f/\Lambda$ as a function of the mass of the excited state. Below 84.6 and 81.3 GeV excited $\tau$ leptons are excluded at any value of the coupling, from the results for pair production followed by either photonic or charged decays. Above these masses single production limits the couplings up to masses close to the center-of-mass energy.

In the context of a composite model there has to be some new interaction at the composite scale binding the preons into quarks and leptons. Remnants of this constituent-binding interaction appear at low energy as an effective interaction, called a contact force. It could manifest itself in small deviations of cross sections and angular distributions from the Standard Model predictions. The reaction $e^+e^- \to \tau^+\tau^-$ and many others have been studied experimentally [74, 819–821]. This offers an alternate possibility for searching for compositeness.

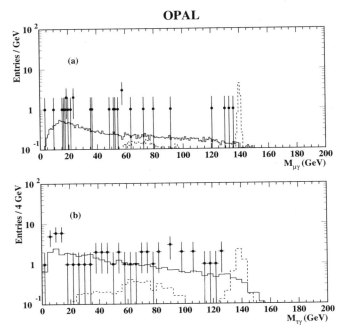

**Fig. 10.11.** Search for single production of excited leptons with photonic decays by the OPAL collaboration [816]: $\ell^\pm\gamma$ invariant-mass distribution after all cuts, from $\mu^+\mu^-\gamma$ events (**a**) and $\tau^+\tau^-\gamma$ (**b**). *Points with error bars*: data. *Solid histogram*: Standard Model background. *Dashed histogram*: Monte Carlo signal with $f/\Lambda = (100 \text{ GeV})^{-1}$. There are two entries per event

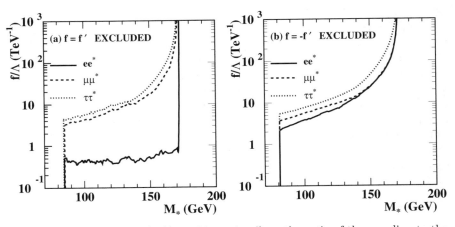

**Fig. 10.12.** Upper limits (95% confidence level) on the ratio of the coupling to the compositeness scale, $f/\Lambda$, as a function of the excited-lepton mass (OPAL [816]). The regions *above* and to the *left* of the *curves* are excluded

## 10.5 New Heavy Leptons

Although this is not truly physics with $\tau$ leptons, a section about searches for even heavier new leptons is added here.

In spite of its remarkable success, the Standard Model leaves many questions unanswered. In particular, it does not explain the number of fermion generations or the fermion mass spectrum. The precision measurements of the $Z^0$ boson at LEP-I and SLC have determined the number of species of light neutrinos to be three [44]. However, this does not exclude a fourth generation, or other more exotic fermions, if all of the new particles have masses greater than $m_Z/2$. (For a review see [822–826].)

There are many models introducing new heavy leptons [41, 702, 827–839]. The most important variations are:

- A new generation of sequential leptons, all heavier than $m_Z/2$, and otherwise identical to the known generations.
- Mirror fermions with chiralities opposite to those of the Standard Model are predicted by left–right symmetric models. There would be a right-handed doublet with a charged lepton and a neutrino plus a left-handed charged singlet.
- Just a singlet neutrino, as predicted in many models of grand unification.

Such new leptons, if they exist, are pair-produced in $e^+e^-$ collisions through a virtual photon or a $Z^0$ boson: $e^+e^- \to L^+L^-$ or $e^+e^- \to L^0\overline{L^0}$. A new charged lepton is denoted by $L^\pm$ and the neutrino by $L^0$. The couplings to the photon and the $Z^0$ are assumed to be identical to those of the Standard Model leptons. This limits the reach of a search to half the center-of-mass energy and therefore the most stringent limits come from the running of LEP-II at the highest energies [816, 840]. Close to the production threshold the cross sections are different for Dirac and Majorana leptons.[4] The threshold behavior for both types is shown in Fig. 10.13. It is in favor of a search for the Dirac type neutrinos.

There are several possible ways for these new leptons to decay:

(a)    $L^0 \to L^\pm W^\pm$,
(b)    $L^0 \to \ell^\pm W^\pm$,
(c)    $L^0 \to \nu_\ell Z^0$;

and, for the charged lepton:

(a)    $L^\pm \to L^0 W^\pm$,
(b)    $L^\pm \to \nu_\ell W^\pm$,
(c)    $L^\pm \to \ell^\pm Z^0$.

---
[4] A new charged lepton, of course, can only be of Dirac type.

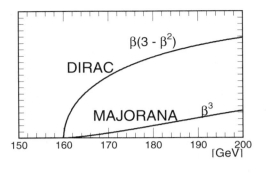

**Fig. 10.13.** The threshold behavior of the cross section for the production of a pair of Dirac or Majorana neutrinos of mass 80 GeV

Decay mode (a) conserves lepton flavor and corresponds to the decays of the known leptons. Either the charged or the neutral lepton can decay through this mode, but not both, depending on which of the two is heavier. The decay modes (b) are introduced through mixing with the known generations and depend on the mixing angles. If mode (a) is inaccessible to the new neutrino and the mixing angle in (b) is very small, then the neutrino becomes stable despite its large mass. In that situation it will be invisible to the experiments.[5] Decay mode (c) is a flavor-changing neutral current and might not be present at all. For new leptons with masses below 80 GeV the $W^{\pm}$ and $Z^0$ bosons are virtual and the decays exhibit three-body kinematics. At higher lepton masses there is enough energy to render the bosons on-shell, resulting in two-body kinematics.

The experiments at LEP-II aimed to cover decay modes (a) and (b) for both charged and neutral new heavy leptons [816, 840]. The search topology was two jets or a lepton from each of the $W^{\pm}$ bosons, plus two isolated leptons or missing momentum. They also searched for pair production of stable charged leptons which left two highly ionizing tracks in the tracking chambers. No signal was found, and new leptons of Dirac type can be excluded up to masses of roughly 80 GeV and a Majorana neutrino can be excluded up to approximately 60 GeV.

---

[5] For a neutrino of mass 80 GeV the decay length exceeds 1 cm for mixing angles below $|U|^2 = \mathcal{O}(10^{-12})$.

# 11. Summary and Outlook

The $\tau$ lepton was discovered in 1974 from a handful of e–$\mu$ events. In the 25 years following the discovery, $\tau$ physics has matured into a field of high precision, testing many aspects of our current understanding of particle physics:

- The coupling of the $\tau$ lepton to the charged weak current has been tested at the level of a few parts per thousand. Its couplings to the weak neutral current have been measured with similar precision.
- Substantial contributions to the precision tests of the electroweak theory at LEP and SLC have been derived from $\tau$ production.
- The coupling to the photon has been investigated. No anomalies have been found so far.
- More than 100 different, mainly hadronic, branching ratios have been measured, allowing tests of QCD and many model predictions.
- The spectral functions have been determined. They allow one to study perturbative QCD at low $Q^2$ and provide one of the most precise measurements of the strong coupling constant $\alpha_s$.
- The structure of the charged current in $\tau$ decays has been examined for deviations from 'V–A'. These investigations are sensitive to many kinds of new physics.
- $\mathcal{CP}$ violation in the production or decay of $\tau$ leptons is a very interesting question which has been analyzed in a search for a deeper understanding of the origin of $\mathcal{CP}$ violation.
- In the light of neutrino oscillations, lepton-flavor-violating $\tau$ decays are expected to occur at some level. The experimental sensitivity has now reached $10^{-6}$ without uncovering a signal.
- The searches for a finite mass of the $\tau$ neutrino are approaching 10 MeV, thereby closing an interesting astrophysical window.

The last few years of $\tau$ physics have been dominated by the results of the LEP experiments and CLEO II. There have also been important contributions from SLD and the new measurement of the $\tau$ mass from BES. But all of these experiments have finished data-taking, at least at energies interesting for $\tau$ physics. The next generation of results will have to come from different experiments.

The CLEO collaboration at present has the largest sample of $\tau$ leptons and they are upgrading to even higher luminosities with CLEO III. The asymmetric b factories PEP-II and KEK-B with the BaBar and BELLE experiments will go into operation this year. Each of them will collect a sample of up to $10^8$ $\tau$ pairs, which will be the basis of $\tau$ physics for the coming years.

$\tau$ leptons will be even more abundant at the LHC, the Large Hadron Collider under construction at CERN. However, it is not obvious how to trigger and analyze these events in the harsh environment of multi-TeV hadron collisions.

A dedicated machine for $\tau$, charm, and two-photon physics has been under discussion for some time. The so-called tau–charm factory would provide $\tau$ samples of similar size as the b factories, but with much improved systematic limitations. Where and whether at all such a machine will be built is still undecided.

In any case there is a rich future of interesting results ahead of us. After all, a big surprise might just be around the corner ...

# References

1. M. L. Perl, Rep. Prog. Phys. **55** (1992) 653.
2. B. C. Barish and R. Stroynowski, Phys. Rep. **157** (1988) 1.
3. A. J. Weinstein and R. Stroynowski, Ann. Rev. Nucl. Part. Sci. **43** (1993) 457.
4. A. Pich, Mod. Phys. Lett. **A5** (1990) 1995.
5. A. Pich, Tau physics, in *Heavy Flavours II*, edited by A. Buras and M. Lindner, World Scientific, Singapore, 1997.
6. A. Golutvin, *Tau Decays*, DESY-94-168, 1994.
7. S. Gentile and M. Pohl, Phys. Rep. **274** (1996) 287.
8. G. Rahal-Callot, Int. J. Mod. Phys. A **13** (1998) 695.
9. M. Davier and B. Jean-Marie, editors, *Proceedings of the Workshop on Tau Lepton Physics*, Orsay, France, 1990.
10. K. K. Ghan, editor, *Proceedings of the 2nd Workshop on Tau Lepton Physics*, Columbus, Ohio, 1992.
11. L. Rolandi, editor, *Proceedings of the 3rd Workshop on Tau Lepton Physics*, volume 40 of Nucl. Phys. B (Proc. Suppl.), Montreux, Switzerland, 1994.
12. J. G. Smith and W. Toki, editors, *Proceedings of the 4th Workshop on Tau Lepton Physics*, volume 55C of Nucl. Phys. B (Proc. Suppl.), Estes Park, Colorado, 1996.
13. A. Pich and A. Ruiz, editors, *Proceedings of the 5th Workshop on Tau Lepton Physics*, volume 76 of Nucl. Phys. B (Proc. Suppl.), Santander, Spain, 1998.
14. W. T. Toner et al., Phys. Lett. B **36** (1972) 251.
15. T. Braunstein et al., Phys. Rev. D **6** (1972) 106.
16. A. Barna et al., Phys. Rev. **173** (1968) 1391.
17. S. Orito et al., Phys. Lett. B **48** (1974) 165.
18. M. L. Perl et al., Phys. Rev. Lett. **35** (1975) 1489.
19. J. E. Augustin et al., Phys. Rev. Lett. **34** (1975) 233.
20. D. Ritson et al., *Proposal for a High-Energy Electron–Positron Colliding-Beam Storage Ring at the Stanford Linear Accelerator Center*, 1964.
21. H. B. Thacker and J. J. Sakurai, Phys. Lett. B **36** (1971) 103.
22. Y.-S. Tsai and A. C. Hearn, Phys. Rev. **140** (1965) B721.
23. Y.-S. Tsai, Phys. Rev. D **4** (1971) 2821, erratum, Phys. Rev. D **13** (1976) 771.
24. R. M. Larsen et al., SLAC proposal SP-2, 1971.
25. M. L. Perl et al., Phys. Lett. B **63** (1976) 466.
26. J. Burmester et al., Phys. Lett. B **68** (1977) 297.
27. J. Burmester et al., Phys. Lett. B **68** (1977) 301.
28. R. Brandelik et al., Phys. Lett. B **70** (1977) 125.
29. G. J. Feldman, The discovery of the tau, in *Proceedings of the International Symposium on Lepton and Photon Interactions at High Energy*, page 39, Stanford, California, 1975.

30. J. J. Aubert et al., Phys. Rev. Lett. **33** (1974) 1404.
31. J. E. Augustin et al., Phys. Rev. Lett. **33** (1974) 1406.
32. M. L. Perl et al., Phys. Lett. B **70** (1977) 487.
33. J. D. Bjorken and S. L. Glashow, Phys. Lett. **11** (1964) 255.
34. C. Bouchiat, J. Iliopoulos, and P. Meyer, Phys. Lett. B **38** (1972) 519.
35. D. J. Gross and R. Jackiw, Phys. Rev. D **6** (1972) 477.
36. S. L. Glashow, J. Iliopoulos, and L. Maiani, Phys. Rev. D **2** (1970) 1285.
37. J. E. Augustin et al., Phys. Rev. Lett. **34** (1975) 764.
38. W. Bacino et al., Phys. Rev. Lett. **41** (1978) 13.
39. C. H. Llewellyn-Smith, Proc. Roy. Soc. A **355** (1977) 585.
40. A. Ali and T. C. Yang, Phys. Rev. D **14** (1976) 3052.
41. J. D. Bjorken and C. H. Llewellyn-Smith, Phys. Rev. D **7** (1973) 887.
42. P. Annis et al., Phys. Lett. B **435** (1998) 458.
43. E. Eskut et al., Phys. Lett. B **434** (1998) 205.
44. C. Caso et al., Euro. Phys. J. C **3** (1998) 1.
45. E872 collaboration, Fermilab, see their web page at http://fn872.fnal.gov/.
46. T. Kafka, E872: the direct observation of the $\nu_\tau$, in *Proceedings of the 5th International Workshop on Topics in Astroparticle and Underground Physics*, edited by A. Bottino, A. di Credico, and P. Monacelli, volume 70 of Nucl. Phys. B (Proc. Suppl.), page 204, Gran Sasso, Italy, 1999.
47. R. Barate et al., Euro. Phys. J. C **4** (1998) 409.
48. R. Barate et al., Euro. Phys. J. C **4** (1998) 29.
49. R. Barate et al., Euro. Phys. J. C **2** (1998) 395.
50. R. Barate et al., Phys. Lett. B **414** (1997) 362.
51. R. Barate et al., Euro. Phys. J. C **1** (1998) 65.
52. R. Barate et al., Phys. Lett. B **405** (1997) 191.
53. R. Barate et al., Z. Phys. C **74** (1997) 387.
54. R. Barate et al., Z. Phys. C **76** (1997) 15.
55. D. Buskulic et al., Z. Phys. C **74** (1997) 263.
56. D. Buskulic et al., Z. Phys. C **70** (1996) 579.
57. D. Buskulic et al., Z. Phys. C **70** (1996) 561.
58. D. Buskulic et al., Z. Phys. C **70** (1996) 549.
59. D. Buskulic et al., Z. Phys. C **69** (1996) 183.
60. D. Buskulic et al., Phys. Lett. B **349** (1995) 585.
61. D. Buskulic et al., Phys. Lett. B **346** (1995) 379.
62. D. Buskulic et al., Phys. Lett. B **346** (1995) 371.
63. D. Buskulic et al., Phys. Lett. B **332** (1994) 219.
64. D. Buskulic et al., Phys. Lett. B **332** (1994) 209.
65. D. Buskulic et al., Z. Phys. C **62** (1994) 539.
66. D. Buskulic et al., Phys. Lett. B **321** (1994) 168.
67. D. Buskulic et al., Z. Phys. C **59** (1993) 369.
68. D. Buskulic et al., Phys. Lett. B **307** (1993) 209.
69. D. Buskulic et al., Phys. Lett. B **297** (1992) 432.
70. D. Buskulic et al., Phys. Lett. B **297** (1992) 459.
71. D. Decamp et al., Phys. Lett. B **279** (1992) 411.
72. D. Decamp et al., Z. Phys. C **54** (1992) 211.
73. D. Decamp et al., Phys. Lett. B **265** (1991) 430.
74. C. Velissaris et al., Phys. Lett. B **331** (1994) 227.
75. A. Bacala et al., Phys. Lett. B **218** (1989) 112.
76. H. Albrecht et al., Phys. Lett. B **431** (1998) 179.
77. H. Albrecht et al., Z. Phys. C **68** (1995) 25.
78. H. Albrecht et al., Phys. Lett. B **349** (1995) 576.
79. H. Albrecht et al., Phys. Lett. B **341** (1995) 441.

80. H. Albrecht et al., Phys. Lett. B **337** (1994) 383.
81. H. Albrecht et al., Phys. Lett. B **316** (1993) 608.
82. H. Albrecht et al., Z. Phys. C **58** (1993) 61.
83. H. Albrecht et al., Z. Phys. C **56** (1992) 339.
84. H. Albrecht et al., Phys. Lett. B **292** (1992) 221.
85. H. Albrecht et al., Z. Phys. C **55** (1992) 179.
86. H. Albrecht et al., Phys. Lett. B **260** (1991) 259.
87. H. Albrecht et al., Phys. Lett. B **250** (1990) 164.
88. H. Albrecht et al., Phys. Lett. B **246** (1990) 278.
89. H. Albrecht et al., Z. Phys. C **41** (1988) 405.
90. H. Albrecht et al., Z. Phys. C **41** (1988) 1.
91. H. Albrecht et al., Phys. Lett. B **202** (1988) 149.
92. H. Albrecht et al., Phys. Lett. B **199** (1987) 580.
93. H. Albrecht et al., Phys. Lett. B **195** (1987) 307.
94. H. Albrecht et al., Phys. Lett. B **185** (1987) 223.
95. H. Albrecht et al., Phys. Lett. B **185** (1987) 228.
96. H. Albrecht et al., Z. Phys. C **33** (1986) 7.
97. H. Albrecht et al., Phys. Lett. B **163** (1985) 404.
98. J. Z. Bai et al., Phys. Rev. D **53** (1996) 20.
99. J. Z. Bai et al., Phys. Rev. Lett. **69** (1992) 3021.
100. D. Antreasyan et al., Phys. Lett. B **259** (1991) 216.
101. H. Janssen et al., Phys. Lett. B **228** (1989) 273.
102. S. Keh et al., Phys. Lett. B **212** (1988) 123.
103. H. J. Behrend et al., Z. Phys. C **46** (1990) 537.
104. H. J. Behrend et al., Phys. Lett. B **222** (1989) 163.
105. H. J. Behrend et al., Phys. Lett. B **200** (1988) 226.
106. H. J. Behrend et al., Z. Phys. C **23** (1984) 103.
107. H. J. Behrend et al., Phys. Lett. B **127** (1983) 270.
108. H. J. Behrend et al., Nucl. Phys. B **211** (1983) 369.
109. H. J. Behrend et al., Phys. Lett. B **114** (1982) 282.
110. S. J. Richichi et al., *Study of Three Prong Hadronic Tau Decays with Charged Kaons*, preprints hep-ex/9810026, CLNS-98-1573, submitted to Phys. Rev. D, 1999.
111. M. Bishai et al., Phys. Rev. Lett. **82** (1999) 281.
112. S. Anderson et al., Phys. Rev. Lett. **81** (1998) 3823.
113. R. Ammar et al., Phys. Lett. B **431** (1998) 209.
114. D. W. Bliss et al., Phys. Rev. D **57** (1998) 5903.
115. K. W. Edwards et al., Phys. Rev. D **56** (1997) 5297.
116. S. Anderson et al., Phys. Rev. Lett. **79** (1997) 3814.
117. T. Bergfeld et al., Phys. Rev. Lett. **79** (1997) 2406.
118. J. P. Alexander et al., Phys. Rev. D **56** (1997) 5320.
119. G. Bonvicini et al., Phys. Rev. Lett. **79** (1997) 1221.
120. T. E. Coan et al., Phys. Rev. D **55** (1997) 7291.
121. R. Ammar et al., Phys. Rev. Lett. **78** (1997) 4686.
122. K. W. Edwards et al., Phys. Rev. D **55** (1997) 3919.
123. A. Anastassov et al., Phys. Rev. D **55** (1997) 2559.
124. P. Avery et al., Phys. Rev. D **55** (1997) 1119.
125. R. Balest et al., Phys. Lett. B **388** (1996) 402.
126. J. Bartelt et al., Phys. Rev. Lett. **76** (1996) 4119.
127. T. E. Coan et al., Phys. Rev. D **53** (1996) 6037.
128. M. S. Alam et al., Phys. Rev. Lett. **76** (1996) 2637.
129. R. Balest et al., Phys. Rev. Lett. **75** (1995) 3809.
130. T. Coan et al., Phys. Lett. B **356** (1995) 580.

131. J. Bartelt et al., Phys. Rev. Lett. **73** (1994) 1890.
132. D. Gibaut et al., Phys. Rev. Lett. **73** (1994) 934.
133. M. Artuso et al., Phys. Rev. Lett. **72** (1994) 3762.
134. M. Battle et al., Phys. Rev. Lett. **73** (1994) 1079.
135. D. Bortoletto et al., Phys. Rev. Lett. **71** (1993) 1791.
136. D. Cinabro et al., Phys. Rev. Lett. **70** (1993) 3700.
137. R. Ballest et al., Phys. Rev. D **47** (1993) 3671.
138. A. Bean et al., Phys. Rev. Lett. **70** (1993) 138.
139. M. Procario et al., Phys. Rev. Lett. **70** (1993) 1207.
140. D. S. Akerib et al., Phys. Rev. Lett. **69** (1992) 3610.
141. M. Battle et al., Phys. Lett. B **291** (1992) 488.
142. M. Artuso et al., Phys. Rev. Lett. **69** (1992) 3278.
143. R. Ammar et al., Phys. Rev. D **45** (1992) 3976.
144. C. Bebek et al., Phys. Rev. D **36** (1987) 690.
145. R. Brandelik et al., Phys. Lett. B **73** (1978) 109.
146. W. Ruckstuhl et al., Phys. Rev. Lett. **56** (1986) 2132.
147. G. B. Mills et al., Phys. Rev. Lett. **54** (1985) 624.
148. G. B. Mills et al., Phys. Rev. Lett. **52** (1984) 1944.
149. W. Bacino et al., Phys. Rev. Lett. **42** (1979) 749.
150. P. Abreu et al., Phys. Lett. B **426** (1998) 411.
151. P. Abreu et al., Phys. Lett. B **404** (1997) 194.
152. P. Abreu et al., Phys. Lett. B **365** (1996) 448.
153. P. Abreu et al., Phys. Lett. B **359** (1995) 411.
154. P. Abreu et al., Phys. Lett. B **357** (1995) 715.
155. P. Abreu et al., Z. Phys. C **67** (1995) 183.
156. P. Abreu et al., Nucl. Phys. B **418** (1994) 403.
157. P. Abreu et al., Phys. Lett. B **302** (1993) 356.
158. P. Abreu et al., Z. Phys. C **55** (1992) 555.
159. P. Abreu et al., Phys. Lett. B **267** (1991) 422.
160. S. Abachi et al., Phys. Lett. B **226** (1989) 405.
161. S. Abachi et al., Phys. Rev. D **40** (1989) 902.
162. K. Sugano, Nucl. Phys. A **478** (1988) 729C.
163. S. Abachi et al., Phys. Rev. Lett. **59** (1987) 2519.
164. S. Abachi et al., Phys. Rev. Lett. **56** (1986) 1039.
165. S. Hegner et al., Z. Phys. C **46** (1990) 547.
166. C. Kleinwort et al., Z. Phys. C **42** (1989) 7.
167. W. Bartel et al., Phys. Lett. B **182** (1986) 216.
168. W. Bartel et al., Z. Phys. C **31** (1986) 359.
169. W. Bartel et al., Phys. Lett. B **161** (1985) 188.
170. M. Acciarri et al., Phys. Lett. B **438** (1998) 405.
171. M. Acciarri et al., Phys. Lett. B **434** (1998) 169.
172. M. Acciarri et al., Phys. Lett. B **426** (1998) 207.
173. M. Acciarri et al., Phys. Lett. B **429** (1998) 387.
174. M. Acciarri et al., Phys. Lett. B **389** (1996) 187.
175. M. Acciarri et al., Phys. Lett. B **377** (1996) 313.
176. M. Acciarri et al., Phys. Lett. B **352** (1995) 487.
177. M. Acciarri et al., Phys. Lett. B **345** (1995) 93.
178. M. Acciarri et al., Phys. Lett. B **341** (1994) 245.
179. O. Adriani et al., Phys. Lett. B **294** (1992) 466.
180. B. Adeva et al., Phys. Lett. B **265** (1991) 451.
181. H. Band et al., Phys. Lett. B **198** (1987) 297.
182. H. Band et al., Phys. Rev. Lett. **59** (1987) 415.
183. E. Fernandez et al., Phys. Rev. Lett. **54** (1985) 1624.

184. D. Y. Wu et al., Phys. Rev. D **41** (1990) 2339.
185. D. Amidei et al., Phys. Rev. D **37** (1988) 1750.
186. K. K. Ghan et al., Phys. Lett. B **197** (1987) 561.
187. K. K. Ghan et al., Phys. Rev. Lett. **59** (1987) 411.
188. P. Burchat et al., Phys. Rev. D **35** (1987) 27.
189. W. B. Schmidke et al., Phys. Rev. Lett. **57** (1986) 527.
190. J. M. Yelton et al., Phys. Rev. Lett. **56** (1986) 812.
191. P. Burchat et al., Phys. Rev. Lett. **54** (1985) 2489.
192. C. Matteuzzi et al., Phys. Rev. D **32** (1985) 800.
193. J. Jaros et al., Phys. Rev. Lett. **51** (1983) 955.
194. C. Matteuzzi et al., Phys. Rev. Lett. **52** (1984) 1869.
195. C. A. Blocker et al., Phys. Rev. Lett. **49** (1982) 1369.
196. C. A. Blocker et al., Phys. Rev. Lett. **48** (1982) 1586.
197. K. G. Hayes et al., Phys. Rev. D **25** (1982) 2869.
198. G. J. Feldman et al., Phys. Rev. Lett. **48** (1982) 66.
199. C. A. Blocker et al., Phys. Lett. B **109** (1982) 119.
200. J. Dorfan et al., Phys. Rev. Lett. **46** (1981) 215.
201. G. S. Abrams et al., Phys. Rev. Lett. **43** (1979) 1555.
202. D. Coffman et al., Phys. Rev. D **36** (1987) 2185.
203. R. M. Baltrusaitis et al., Phys. Rev. Lett. **55** (1985) 1842.
204. B. Adeva et al., Phys. Rev. D **38** (1988) 2665.
205. B. Adeva et al., Phys. Lett. B **179** (1986) 177.
206. B. Adeva et al., Phys. Rep. **109** (1984) 131.
207. D. P. Barber et al., Phys. Rev. Lett. **46** (1981) 1663.
208. D. P. Barber et al., Phys. Lett. B **95** (1980) 149.
209. G. Abbiendi et al., Phys. Lett. B **447** (1999) 134.
210. K. Ackerstaff et al., Euro. Phys. J. C **8** (1999) 3.
211. K. Ackerstaff et al., Euro. Phys. J. C **7** (1999) 571.
212. K. Ackerstaff et al., Euro. Phys. J. C **8** (1999) 183.
213. K. Ackerstaff et al., Euro. Phys. J. C **5** (1998) 229.
214. K. Ackerstaff et al., Phys. Lett. B **431** (1998) 188.
215. K. Ackerstaff et al., Euro. Phys. J. C **4** (1998) 193.
216. K. Ackerstaff et al., Phys. Lett. B **404** (1997) 213.
217. K. Ackerstaff et al., Z. Phys. C **75** (1997) 593.
218. K. Ackerstaff et al., Z. Phys. C **74** (1997) 403.
219. G. Alexander et al., Phys. Lett. B **388** (1996) 437.
220. G. Alexander et al., Z. Phys. C **72** (1996) 365.
221. G. Alexander et al., Z. Phys. C **72** (1996) 231.
222. G. Alexander et al., Phys. Lett. B **374** (1996) 341.
223. G. Alexander et al., Phys. Lett. B **368** (1996) 244.
224. G. Alexander et al., Phys. Lett. B **369** (1996) 163.
225. R. Akers et al., Z. Phys. C **68** (1995) 555.
226. R. Akers et al., Z. Phys. C **67** (1995) 45.
227. R. Akers et al., Z. Phys. C **66** (1995) 543.
228. R. Akers et al., Z. Phys. C **66** (1995) 31.
229. R. Akers et al., Phys. Lett. B **338** (1994) 497.
230. R. Akers et al., Z. Phys. C **65** (1995) 1.
231. R. Akers et al., Z. Phys. C **65** (1995) 183.
232. R. Akers et al., Phys. Lett. B **328** (1994) 207.
233. R. Akers et al., Phys. Lett. B **339** (1994) 278.
234. R. Akers et al., Z. Phys. C **61** (1994) 19.
235. P. D. Acton et al., Z. Phys. C **59** (1993) 183.
236. P. D. Acton et al., Phys. Lett. B **288** (1992) 373.

237. P. D. Acton et al., Phys. Lett. B **281** (1992) 405.
238. P. D. Acton et al., Phys. Lett. B **273** (1991) 355.
239. G. Alexander et al., Phys. Lett. B **266** (1991) 201.
240. C. Berger et al., Z. Phys. C **C28** (1985) 1.
241. W. Wagner et al., Z. Phys. C **3** (1980) 193.
242. G. Alexander et al., Phys. Lett. B **81** (1979) 84.
243. G. Alexander et al., Phys. Lett. B **78** (1978) 162.
244. G. Alexander et al., Phys. Lett. B **73** (1978) 99.
245. K. Abe et al., Phys. Rev. Lett. **79** (1997) 804.
246. K. Abe et al., Phys. Rev. Lett. **78** (1997) 4691.
247. K. Abe et al., Phys. Rev. D **52** (1995) 4828.
248. W. Braunschweig et al., Z. Phys. C **43** (1989) 549.
249. W. Braunschweig et al., Z. Phys. C **39** (1988) 331.
250. M. Althoff et al., Z. Phys. C **26** (1985) 521.
251. M. Althoff et al., Phys. Lett. B **141** (1984) 264.
252. R. Brandelik et al., Phys. Lett. B **110** (1982) 173.
253. R. Brandelik et al., Phys. Lett. B **92** (1980) 199.
254. B. Howell et al., Phys. Lett. B **291** (1992) 206.
255. I. Adachi et al., Phys. Lett. B **208** (1988) 319.
256. D. A. Bauer et al., Phys. Rev. D **50** (1994) R13.
257. H. Aihara et al., Phys. Rev. Lett. **59** (1987) 751.
258. H. Aihara et al., Phys. Rev. D **35** (1987) 1553.
259. H. Aihara et al., Phys. Rev. Lett. **57** (1986) 1836.
260. H. Aihara et al., Phys. Rev. D **30** (1984) 2436.
261. H. Hanai et al., Phys. Lett. B **403** (1997) 155.
262. K. Abe et al., Z. Phys. C **48** (1990) 13.
263. J. H. Kühn, Phys. Lett. B **313** (1993) 458.
264. D. Buskulic et al., Measurement of the tau polarisation at LEP I, preliminary result, *International Conference on High Energy Physics*, July 1998, Vancouver, Canada (Abstract 939).
265. A. Stahl, *Untersuchung myonischer $\tau$ Zerfälle mit dem ALEPH Detektor am LEP*, PhD thesis, Institut für Hochenergiephysik, Universität Heidelberg, 1991, EX HD IHEP 91-10.
266. G. G. G. Massaro, Test of electroweak interactions at LEP, in *Proceedings of the 2nd Workshop on Tau Lepton Physics*, edited by K. K. Ghan, pages 387–400, Columbus, Ohio, 1992.
267. J. Alcaraz, Tau leptonic branching fractions from L3, in *Proceedings of the 3rd Workshop on Tau Lepton Physics*, edited by L. Rolandi, volume 40 of Nucl. Phys. B (Proc. Suppl.), pages 237–246, Montreux, Switzerland, 1994.
268. P. Watkins, Measurement of the tau leptonic branching ratios by the OPAL experiment, in *Proceedings of the 3rd Workshop on Tau Lepton Physics*, edited by L. Rolandi, volume 40 of Nucl. Phys. B (Proc. Suppl.), pages 247–254, Montreux, Switzerland, 1994.
269. C. Kiesling, $\tau$ decays: the solution to a long standing problem?, in *Proceedings of the 24th Rencontre de Moriond: Electroweak Interactions and Unified Theories*, edited by J. Trah Thanh Van, pages 323–338, Les Arcs, France, 1989.
270. D. Buskulic et al., Nucl. Instrum. Meth. **A360** (1995) 481.
271. D. Karlen, Comp. in Phys. **12** (1998) 380, preprint physics/9805018.
272. L. Lonnblad, C. Peterson, and T. Rognvaldsson, Comp. Phys. Comm. **70** (1992) 167.

273. C. Peterson and T. Rognvaldsson, *An Introduction to Artificial Neural Networks*, lectures given at 1991 CERN School of Computing, Ystad, Sweden, Aug. 23 – Sep. 2, 1991.
274. B. Müller, M. T. Strickland, and J. E. Reinhardt, *Neural Networks: An Introduction*, Springer, Berlin, Heidelberg, 2nd edition, 1995.
275. P. D. Wasserman, *Neural Computing, Theory and Practice*, Van Nostrand Reinhold, New York, 1989.
276. J. Hertz, A. Krogh, and R. Palmer, *Introduction to the Theory of Neural Computation*, Addison-Wesley, Reading, Massachusetts, 1991.
277. C. M. Bishop, *Neural Networks for Pattern Recognition*, Clarendon Press, Oxford, 1995.
278. C. Peterson, T. Rognvaldsson, and L. Lonnblad, Comp. Phys. Comm. **81** (1994) 185.
279. R. Odorico, Comp. Phys. Comm. **96** (1996) 314.
280. V. Innocente, Y. F. Wang, and Z. P. Zhang, Nucl. Instrum. Meth. A **323** (1992) 647.
281. M. T. Ronan, Strange decays of the $\tau$ lepton, in *Proceedings of the 2nd Workshop on Tau Lepton Physics*, edited by K. K. Ghan, pages 225–237, Columbus, Ohio, 1992.
282. W. Ruckstuhl, Tau decays into kaons in DELPHI, in *Proceedings of the 2nd Workshop on Tau Lepton Physics*, edited by K. K. Ghan, volume 40 of Nucl. Phys. B (Proc. Suppl.), pages 371–380, Columbus, Ohio, 1994.
283. S. Jadach and Z. Wąs, Comp. Phys. Comm. **36** (1985) 191.
284. S. Jadach, B. F. L. Ward, and Z. Wąs, Comp. Phys. Comm. **66** (1991) 276.
285. S. Jadach, B. F. L. Ward, and Z. Wąs, Comp. Phys. Comm. **79** (1994) 503.
286. S. Jadach, M. Skrzypek, and B. F. L. Ward, Phys. Rev. D **55** (1997) 1206.
287. S. Jadach, E. Richter-Wąs, B. F. L. Ward, and Z. Wąs, Phys. Lett. B **353** (1995) 362.
288. B. F. L. Ward et al., Acta Phys. Polon. B **25** (1994) 245.
289. S. Jadach and B. F. L. Ward, Acta Phys. Polon. B **22** (1991) 229.
290. E. Barberio and Z. Wąs, Comp. Phys. Comm. **79** (1994) 291.
291. E. Barberio, B. van Eijk, and Z. Wąs, Comp. Phys. Comm. **66** (1991) 115.
292. R. K. F. A. Berends and S. Jadach, Comp. Phys. Comm. **29** (1983) 185.
293. D. Y. Bardin, M. S. Bilenkii, T. Riemann, M. Sachwitz, and H. Vogt, Comp. Phys. Comm. **59** (1990) 303.
294. B. A. Kniehl and R. G. Stuart, Comp. Phys. Comm. **72** (1992) 175.
295. S. Jadach, J. H. Kühn, and Z. Wąs, Comp. Phys. Comm. **64** (1991) 275.
296. M. Jezabek, Z. Wąs, S. Jadach, and J. H. Kühn, Comp. Phys. Comm. **70** (1992) 69.
297. S. Jadach, Z. Wąs, R. Decker, and J. H. Kühn, Comp. Phys. Comm. **76** (1993) 361.
298. F. A. Berends, R. Kleiss, S. Jadach, and Z. Wąs, Acta Phys. Polon. B **14** (1983) 413.
299. Z. Wąs and S. Jadach, TAUOLA: Monte Carlo for $\tau$ decays. A question of systematic errors, in *Proceedings of the 2nd Workshop on Tau Lepton Physics*, edited by K. K. Ghan, pages 37–58, Columbus, Ohio, 1992.
300. J. H. Kühn and A. Santamaria, Z. Phys. C **48** (1990) 445.
301. R. Fischer, J. Wess, and F. Wagner, Z. Phys. C **3** (1979) 313.
302. J. H. Kühn and F. Wagner, Nucl. Phys. B **236** (1984) 16.
303. R. Decker, E. Mirkes, R. Sauer, and Z. Wąs, Z. Phys. C **58** (1993) 445.
304. G. Kramer, W. F. Palmer, and S. S. Pinsky, Phys. Rev. D **30** (1984) 89.
305. G. Kramer and W. F. Palmer, Z. Phys. C **25** (1984) 195.
306. T. N. Pham, C. Roiesnel, and T. N. Truong, Phys. Lett. B **78** (1978) 623.

307. W. Bartel et al., Phys. Lett. B **77** (1978) 331.
308. J. M. Wu and P. Y. Zhao, Chinese J. High Energy Nucl. Phys. **17** (1993) 379.
309. J. P. Alexander, G. Bonvicini, P. Drell, R. Frey, and V. Luth, Nucl. Phys. B **320** (1989) 45.
310. M. Perrottet, An improved calculation of $\sigma(e^+e^- \to \tau^+\tau^-)$ near threshold, in *Proceedings of the 3rd Workshop on the Tau–Charm Factory*, edited by J. Kirkby and R. Kirkby, pages 89–95, Marbella, Spain, 1993.
311. M. B. Voloshin, Topics in tau physics at a tau charm factory, invited paper, *Tau Charm Factory Workshop*, SLAC, Stanford, California, May 1989; University of Minnesota, TPI-MINN-89/33-T.
312. F. A. Berends and G. J. Komen, Phys. Lett. B **63** (1976) 432.
313. E. A. Kuraev and V. S. Fadin, Sov. J. Nucl. Phys. **41** (1985) 466.
314. E. A. Kuraev and V. S. Fadin, *Calculation of Radiative Corrections to the Cross Section of One Photon Annihilation by Means of Structure Functions*, DESY L-Trans-297, 1985.
315. E. A. Kuraev and V. S. Fadin, Yad. Fiz. **41** (1985) 733.
316. K. Hikasa et al., Phys. Rev. D **45** (1992) S1.
317. P. Weber, Review of $\tau$ lifetime and leptonic branching ratio measurements, in *Proceedings of the 4th Workshop on Tau Lepton Physics*, edited by J. G. Smith and W. Toki, volume 55C of Nucl. Phys. B (Proc. Suppl.), pages 107–119, Estes Park, Colorado, 1996.
318. S. R. Wasserbach, Review of $\tau$ lifetime measurements, in *Proceedings of the 5th Workshop on Tau Lepton Physics*, edited by A. Pich and A. Ruiz, volume 76 of Nucl. Phys. B (Proc. Suppl.), pages 107–116, Santander, Spain, 1998.
319. A. P. Colijn, L3 measurement of the tau lepton lifetime, in *Proceedings of the 5th Workshop on Tau Lepton Physics*, edited by A. Pich and A. Ruiz, volume 76 of Nucl. Phys. B (Proc. Suppl.), pages 101–106, Santander, Spain, 1998.
320. K. Abe et al., Updated measurement of the tau lifetime at SLD, preliminary result, *International Conference on High Energy Physics*, July 1996, Warsaw, Poland (Abstract PA07-064), SLAC-PUB 7213.
321. A. Andreazza et al., A measurement of the tau lifetime, preliminary result, *International Conference on High Energy Physics*, August 1997, Jerusalem, Israel (Abstract 317; pa5,7 pl6).
322. A. Andreazza et al., An updated measurement of the tau lifetime, preliminary result, *5th Workshop on Tau Lepton Physics*, September 1998, Santander, Spain. DELPHI 98-153 CONF 207.
323. M. Aguilar-Benitez et al., Phys. Lett. B **170** (1986) 1.
324. G. P. Yost et al., Phys. Lett. B **204** (1988) 1.
325. J. J. Hernandez et al., Phys. Lett. B **239** (1990) 1.
326. L. Montanet et al., Phys. Rev. D **50** (1994) 1173.
327. R. M. Barnett et al., Phys. Rev. D **54** (1996) 1.
328. S. R. Wasserbaech, Phys. Rev. D **48** (1993) 4216.
329. W. Grimus, Fortschr. Phys. **36** (1988) 201.
330. W. Bernreuther and O. Nachtmann, Z. Phys. C **73** (1997) 647.
331. T. G. Rizzo, Phys. Rev. D **56** (1997) 3074.
332. M. T. Dova, J. Swain, and L. Taylor, Constraints on anomalous charged current couplings, tau neutrino mass and fourth generation mixing from tau leptonic branching fractions, in *Proceedings of the 5th Workshop on Tau Lepton Physics*, edited by A. Pich and A. Ruiz, volume 76 of Nucl. Phys. B (Proc. Suppl.), pages 133–138, Santander, Spain, 1998, eprint hep-ph/9811209.
333. M.-T. Dova, J. Swain, and L. Taylor, Phys. Rev. D **58** (1998) 015005.

334. M. A. Samuel, Guo-wen Li, and R. Mendel, Phys. Rev. Lett. **67** (1991) 668, erratum, Phys. Rev. Lett. **69** (1992) 995.
335. F. Hamzeh and N. F. Nasrallah, Phys. Lett. B **373** (1996) 211.
336. J. Bernabéu, G. González-Sprinberg, M. Tung, and J. Vidal, Nucl. Phys. B **436** (1995) 474.
337. I. B. Khriplovich, *Nuclear Anapole Moments*, Fundamental Symmetries in Nuclei and Particles, World Scientific, Singapore, 1989.
338. I. B. Khriplovich, *Parity Nonconservation in Atomic Phenomena*, Gordon and Breach, Philadelphia, 1991.
339. W. Bernreuther, U. Löw, J. P. Ma, and O. Nachtmann, Z. Phys. C **43** (1989) 117.
340. W. Bernreuther and O. Nachtmann, Phys. Lett. B **268** (1991) 424.
341. R. Escribano and E. Massó, Phys. Lett. B **301** (1993) 419.
342. J. A. Grifols and A. Méndez, Phys. Lett. B **255** (1991) 611, erratum, Phys. Lett. B **259** (1991) 512.
343. L. Taylor, Anomalous magnetic and electric dipole moments of the tau, in *Proceedings of the 5th Workshop on Tau Lepton Physics*, edited by A. Pich and A. Ruiz, volume 76 of Nucl. Phys. B (Proc. Suppl.), pages 237–244, Santander, Spain, 1998, eprint hep-ph/9810463.
344. D. Zeppenfeld, private communication with OPAL.
345. J. Biebel and T. Riemann, Z. Phys. C **76** (1997) 53.
346. S. S. Gau, T. Paul, J. Swain, and L. Taylor, Nucl. Phys. B **523** (1998) 439.
347. T. Paul and Z. Wąs, *Inclusion of Tau Anomalous Magnetic and Electric Dipole Moments in the KORALZ Monte Carlo*, L3 internal note 2184 and eprint hep-ph/9801301, 1998.
348. T. Paul, J. Swain, and Z. Was, *Library of Anomalous $\tau\tau\gamma$ Couplings for $\tau^+\tau^-(n\gamma)$ Monte Carlo Programs*, eprint hep-ph/9905207, 1999.
349. R. Escribano and E. Massó, Phys. Lett. B **395** (1997) 369.
350. J. Bernabéu, G. A. González-Sprinberg, and J. Vidal, Phys. Lett. B **326** (1994) 168.
351. U. Stiegler, Z. Phys. C **57** (1993) 511.
352. G. L. Shaw and F. Daghighian, Phys. Rev. D **26** (1982) 1798.
353. D. J. Silverman and G. L. Shaw, Phys. Rev. D **27** (1983) 1196.
354. J. Reid, M. Samuel, K. A. Milton, and T. G. Rizzo, Phys. Rev. D **30** (1984) 245.
355. G. Domokos, S. Kovesi-Domokos, C. Vaz, and D. Wurmser, Phys. Rev. D **32** (1985) 247.
356. G. Kopp, D. Schaile, M. Spira, and P. M. Zerwas, Z. Phys. C **65** (1995) 545.
357. A. Stahl, Michel parameters: averages and interpretation, in *Proceedings of the 5th Workshop on Tau Lepton Physics*, edited by A. Pich and A. Ruiz, volume 76 of Nucl. Phys. B (Proc. Suppl.), pages 173–181, Santander, Spain, 1998.
358. S. M. Berman, Phys. Rev. **112** (1958) 267.
359. T. Kinoshita and A. Sirlin, Phys. Rev. **113** (1959) 1652.
360. G. Källén, *Radiative Corrections in Elementary Particle Physics*, volume 46 of Springer Tracts in Mod. Phys. , Springer Verlag, Heidelberg, 1968.
361. M. Roos and A. Sirlin, Nucl. Phys. B **29** (1971) 296.
362. A. Sirlin, Rev. Mod. Phys. **50** (1978) 573.
363. W. J. Marciano and A. Sirlin, Phys. Rev. Lett. **61** (1988) 1815.
364. T. D. Lee and C. N. Yang, Phys. Rev. **108** (1957) 1611.
365. A. Stahl, Phys. Lett. B **324** (1994) 121.
366. J. Swain and L. Taylor, Phys. Rev. D **55** (1997) 1R.

## References

367. S. I. Eidelman and V. N. Ivanchenko, Present status of $e^+e^- \to$ hadrons, in *Proceedings of the 5th Workshop on Tau Lepton Physics*, edited by A. Pich and A. Ruiz, volume 76 of Nucl. Phys. B (Proc. Suppl.), pages 319–326, Santander, Spain, 1998.
368. S. Weinberg, Phys. Rev. Lett. **18** (1967) 507.
369. M. G. Bowler, Phys. Lett. B **182** (1986) 400.
370. M. G. Bowler, Phys. Lett. B **209** (1988) 99.
371. N. A. Törnqvist, Z. Phys. C **36** (1987) 695, erratum, Z. Phys. C **40** (1988) 632.
372. N. Isgur, C. Morningstar, and C. Reader, Phys. Rev. D **39** (1989) 1357.
373. V. S. M. T. Das and S. Okubo, Phys. Rev. Lett. **18** (1967) 761.
374. S. Matsuda and S. Oneda, Phys. Rev. Lett. **171** (1968) 1743.
375. T. N. Truong, Phys. Rev. D **30** (1984) 1509.
376. K. G. Hayes and M. L. Perl, Phys. Rev. D **38** (1988) 3351.
377. K. G. Hayes, M. L. Perl, and B. Efron, Phys. Rev. D **39** (1989) 274.
378. M. L. Perl, Phys. Rev. D **38** (1988) 845.
379. C. O. Escobar, O. L. G. Peres, V. Pleitez, and R. Z. Funchal, Europhys. Lett. **21** (1993) 169.
380. C. K. Jung, Phys. Rev. D **47** (1993) 3994.
381. W. J. Marciano, Phys. Rev. D **45** (1992) 721.
382. C. Kiesling, $\tau$ decays: an experimental review, in *Proceedings of the 1st Workshop on Tau Lepton Physics*, edited by M. Davier and B. Jean-Marie, pages 127–149, Orsay, France, 1990.
383. Z. Zhang, Measurement of $\tau$ branching ratios from ALEPH, in *Proceedings of the 1st Workshop on Tau Lepton Physics*, edited by M. Davier and B. Jean-Marie, pages 151–160, Orsay, France, 1990.
384. J. J. Hernández, Phys. Lett. B **239** (1990) 1.
385. D. Karlen, Experimental status of the Standard Model, in *Proceedings of the International Conference on High Energy Physics (ICHEP98)*, Vancouver, Canada, 1998.
386. W. Hollik, Standard Model theory, in *Proceedings of the International Conference on High Energy Physics (ICHEP98)*, Vancouver, Canada, 1998, eprint hep-ph/9811313.
387. P. Langacker, editor, *Precision Tests of the Standard Electroweak Model*, volume 14 of Advanced Series on Directions in High Energy Physics, World Scientific, Singapore, 1995.
388. M. Martinez, R. Miquel, L. Rolandi, and R. Tenchini, Rev. Mod. Phys. **71** (1999) 575.
389. W. Hollik, J. Phys. G **23** (1997) 1503.
390. J. Mnich, Phys. Rep. **271** (1996) 181.
391. D. Schaile, Fortschr. Phys. **42** (1994) 429.
392. F. Jegerlehner, Prog. Nucl. Part. Phys. **27** (1991) 1.
393. W. F. L. Hollik, Fortschr. Phys. **38** (1990) 165.
394. G. Altarelli, *The Standard Electroweak Theory and Beyond*, eprint hep-ph/9811456, 1998.
395. G. Altarelli, R. Barbieri, and F. Caravaglios, Int. J. Mod. Phys. A **13** (1998) 1031.
396. G. Altarelli, *The Status of the Standard Model*, eprint hep-ph/9710434, 1997.
397. G. Altarelli, *Status of Precision Tests of the Standard Model*, eprint hep-ph/9611239, 1996.
398. D. Bardin and G. Passarino, *Upgrading of Precision Calculations for Electroweak Observables*, eprint hep-ph/9803425, 1998.

399. P. B. Renton, *Precision Electroweak Data: Present Status and Future Prospects*, University of Oxford, OUNP-98-04, 1998.
400. P. B. Renton, Int. J. Mod. Phys. A **12** (1997) 4109.
401. A. Blondel, Nuovo Cim. A **109** (1996) 771.
402. D. Strom, LEP Precision Electroweak Measurements from the $Z^0$ Resonance, invited talk, *23rd Annual SLAC Summer Institute on Particle Physics: The Top Quark and the Electroweak Interaction (SSI 95)*, Stanford, California, 1995.
403. J. L. Rosner, Comments Nucl. Part. Phys. **22** (1998) 205.
404. A. Pich, *Electroweak Precision Physics*, eprint hep-ph/9802257, 1997.
405. A. Pich, *Precision Tests of the Standard Model*, eprint hep-ph/9711279, 1997.
406. J. Sola, editor, *Proceedings of the 4th Interantional Symposium on Radiative Corrections (RADCOR 98)*, Barcelona, Spain, 1998.
407. D. Bardin et al., *Electroweak Working Group Report*, eprint hep-ph/9709229, 1997.
408. D. Abbaneo et al., *A Combination of Preliminary Electroweak Measurements and Constraints on the Standard Model*, preprint CERN-EP/99-015, 1999.
409. T. Adams et al., Measurement of $\sin^2\theta_W$ from $\nu N$ scattering at NuTeV, preliminary result, *International Conference on High Energy Physics*, July 1998, Vancouver, Canada.
410. F. Abe et al., Phys. Rev. Lett. **80** (1998) 2767.
411. F. Abe et al., Phys. Rev. Lett. **80** (1998) 2779.
412. F. Abe et al., Phys. Rev. Lett. **80** (1998) 5720.
413. B. Abbott et al., Direct measurement of the top quark mass at DØ, preliminary result, *International Conference on High Energy Physics*, July 1998, Vancouver, Canada (Abstract 602).
414. B. Abbott et al., Measurement of the W boson mass at DØ, preliminary result, *International Conference on High Energy Physics*, July 1998, Vancouver, Canada (Abstract 461).
415. P. McNamara, Standard Model Higgs at LEP, *International Conference on High Energy Physics*, July 1998, Vancouver, Canada (ID 1001).
416. S. Jadach and Z. Wąs, The $\tau$ polarization measurement, in *Z Physics at LEP 1*, edited by G. Altarelli et al., pages 235–265, CERN Yellow Report 89-08, 1989.
417. M. Böhm and W. Hollik, Forward–backward asymmetries, in *Z Physics at LEP 1*, edited by G. Altarelli et al., pages 203–234, CERN Yellow Report 89-08, 1989.
418. D. Abbaneo et al., *A Combination of Preliminary Electroweak Measurements and Constraints on the Standard Model*, preprint CERN-PPE/97-154, 1997.
419. T. Riemann, ZFITTER: an analytical program for fermion pair production, in *Proceedings of the International Conference on High Energy Physics (ICHEP92)*, Dallas, Texas, 1992.
420. D. Bardin et al., *ZFITTER: An Analytical Program for Fermion Pair Production in $e^+e^-$ Annihilation*, CERN preprint CERN-TH.6442/92 and eprint hep-ph/9412201, 1992.
421. P. Abreu et al., Results on fermion-pair production at LEP running near 189 GeV, preliminary result, *International Conference on High Energy Physics*, July 1998, Vancouver, Canada (Abstract 643).
422. P. Abreu et al., Cross sections and leptonic forward–backward asymmetries from the $Z^0$ running of LEP, preliminary result, *International Conference on High Energy Physics*, July 1998, Vancouver, Canada (Abstract 438).
423. K. Abe et al., Phys. Rev. Lett. **73** (1994) 25.

424. M. Woods, *The Scanning Compton Polarimeter for the SLD Experiment*, SLAC-PUB-7319, eprint hep-ex/9611005, 1996.
425. K. Abe et al., Phys. Rev. Lett. **78** (1997) 2075.
426. M. Woods, *The Polarized Electron Beam for the SLAC Linear Collider*, SLAC-PUB-7320, eprint hep-ex/9611006, 1996.
427. M. Martinez and R. Miquel, Z. Phys. C **53** (1992) 115.
428. K. G. Baird, Measurements of A(LR) and A(lepton) from SLD, in *Proceedings of the International Conference on High Energy Physics (ICHEP98)*, Vancouver, Canada, 1998, SLAC-PUB-8017, eprint hep-ex/9812008.
429. S. Kawasaki, T. Shirafuji, and Y. Tsai, Prog. Theor. Phys. **49** (1973) 1656.
430. S.-Y. Pi and A. I. Sanda, Ann. Phys. **106** (1977) 171.
431. C. Cohen-Tannoudji, B. Diu, and F. Laloe, *Quantum Mechanics*, Wiley, New York, 1977, chapter 9.
432. A. Rougé, Tau decays as polarization analysers, in *Proceedings of the Workshop on Tau Lepton Physics*, edited by M. Davier and B. Jean-Marie, pages 213–222, Orsay, France, 1990.
433. K. Hagiwara, A. D. Martin, and D. Zeppenfeld, Phys. Lett. B **235** (1990) 198.
434. M. Davier, L. Duflot, F. L. Diberder, and A. Rouge, Phys. Lett. B **306** (1993) 411.
435. A. Rougé, Z. Phys. C **48** (1990) 75.
436. J. H. Kühn, Phys. Rev. D **52** (1995) 3128.
437. J. H. Kühn and E. Mirkes, Z. Phys. C **56** (1992) 661.
438. P. Overmann, *A New Method to Measure the Tau Polarization at the Z Peak*, University of Dortmund: DO-TH-93-24, 1993.
439. S. Amato et al., Measurement of the tau polarisation, preliminary result, International Conference on High Energy Physics, July 1998, Vancouver, Canada (Abstract 243).
440. W. T. Ford et al., Phys. Rev. D **36** (1987) 1971.
441. R. Alemany, N. Rius, J. Bernabéu, J. J. Gomez-Cadenas, and A. Pich, Nucl. Phys. B **379** (1992) 3.
442. C. A. Nelson, Phys. Rev. D **40** (1989) 123.
443. C. A. Nelson, Phys. Rev. D **43** (1991) 1465.
444. J. Bernabéu, N. Rius, and A. Pich, Phys. Lett. B **257** (1991) 219.
445. J. M. Roney, Combining the LEP tau polarization measurements, in *Proceedings of the 5th Workshop on Tau Lepton Physics*, edited by A. Pich and A. Ruiz, volume 76 of Nucl. Phys. B (Proc. Suppl.), pages 69–74, Santander, Spain, 1998.
446. E. L. Berger and H. J. Lipkin, Phys. Lett. B **189** (1987) 226.
447. E. L. Berger and H. J. Lipkin, Phys. Rev. Lett. **59** (1987) 1394.
448. M. Kobayashi and T. Maskawa, Prog. Theor. Phys. **49** (1973) 652.
449. S. Weinberg, Phys. Rev. **112** (1958) 1375.
450. K. Sugimoto et al., Phys. Rev. Lett. **34** (1975) 1533.
451. F. P. Calaprice et al., Phys. Rev. Lett. **35** (1975) 1566.
452. M. Derrick et al., Phys. Lett. B **189** (1987) 260.
453. J. H. Kühn and E. Mirkes, Phys. Lett. B **286** (1992) 381.
454. R. Decker and E. Mirkes, Z. Phys. C **57** (1993) 495.
455. M. Finkemeier and E. Mirkes, Z. Phys. C **69** (1996) 243.
456. R. Decker, M. Finkemeier, P. Heiliger, and H. H. Jonsson, Z. Phys. C **70** (1996) 247.
457. J. Gasser and H. Leutwyler, Ann. Phys. **158** (1984) 142.
458. A. Pich, *Introduction to Chiral Perturbation Theory*, eprint hep-ph/9308351, 1993.

459. G. Colangelo, M. Finkemeier, and R. Urech, Phys. Rev. D **54** (1996) 4403.
460. A. J. Weinstein, A precision measurement of the branching fraction $B\left(\tau^{\pm} \to h^{\pm}\pi^0\nu_\tau\right)$ from CLEO II, in *Proceedings of the 3rd Workshop on Tau Lepton Physics*, edited by L. Rolandi, volume 40 of Nucl. Phys. B (Proc. Suppl.), page 163, Montreux, Switzerland, 1994.
461. S. L. Adler, Phys. Rev. **177** (1969) 2426.
462. W. A. Bardeen, Phys. Rev. **184** (1969) 1848.
463. J. Wess and B. Zumino, Phys. Lett. B **37** (1971) 95.
464. E. Witten, Nucl. Phys. B **223** (1983) 422.
465. G. J. Gounaris and J. J. Sakurai, Phys. Rev. Lett. **21** (1968) 244.
466. B. C. de Beauregard, T. N. Pham, B. Pire, and T. N. Truong, Phys. Lett. B **67** (1977) 213.
467. A. Quenzer et al., Phys. Lett. B **76** (1978) 512.
468. M. Feindt, Z. Phys. C **48** (1990) 681.
469. R. Kokoski and N. Isgur, Phys. Rev. D **35** (1987) 907.
470. E. Braaten, R. J. Oakes, and S.-M. Tse, Int. J. Mod. Phys. A **5** (1990) 2737.
471. H. Davoudiasl and M. B. Wise, Phys. Rev. D **53** (1996) 2523.
472. Y. P. Ivanov, A. A. Osipov, and M. K. Volkov, Z. Phys. C **49** (1991) 563.
473. T. N. Pham, C. Roiesnel, and T. N. Truong, Phys. Rev. Lett. **41** (1978) 371.
474. T. N. Pham, C. Roiesnel, and T. N. Truong, Phys. Lett. B **80** (1978) 119.
475. L. Beldjoudi and T. N. Truong, Phys. Lett. B **344** (1995) 419.
476. L. Beldjoudi and T. N. Truong, Phys. Lett. B **351** (1995) 357.
477. A. Pais, Ann. Phys. **9** (1960) 548.
478. F. J. Gilman and S. H. Rhie, Phys. Rev. D **31** (1985) 1066.
479. A. Rougé, Z. Phys. C **70** (1996) 65.
480. F. J. Gilman and D. H. Miller, Phys. Rev. D **17** (1978) 1846.
481. R. P. Feynman and M. Gell-Mann, Phys. Rev. **109** (1958) 193.
482. S. I. Eidelman and V. N. Ivanchenko, Phys. Lett. B **257** (1991) 437.
483. S. I. Eidelman and V. N. Ivanchenko, Tau decays and CVC, in *Proceedings of the 3rd Workshop on Tau Lepton Physics*, edited by L. Rolandi, volume 40 of Nucl. Phys. B (Proc. Suppl.), pages 131–138, Montreux, Switzerland, 1994.
484. R. J. Sobie, Z. Phys. C **65** (1995) 79.
485. L. M. Barkov et al., Nucl. Phys. B **256** (1985) 365.
486. Z. Y. Fang, G. L. Castro, and J. Pestieau, Nuovo Cim. A **100** (1988) 155.
487. V. Chabaud et al., Nucl. Phys. B **223** (1983) 1.
488. D. Amelin et al., Phys. Lett. B **356** (1995) 595.
489. M. Schmidtler, The hadronic structure in $\tau \to 3\pi\nu$, in *Proceedings of the 5th Workshop on Tau Lepton Physics*, edited by A. Pich and A. Ruiz, volume 76 of Nucl. Phys. B (Proc. Suppl.), pages 271–282, Santander, Spain, 1998.
490. T. E. Coan et al., Hadronic structure in the decay $\tau^- \to \nu_\tau\pi^-\pi^0\pi^0$ and the sign of the tau neutrino helicity, preliminary result, *International Conference on High Energy Physics*, July 1998, Vancouver, Canada (Abstract 976), CLEO-CONF 98-19.
491. D. Asner et al., *Hadronic Structure in the Decay $\tau^- \to \nu_\tau\pi^-\pi^0\pi^0$ and the Sign of the Tau Neutrino Helicity*, preprints CLNS 99/1601, hep-ex/9902022, submitted to Phys. Rev. D, 1999.
492. E. C. Poggio, H. R. Quinn, and S. Weinberg, Phys. Rev. D **13** (1976) 1958.
493. J. J. Sakurai, Phys. Lett. B **46** (1973) 207.
494. S. Narison and A. Pich, Phys. Lett. B **211** (1988) 183.
495. E. Braaten, S. Narison, and A. Pich, Nucl. Phys. B **373** (1992) 581.
496. K. Schilcher and M. D. Tran, Phys. Rev. D **29** (1984) 570.

497. A. Pich, QCD predictions for the $\tau$ hadronic width and determination of $\alpha_s\left(m_\tau^2\right)$, in *Proceedings of the 2nd Workshop on Tau Lepton Physics*, edited by K. K. Ghan, Columbus, Ohio, 1992.
498. A. Pich, QCD predictions for the $\tau$ hadronic width: determination of $\alpha_s\left(m_\tau^2\right)$, in *Proceedings of the QCD 94 Workshop*, edited by S. Narison, volume 39B,C of Nucl. Phys. B (Proc. Suppl.), page 326, Montpellier, France, 1994.
499. E. Braaten, Phys. Rev. Lett. **60** (1988) 1606.
500. E. Braaten, Phys. Rev. D **39** (1989) 1458.
501. M. Neubert, Nucl. Phys. B **463** (1996) 511.
502. F. L. Diberder and A. Pich, Phys. Lett. B **286** (1992) 147.
503. K. G. Chetyrkin, A. L. Kataev, and F. V. Tkachev, Phys. Lett. B **85** (1979) 277.
504. M. Dine and J. Sapirstein, Phys. Rev. Lett. **43** (1979) 668.
505. W. Celmaster and R. J. Gonsalves, Phys. Rev. Lett. **44** (1980) 560.
506. S. G. Gorishnii, A. L. Kataev, and S. A. Larin, Phys. Lett. B **259** (1991) 144.
507. L. R. Surguladze and M. A. Samuel, Phys. Rev. Lett. **66** (1991) 560.
508. W. E. Caswell, Phys. Rev. Lett. **33** (1974) 244.
509. D. R. T. Jones, Nucl. Phys. B **75** (1974) 531.
510. E. Egorian and O. V. Tarasov, Theor. Math. Phys. **41** (1979) 863.
511. O. V. Tarasov, A. A. Vladimirov, and A. Y. Zharkov, Phys. Lett. B **93** (1980) 429.
512. S. A. Larin and J. A. M. Vermaseren, Phys. Lett. B **303** (1993) 334.
513. T. van Ritbergen, J. A. M. Vermaseren, and S. A. Larin, Phys. Lett. B **400** (1997) 379.
514. P. A. Raczka, Phys. Rev. D **57** (1998) 6862.
515. P. A. Raczka and A. Szymacha, Z. Phys. C **70** (1996) 125.
516. P. M. Stevenson, Phys. Rev. D **23** (1981) 2916.
517. A. L. Kataev and V. V. Starshenko, Mod. Phys. Lett. A **10** (1995) 235.
518. G. Grunberg, Phys. Rev. D **29** (1984) 2315.
519. D. J. Broadhurst and A. L. Kataev, Phys. Lett. B **315** (1993) 179.
520. A. Pich, QCD tests from tau decays, in *Proceedings of the 20th John Hopkins Workshop on Current Problems in Particle Theory*, edited by M. Jamin et al., Heidelberg, Germany, 1996.
521. F. L. Diberder, Experimental estimates of higher order perturbative corrections, in *Proceedings of the QCD 94 Workshop*, edited by S. Narison, volume 39B,C of Nucl. Phys. B (Proc. Suppl.), page 318, Montpellier, France, 1994.
522. K. Symanzik, Comm. Math. Phys. **23** (1971) 49.
523. C. Callan, Phys. Rev. D **5** (1972) 3302.
524. C. Callan, Fortschr. Phys. **18** (1970) 249.
525. M. A. Shifman, A. I. Vainshtein, and V. I. Zakharov, Nucl. Phys. B **147** (1979) 448.
526. M. A. Shifman, A. I. Vainshtein, and V. I. Zakharov, Nucl. Phys. B **147** (1979) 385.
527. M. A. Shifman, A. I. Vainshtein, and V. I. Zakharov, Nucl. Phys. B **147** (1979) 519.
528. K. G. Wilson, Phys. Rev. **179** (1969) 1499.
529. K. G. Chetyrkin and A. Kwiatkowski, Z. Phys. C **59** (1993) 525, erratum, eprint hep-ph/9805232, 1998.
530. C. Becchi, S. Narison, E. de Rafael, and F. J. Yndurain, Z. Phys. C **8** (1981) 335.
531. D. J. Broadhurst, Phys. Lett. B **101** (1981) 423.
532. S. C. Generalis, J. Phys. G **15** (1989) L225.

533. K. G. Chetyrkin, V. P. Spiridonov, and S. G. Gorishnii, Phys. Lett. B **160** (1985) 149.
534. W. Hubschmid and S. Mallik, Nucl. Phys. B **207** (1982) 29.
535. D. A. Ross, Nucl. Phys. B **188** (1981) 109.
536. L. V. Larin, V. P. Spiridonov, and K. G. Chetyrkin, Sov. J. Nucl. Phys. **44** (1986) 892.
537. D. J. Broadhurst and S. C. Generalis, Phys. Lett. B **165** (1985) 175.
538. F. L. Diberder and A. Pich, Phys. Lett. B **289** (1992) 165.
539. S. Narison, *QCD Spectral Sum Rules*, volume 26 of Lecture Notes in Physics, World Scientific, Singapore, 1989.
540. E. Braaten and C.-S. Li, Phys. Rev. D **42** (1990) 3888.
541. A. Sirlin, Nucl. Phys. B **196** (1982) 83.
542. A. Sirlin, Phys. Rev. D **22** (1980) 971.
543. W. J. Marciano and A. Sirlin, Phys. Rev. D **22** (1980) 2695.
544. M. Davier, $\tau$ decays into strange particles and QCD, in *Proceedings of the 4th Workshop on Tau Lepton Physics*, edited by J. G. Smith and W. Toki, volume 55C of Nucl. Phys. B (Proc. Suppl.), page 395, Estes Park, Colorado, 1996.
545. S. Chen, Measurement of $\tau$ decays with kaons from ALEPH and $m_s$ determination, in *Proceedings of the QCD 97 Workshop*, edited by S. Narison, volume 64 of Nucl. Phys. B (Proc. Suppl.), page 265, Montpellier, France, 1997.
546. S. Narison, $\alpha_s$ *from Tau Decays*, eprint hep-ph/9412295, 1994.
547. S. Bethke, QCD tests at $e^+e^-$ colliders, in *Proceedings of the QCD 97 Workshop*, edited by S. Narison, volume 64 of Nucl. Phys. B (Proc. Suppl.), page 54, Montpellier, France, 1997.
548. S. Narison and A. Pich, Phys. Lett. B **304** (1993) 359.
549. M. Girone and M. Neubert, Phys. Rev. Lett. **76** (1996) 3061.
550. V. S. M. T. Das and S. Okubo, Phys. Rev. Lett. **19** (1967) 895.
551. S. R. Amendolia et al., Nucl. Phys. B **277** (1986) 168.
552. T. Das, G. S. Guralnik, V. S. Mathur, F. E. Low, and J. E. Young, Phys. Rev. Lett. **18** (1967) 759.
553. J. S. Conway, Physics with taus at CDF, in *Proceedings of the 4th Workshop on Tau Lepton Physics*, edited by J. G. Smith and W. Toki, volume 55C of Nucl. Phys. B (Proc. Suppl.), pages 409–415, Estes Park, Colorado, 1996.
554. M. Gallinaro, $\tau$ physics at $p\bar{p}$ colliders, in *Proceedings of the 5th Topical Seminar on the Irresistible Rise of the Standard Model*, edited by F.-L. Navarria and P. G. Pelfer, volume 65 of Nucl. Phys. B (Proc. Suppl.), pages 147–151, San Miniato al Todesco, Italy, 1997.
555. M. Gallinaro, Searches with taus at the Tevatron, in *Proceedings of the 5th Workshop on Tau Lepton Physics*, edited by A. Pich and A. Ruiz, volume 76 of Nucl. Phys. B (Proc. Suppl.), pages 207–214, Santander, Spain, 1998.
556. S. Protopopescu, $W \to \tau \nu_\tau$ at the Tevatron, in *Proceedings of the 5th Workshop on Tau Lepton Physics*, edited by A. Pich and A. Ruiz, volume 76 of Nucl. Phys. B (Proc. Suppl.), pages 91–97, Santander, Spain, 1998.
557. F. Abe et al., Phys. Rev. Lett. **79** (1997) 3585.
558. F. Abe et al., Phys. Rev. D **54** (1996) 735.
559. F. Abe et al., Phys. Rev. Lett. **79** (1997) 357.
560. J. C. Pati and A. Salam, Phys. Rev. D **8** (1973) 1240.
561. J. C. Pati and A. Salam, Phys. Rev. Lett. **31** (1973) 661.
562. J. C. Pati and A. Salam, Phys. Rev. D **10** (1974) 275.
563. H. Georgi and S. L. Glashow, Phys. Rev. Lett. **32** (1974) 438.
564. L. F. Abbott and E. Farhi, Phys. Lett. B **101** (1981) 69.

565. B. Schrempp and F. Schrempp, Phys. Lett. B **153** (1985) 101.
566. J. C. Pati, Phys. Lett. B **228** (1989) 228.
567. K. Lane and M. V. Ramana, Phys. Rev. D **44** (1991) 2678.
568. S. F. King, Rep. Prog. Phys. **58** (1995) 263.
569. F. Abe et al., Phys. Rev. Lett. **75** (1995) 1012.
570. S. Abachi et al., Phys. Rev. Lett. **72** (1994) 965.
571. F. Abe et al., Phys. Rev. Lett. **78** (1997) 2906.
572. F. Abe et al., Phys. Rev. Lett. **82** (1999) 3206.
573. E. Eichten, I. Hinchliffe, K. D. Lane, and C. Quigg, Rev. Mod. Phys. **56** (1984) 579.
574. E. Eichten, I. Hinchliffe, K. D. Lane, and C. Quigg, Phys. Rev. D **34** (1986) 1547.
575. C. Albajar et al., Z. Phys. C **44** (1989) 15.
576. C. Albajar et al., Phys. Lett. B **253** (1991) 503.
577. J. Alitti et al., Phys. Lett. B **280** (1992) 137.
578. B. Abbott et al., A measurement of the production cross section times $\tau$ branching ratio for W bosons in $p\bar{p}$ collisions at $\sqrt{s} = 1.8$ TeV, preliminary result, *International Conference on High Energy Physics*, July 1998, Vancouver, Canada (Abstract 463).
579. F. Abe et al., Phys. Rev. Lett. **68** (1992) 3398.
580. D. P. Roy, *Neutrino Mass and Oscillation: An Introductory Review*, eprint hep-ph/9903506, 1998.
581. D. Suematsu, Prog. Theor. Phys. **99** (1998) 483.
582. S. Lola and J. D. Vergados, Prog. Nucl. Part. Phys. **40** (1998) 71.
583. J. Brunner, Fortschr. Phys. **45** (1997) 343.
584. G. Gelmini and E. Roulet, Rep. Prog. Phys. **58** (1995) 1207.
585. L. Oberauer and F. von Feilitzsch, Rep. Prog. Phys. **55** (1992) 1093.
586. J. W. F. Valle, Prog. Nucl. Part. Phys. **26** (1991) 91.
587. M. Schwartz, Rev. Mod. Phys. **61** (1989) 527.
588. J. Steinberger, Rev. Mod. Phys. **61** (1989) 533.
589. L. M. Lederman, Rev. Mod. Phys. **61** (1989) 547.
590. S. M. Bilenkii and S. T. Petcov, Rev. Mod. Phys. **59** (1987) 671.
591. J. L. Vuilleumier, Rep. Prog. Phys. **49** (1986) 1293.
592. F. Eisele, Rep. Prog. Phys. **49** (1986) 233.
593. A. Bottino, A. di Credico, and P. Monacelli, editors, *Proceedings of the 5th International Workshop on Topics in Astroparticle and Underground Physics*, volume 70 of Nucl. Phys. B (Proc. Suppl.), Gran Sasso, Italy, 1999.
594. A. Dar, G. Eilam, and M. Gronau, editors, *Proceedings of the 16th International Conference on Neutrino Physics and Astrophysics*, volume 38 of Nucl. Phys. B (Proc. Suppl.), Eilat, Israel, 1994.
595. Y. Fukuda et al., Phys. Lett. B **433** (1998) 9.
596. Y. Fukuda et al., Phys. Lett. B **436** (1998) 33.
597. Y. Fukuda et al., Phys. Rev. Lett. **81** (1998) 1562.
598. S. E. Csorna et al., Phys. Rev. D **35** (1987) 2747.
599. J. J. G. Cadenas, A. Seiden, M. C. Gonzalez-Garcia, D. H. Coward, and R. H. Schindler, Phys. Rev. D **41** (1990) 2179.
600. B. D. Fields, Astroparticle Phys. **6** (1997) 169.
601. R. Mayle, D. N. Schramm, M. S. Turner, and J. R. Wilson, Phys. Lett. B **317** (1993) 119.
602. H. Harari, Intrinsic $\nu_\tau$ properties, *3rd Workshop on Tau Lepton Physics*, Montreux, Switzerland, Sept. 1994.
603. R. Cowsik and J. Mcclelland, Phys. Rev. Lett. **29** (1972) 669.
604. M. Kawasaki et al., Nucl. Phys. B **419** (1994) 105.

605. K. Fujikawa and R. Shrock, Phys. Rev. Lett. **45** (1980) 963.
606. E. Ma and J. Okada, Phys. Rev. Lett. **41** (1978) 287.
607. K. J. F. Gaemers, R. Gastmans, and F. M. Renard, Phys. Rev. D **19** (1979) 1605.
608. M. Acciarri et al., Phys. Lett. B **412** (1997) 201.
609. P. Abreu et al., Z. Phys. C **74** (1997) 577.
610. T. M. Gould and I. Z. Rothstein, Phys. Lett. B **333** (1994) 545.
611. N. G. Deshpande and K. V. L. Sharma, Phys. Rev. D **43** (1991) 943.
612. H. Grotch and R. W. Robinett, Z. Phys. C **39** (1988) 553.
613. A. M. Cooper-Sarkar et al., Phys. Lett. B **280** (1992) 153.
614. H. Grassler et al., Nucl. Phys. B **273** (1986) 253.
615. A. V. Kyuldjiev, Nucl. Phys. B **243** (1984) 387.
616. L. Bergström and H. R. Rubinstein, Phys. Lett. B **253** (1991) 168.
617. F. Scheck, Phys. Rep. **44** (1978) 187.
618. K. Mursula, M. Roos, and F. Scheck, Nucl. Phys. B **219** (1983) 321.
619. W. Fetscher, Phys. Rev. D **42** (1990) 1544.
620. W. Fetscher, H. J. Gerber, and K. F. Johnson, Phys. Lett. B **173** (1986) 102.
621. W. Fetscher and H. J. Gerber, Precision measurements in muon and tau decays, in *Precision Tests of the Standard Model*, edited by P. Langacker, World Scientific, Singapore, 1993.
622. M. Fierz, Z. Phys. **101** (1937) 553.
623. P. Langacker and D. London, Phys. Rev. D **39** (1989) 266.
624. A. A. Poblaguev, Phys. Lett. B **238** (1990) 108.
625. M. V. Chizhov, *On the Muon Decay Parameters*, eprint hep-ph/9612399, 1996.
626. T. Barakat, *The $K^+ \to \pi^+ \nu \bar{\nu}$ Rare Decay in Two Higgs Doublet Model*, Lefkosa-Mersin Near East University CV-HEP-98-03, and eprint hep-ph/9807317, 1998.
627. V. N. Bolotov et al., Phys. Lett. B **243** (1990) 308.
628. S. A. Akimenko et al., Phys. Lett. B **259** (1991) 225.
629. P. Seager, Michel parameters and limits on tensor couplings from DELPHI, in *Proceedings of the 5th Workshop on Tau Lepton Physics*, edited by A. Pich and A. Ruiz, volume 76 of Nucl. Phys. B (Proc. Suppl.), pages 141–146, Santander, Spain, 1998.
630. C. Greub, D. Wyler, and W. Fetscher, Phys. Lett. B **324** (1994) 109.
631. L. Michel, Proc. Phys. Soc. A **63** (1950) 514.
632. C. Bouchiat and L. Michel, Phys. Rev. **106** (1957) 170.
633. A. Rougé, Results on tau neutrino from colliders, in *Proceedings of the 30th Rencontres de Moriond: Euroconferences: Dark Matter in Cosmology, Clocks and Tests of Fundamental Laws*, edited by D. H. B. Guiderdoni, G. Greene and J. T. Thanh, Les Arcs, France, 1995.
634. A. Stahl and H. Voss, Z. Phys. C **74** (1997) 73.
635. C. Jarlskog, Nucl. Phys. **75** (1966) 659.
636. R. Bartoldus, *Measurements of the Michel Parameters in Leptonic Tau Decays with the OPAL Detector at LEP*, PhD thesis, Physikalisches Institut, Universität Bonn, 1998.
637. R. Bartoldus, Michel parameters from OPAL, in *Proceedings of the 5th Workshop on Tau Lepton Physics*, edited by A. Pich and A. Ruiz, volume 76 of Nucl. Phys. B (Proc. Suppl.), pages 147–157, Santander, Spain, 1998.
638. H. Thurn and H. Kolanoski, Z. Phys. C **60** (1993) 277.
639. W. T. Ford, Phys. Rev. D **36** (1987) 1971.
640. S. Behrends et al., Phys. Rev. D **32** (1985) 2468.
641. C. A. Nelson, Phys. Lett. B **355** (1995) 561.

642. P. H. Eberhard et al., The $\tau$ polarization measurement at LEP, in *Proceedings of the Workshop on Z Physics at LEP*, edited by G. Altarelli, R. Kleiss, and V. Verzegnassi, volume 1-3, 1989, CERN-89-08.
643. R. Barate et al., Measurement of the Michel parameters and the neutrino helicity of the tau lepton, preliminary result, *International Conference on High Energy Physics*, July 1998, Vancouver, Canada (Abstract 981).
644. U. Stiegler, Z. Phys. C **58** (1993) 601.
645. T. Hebbeker and W. Lohmann, Z. Phys. C **74** (1997) 399.
646. A. Pich and J. P. Silva, Phys. Rev. D **52** (1995) 4006.
647. H. E. Haber, Can the Higgs sector be probed in $\tau$ lepton decays?, in *Proceedings of the Tau Charm Factory Workshop*, 1989, SLAC-Report-343.
648. B. McWilliams and L.-F. Li, Nucl. Phys. B **179** (1981) 62.
649. P. Abreu et al., Upper limit for the decay $B^- \to \tau^- \bar{\nu}_\tau$ and measurement of the branching ratio $b \to \tau \nu_\tau X$, preliminary result, *International Conference on High Energy Physics*, July 1998, Vancouver, Canada (Abstract 242).
650. R. Barate et al., Measurements of BR(b $\to \tau \nu X$) and BR(b $\to \tau \nu D^{*\pm}(X)$) and upper limits on BR(B$^- \to \tau \nu$) and BR(b $\to s \nu \nu$), preliminary result, *International Conference on High Energy Physics*, July 1998, Vancouver, Canada (Abstract 982).
651. J. H. Christenson, J. W. Cronin, V. L. Fitch, and R. Turlay, Phys. Rev. Lett. **13** (1964) 138.
652. C. Jarlskog, editor, *CP Violation*, volume 3 of Advanced Series on Directions in High Energy Physics, World Scientific, Singapore, 1989.
653. L. Wolfenstein, editor, *CP Violation*, volume 5 of Current Physics–Sources and Comments, North-Holland, Amsterdam, Netherlands, 1989.
654. W. Bernreuther, *CP Violation*, eprint hep-ph/9808453, 1998.
655. A. Pich, Weak decays, quark mixing and CP violation: theory overview, in *Proceedings of the 16th Workshop on Weak Interactions and Neutrinos*, edited by G. Fiorillo, V. Palldino, and P. Strolin, volume 66 of Nucl. Phys. B (Proc. Suppl.), page 456, Capri, Italy, 1998, eprint hep-ph/9709441.
656. H. Quinn, CP violation, in *Proceedings of the Conference on Production and Decay of Hyperons, Charm and Beauty Hadrons*, edited by J.-P. Engel, A. Fridman, and P. Roudeau, volume 50 of Nucl. Phys. B (Proc. Suppl.), pages 17–23, Strasbourg, France, 1995.
657. K. Kojima, W. Sugiyama, and S. Y. Tsai, Prog. Theor. Phys. **95** (1996) 913.
658. B. Winstein and L. Wolfenstein, Rev. Mod. Phys. **65** (1993) 1113.
659. Y. Nir and H. R. Quinn, Ann. Rev. Nucl. Part. Sci. **42** (1992) 211.
660. K. Kleinknecht, Prog. Nucl. Part. Phys. **25** (1990) 81.
661. L. Wolfenstein, Ann. Rev. Nucl. Part. Sci. **36** (1986) 137.
662. W. Bernreuther, A. Brandenburg, and P. Overmann, Phys. Lett. B **391** (1997) 413.
663. G. C. Branco, CP violation in the Standard Model and beyond, Lectures given at *5th Mexican School of Particles and Fields*, Guanajuato, Mexico, 30 Nov. – 11 Dec. 1992, CERN-TH.7176/94.
664. W. Bernreuther and M. Suzuki, Rev. Mod. Phys. **63** (1991) 313.
665. M. B. Gavela, P. Hernandez, J. Orloff, and O. Pene, Mod. Phys. Lett. A **9** (1994) 795.
666. G. R. Farrar and M. E. Shaposhnikov, Phys. Rev. D **50** (1994) 774.
667. A. G. Cohen, D. B. Kaplan, and A. E. Nelson, Ann. Rev. Nucl. Part. Sci. **43** (1993) 27.
668. F. Hoogeveen, Nucl. Phys. B **341** (1990) 322.
669. J. F. Donoghue, Phys. Rev. D **18** (1978) 1632.

670. S. M. Barr and W. J. Marciano, Electric dipole moments, in *CP Violation*, edited by C. Jarlskog, page 455, World Scientific, Singapore, 1989.
671. Y. A. Golfand and E. P. Likhtman, J. Exp. Theor. Phys. Lett. **13** (1971) 323.
672. D. V. Volkov and V. P. Akulov, Phys. Lett. B **46** (1973) 109.
673. J. Wess and B. Zumino, Nucl. Phys. B **70** (1974) 39.
674. H. E. Haber and Y. Nir, Nucl. Phys. B **335** (1990) 363.
675. H. P. Nilles, Phys. Rep. **110** (1984) 1.
676. H. E. Haber and G. L. Kane, Phys. Rep. **117** (1985) 75.
677. R. Barbieri, Supersymmetry searches, in *Z Physics at LEP 1*, edited by G. Altarelli et al., volume 2, page 121, CERN Yellow Report 89-08, 1989.
678. J. M. Frère and G. L. Kane, Nucl. Phys. **B223** (1983) 331.
679. J. Ellis, A. B. Lahanas, D. V. Nanopoulos, and K. Tamvakis, Phys. Lett. B **134** (1984) 429.
680. J. Ellis, J. S. Hagelin, D. V. Nanopoulos, and M. Srednicki, Phys. Lett. B **127** (1983) 233.
681. M. Dugan, B. Grinstein, and L. Hall, Nucl. Phys. B **255** (1985) 413.
682. S. M. Barr and A. Masiero, Phys. Rev. D **38** (1988) 366.
683. J. Ellis, S. Ferrara, and D. V. Nanopoulos, Phys. Lett. B **114** (1982) 231.
684. W. Buchmüller and D. Wyler, Phys. Lett. B **121** (1983) 321.
685. F. del Aguila, M. B. Gavela, J. A. Grifols, and A. Mendez, Phys. Lett. B **126** (1983) 71.
686. J. M. Frère and M. B. Gavela, Phys. Lett. B **132** (1983) 107.
687. J. Polchinski and M. B. Wise, Phys. Lett. B **125** (1983) 393.
688. E. Franco and M. Mangano, Phys. Lett. B **135** (1984) 445.
689. J. M. Gerard, W. Grimus, A. Raychaudhuri, and G. Zoupanos, Phys. Lett. B **140** (1984) 349.
690. S. T. Petcov, Phys. Lett. B **178** (1986) 57.
691. T. D. Lee, Phys. Rev. D **8** (1973) 1226.
692. T. D. Lee, Phys. Rep. **9** (1974) 143.
693. P. Sikivie, Phys. Lett. B **65** (1976) 141.
694. A. B. Lahanas and C. E. Vayonakis, Phys. Rev. D **19** (1979) 2158.
695. G. C. Branco, A. J. Buras, and J. M. Gerard, Nucl. Phys. B **259** (1985) 306.
696. S. Weinberg, Phys. Rev. Lett. **37** (1976) 657.
697. I. I. Bigi and A. I. Sanda, On spontaneous CP violation triggered by scalar bosons, in *CP Violation*, edited by C. Jarlskog, pages 362–383, World Scientific, Singapore, 1989.
698. S. M. Barr and A. Zee, Phys. Rev. Lett. **65** (1990) 21.
699. J. C. Pati and A. Salam, Phys. Rev. D **10** (1974) 275.
700. R. N. Mohapatra, CP violation and left–right symmetry, in *CP Violation*, edited by C. Jarlskog, pages 384–435, World Scientific, Singapore, 1989.
701. M. Gell-Mann, P. Ramond, and R. Slansky, Complex spinors and unified theories, in *Supergravity*, edited by D. Friedman and P. van Nieuwhuisen, page 315, North-Holland, Amsterdam, 1979.
702. R. N. Mohapatra and G. Senjanovic, Phys. Rev. Lett. **44** (1980) 912.
703. K. S. Babu and R. N. Mohapatra, Phys. Rev. Lett. **62** (1989) 1079.
704. J. P. Ma and A. Brandenburg, Z. Phys. C **56** (1992) 97.
705. L. J. Hall and L. J. Randall, Nucl. Phys. B **274** (1986) 157.
706. S. M. Barr and A. Masiero, Phys. Rev. Lett. **58** (1987) 187.
707. A. Barroso and J. Maalampi, Phys. Lett. B **187** (1987) 85.
708. C. Q. Geng, Z. Phys. C **48** (1990) 279.
709. C. Q. Geng and J. N. Ng, Phys. Rev. D **42** (1990) 1509.
710. J. F. Nieves, Phys. Lett. B **164** (1985) 85.
711. S. M. Barr, Phys. Rev. D **34** (1986) 1567.

712. W. Bernreuther and O. Nachtmann, Phys. Rev. Lett. **63** (1989) 2787, erratum, Phys. Rev. Lett. **64** (1990) 1072.
713. W. Bernreuther, G. W. Botz, O. Nachtmann, and P. Overmann, Z. Phys. C **52** (1991) 567.
714. C. A. Nelson, Phys. Rev. D **41** (1990) 2805.
715. S. Goozovat and C. A. Nelson, Phys. Lett. B **267** (1991) 128.
716. C. A. Nelson, *Testing CP, T and (V–A) Symmetry through Tau Leptons*, preprint SUNY BING 9/29/96, hep-ph/9610216, 1996.
717. M.-C. Chen et al., Test of $\mathcal{CP}$ violation in $e^+e^- \to Z^0 \to \tau^+\tau^-$ and upper limit on the weak dipole moment of the tau lepton, preliminary result, *International Conference on High Energy Physics*, August 1997, Jerusalem, Israel (Abstract pa7 pl6).
718. D. Atwood and A. Soni, Phys. Rev. D **45** (1992) 2405.
719. M. Diehl and O. Nachtmann, Z. Phys. C **62** (1994) 397.
720. M. Wunsch, A test of CP-invariance in $Z^0 \to \tau^+\tau^-$, preliminary result, *International Conference on High Energy Physics*, July 1996, Warsaw, Poland (Abstract PA08-030).
721. T. Barklow, SLD measurement of dipole moments, in *Proceedings of the 5th Workshop on Tau Lepton Physics*, edited by A. Pich and A. Ruiz, volume 76 of Nucl. Phys. B (Proc. Suppl.), Santander, Spain, 1998, eprint hep-ph/9811209.
722. M. Schumacher, CP tests and the weak dipole moment in tau pair production at LEP, in *Proceedings of the International Conference on High Energy Physics*, edited by D. Lellouch, G. Mikenberg, and E. Rabinovici, page 773, Jerusalem, Israel, 1997, University of Bonn, BONN-HE-98-01.
723. A. Zalite, LEP summary on weak dipole moments, in *Proceedings of the 5th Workshop on Tau Lepton Physics*, edited by A. Pich and A. Ruiz, pages 229–236, Santander, Spain, 1998, eprint hep-ph/9811209.
724. B. Ananthanarayan and S. D. Rindani, Phys. Rev. D **50** (1994) 4447.
725. B. Ananthanarayan and S. D. Rindani, Phys. Rev. D **51** (1995) 5996.
726. M. Perl, Perspectives on tau physics, in *Proceedings of the 5th Workshop on Tau Lepton Physics*, edited by A. Pich and A. Ruiz, volume 76 of Nucl. Phys. B (Proc. Suppl.), pages 3–19, Santander, Spain, 1998.
727. Y. S. Tsai, CP violation signals in tau decays, in *Proceedings of the 5th Workshop on Tau Lepton Physics*, edited by A. Pich and A. Ruiz, Nucl. Phys. B (Proc. Suppl.), Santander, Spain, 1998.
728. Y. S. Tsai, Test of $\mathcal{T}$ and $\mathcal{CP}$ violation in leptonic decay of $\tau^\pm$, *Workshop on Tau/Charm Factory*, June 1995, Argonne National Laboratory, eprint hep-ph/9506252, SLAC-PUB-95-6916, 1995.
729. Y. S. Tsai, Phys. Rev. D **51** (1995) 3172.
730. J. H. Kühn and E. Mirkes, Phys. Lett. B **398** (1997) 407.
731. U. Kilian, J. G. Korner, K. Schilcher, and Y. L. Wu, Z. Phys. C **62** (1994) 413.
732. S. Y. Choi, K. Hagiwara, and M. Tanabashi, Phys. Rev. D **52** (1995) 1614.
733. S. Y. Choi, J. Lee, and J. Song, Phys. Lett. B **437** (1998) 191.
734. C. A. Nelson, *Tests for Leptonic $\mathcal{CP}$ Violation in Tau $\to$ Rho Neutrino Decay*, preprint SUNY-BING-7-19-92, 1992.
735. C. A. Nelson et al., Phys. Rev. D **50** (1994) 4544.
736. N. V. Samsonenko and A. D. Nevskii, Bull. Russ. Acad. Sci. Phys. **57** (1993) 1545.
737. N. V. Samsonenko, C. L. Katkhat, and A. D. Nevskii, Bull. Russ. Acad. Sci. Phys. **57** (1993) 1.
738. C. L. Katkhat, Bull. Russ. Acad. Sci. Phys. **53** (1989) 100.
739. C. L. Katkhat and N. V. Samsonenko, Nucl. Phys. A **500** (1989) 669.

740. C. Leroy and J. Pestieau, Phys. Lett. B **72** (1978) 398.
741. S. N. Biswas, S. R. Choudhury, A. Goyal, and J. N. Passi, Phys. Lett. B **80** (1979) 393.
742. A. Pich, Phys. Lett. B **196** (1987) 561.
743. S. Tisserant and T. N. Truong, Phys. Lett. B **115** (1982) 264.
744. H. Neufeld and H. Rupertsberger, Z. Phys. C **68** (1995) 91.
745. A. Bramon, S. Narison, and A. Pich, Phys. Lett. B **196** (1987) 543.
746. F. Scheck and R. Tegen, Z. Phys. C **7** (1981) 111.
747. R. Tegen, Z. Phys. C **7** (1981) 121.
748. C. K. Zachos and Y. Meurice, Mod. Phys. Lett. A **A2** (1987) 247.
749. N. Paver and D. Treleani, Nuovo Cim. Lett. **31** (1981) 364.
750. C. Dominguez, Phys. Rev. D **20** (1979) 802.
751. V. P. Barannik, A. P. Korzh, and M. P. Rekalo, Acta Phys. Polon. B **13** (1982) 835.
752. P. Baringer et al., Phys. Rev. Lett. **59** (1987) 1993.
753. M. C. González-Garcia and J. W. F. Valle, Mod. Phys. Lett. A **7** (1992) 477.
754. G.-G. Wong and W.-S. Hou, Phys. Rev. D **50** (1994) 2962.
755. A. Pilaftsis, Mod. Phys. Lett. A **9** (1994) 3595.
756. J. Hisano, T. Moroi, K. Tobe, M. Yamaguchi, and T. Yanagida, Phys. Lett. B **357** (1995) 579.
757. A. Ilakovac and A. Pilaftsis, Nucl. Phys. B **437** (1995) 491.
758. A. Ilakovac, B. A. Kniehl, and A. Pilaftsis, Phys. Rev. D **52** (1995) 3993.
759. A. Ilakovac, Phys. Rev. D **54** (1996) 5653.
760. R. N. Mohapatra, Phys. Rev. D **46** (1992) 2990.
761. R. N. Mohapatra, S. Nussinov, and X. Zhang, Phys. Rev. D **49** (1994) 2410.
762. S. Pastor, S. D. Rindani, and J. W. Valle, *Lepton Flavor Violation in a Left–Right Symmetric Model*, eprint hep-ph/9705394, 1997.
763. J. C. Romao, N. Rius, and J. W. F. Valle, Nucl. Phys. B **363** (1991) 369.
764. G. Bhattacharyya and D. Choudhury, Mod. Phys. Lett. A **10** (1995) 1699.
765. D. Choudhury and P. Roy, Phys. Lett. B **378** (1996) 153.
766. J. E. Kim, P. Ko, and D.-G. Lee, Phys. Rev. D **56** (1997) 100.
767. M. E. Gomez, G. K. Leontaris, S. Lola, and J. D. Vergados, Phys. Rev. D **59** (1999) 116009.
768. G. K. Leontaris and N. D. Tracas, Phys. Lett. B **431** (1998) 90.
769. J. Hisano, T. Moroi, K. Tobe, and M. Yamaguchi, Phys. Lett. B **391** (1997) 341.
770. R. Barbieri and L. J. Hall, Phys. Lett. B **338** (1994) 212.
771. S. Kelley, J. L. Lopez, D. V. Nanopoulos, and H. Pois, Nucl. Phys. B **358** (1991) 27.
772. R. Arnowitt and P. Nath, Phys. Rev. Lett. **66** (1991) 2708.
773. Ji-zhi Wu, S. Urano, and R. Arnowitt, Phys. Rev. D **47** (1993) 4006.
774. S. F. King and M. Oliveira, *Lepton Flavor Violation in String Inspired Models*, CERN-TH/98-28, eprint hep-ph/9804283, 1998.
775. P. Depommier and C. Leroy, Rep. Prog. Phys. **58** (1995) 61.
776. J. W. F. Valle, Prog. Nucl. Part. Phys. **26** (1991) 91.
777. J. D. Vergados, Phys. Rep. **133** (1986) 1.
778. S. Pakvasa, Flavor changing neutral currents: then and now, in *Proceedings of the International Symposium on 30 Years of Neutral Currents: From Weak Neutral Currents to the W / Z and Beyond*, edited by A. Mann and D. Cline, Santa Monica, CA, 1993.

779. M. L. Perl, Searches for new particles at a tau charm factory, in *Proceedings of the Trieste Workshop on the Search for New Elementary Particles: Status and Prospects*, edited by G. Herten, L. Beers, and M. Perl, volume 3B of Int. J. Mod. Phys. A (Proc. Suppl.), pages 188–210, Trieste, Italy, 1992.
780. X.-Y. Pham, *Lepton Flavor Changing in Neutrinoless Tau Decays*, eprint hep-ph/9810484, 1998.
781. X.-Y. Pham, *Is the Lepton Flavor Changing Observable in $Z^0 \to \mu^\pm + \tau^\pm$ Decay?*, eprint hep-ph/9809322, 1998.
782. J. Hisano and D. Nomura, Phys. Rev. D **59** (1999) 116005.
783. T.-P. Cheng and L.-F. Li, Phys. Rev. D **16** (1977) 1425.
784. B. W. Lee and R. E. Shrock, Phys. Rev. D **16** (1977) 1444.
785. W. J. Marciano and A. I. Sanda, Phys. Lett. B **67** (1977) 303.
786. S. T. Petcov, Sov. J. Nucl. Phys. **25** (1977) 340.
787. M. J. S. Levine, Phys. Rev. D **36** (1987) 1329.
788. E. W. N. Glover and J. J. van der Bij, Rare Z-decays, in *Z-Physics at LEP I*, edited by G. Altarelli, R. Kleiss, and C. Verzegnassi, 1989, Yellow Report CERN 89-08, Vol. II.
789. T. K. Kuo and N. Nakagawa, Phys. Rev. D **32** (1985) 306.
790. G. Eilam and T. G. Rizzo, Phys. Lett. B **188** (1987) 91.
791. J. Bernabéu et al., Phys. Lett. B **187** (1987) 303.
792. J. Bernabéu and A. Santamaria, Phys. Lett. B **197** (1987) 418.
793. J. G. Körner, A. Pilaftsis, and K. Schilcher, Phys. Lett. B **300** (1993) 381.
794. M. Frank and H. Hamidian, Phys. Rev. D **54** (1996) 6790.
795. F. Gabbiani, J. H. Kim, and A. Masiero, Phys. Lett. B **214** (1988) 398.
796. A. Mendez and L. M. Mir, Phys. Rev. D **40** (1989) 251.
797. M. A. Doncheski et al., Phys. Rev. D **40** (1989) 2301.
798. R. Akers et al., Z. Phys. C **67** (1995) 555.
799. D. Decamp et al., Phys. Rep. **216** (1992) 253.
800. P. Abreu et al., Z. Phys. C **73** (1997) 243.
801. O. Adriani et al., Phys. Lett. B **316** (1993) 427.
802. J. J. G. Cadenas et al., Phys. Rev. Lett. **66** (1991) 1007.
803. K. Hagiwara, D. Zeppenfeld, and S. Komamiya, Z. Phys. C **29** (1985) 115.
804. F. Boudjema and A. Djouadi, Phys. Lett. B **240** (1990) 485.
805. F. Boudjema, A. Djouadi, and J. L. Kneur, Z. Phys. C **57** (1993) 425.
806. L. B. Okun, M. B. Voloshin, and M. I. Vysotsky, Sov. Phys. JETP **64** (1986) 446.
807. M. B. Voloshin, Phys. Lett. B **209** (1988) 360.
808. F. Boudjema and F. M. Renard, Compositeness, in *Z-Physics at LEP I*, edited by G. Altarelli, R. Kleiss, and C. Verzegnassi, 1989, Yellow Report CERN 89-08, Vol. II.
809. G. F. Guidice et al., Searches for new physics, in *Physics at LEP2*, edited by G. Altarelli, T. Sjöstrand, and F. Zwirner, 1996, Yellow Report CERN 06-01.
810. P. Abreu et al., Z. Phys. C **53** (1992) 41.
811. O. Adriani et al., Phys. Rep. **236** (1993) 1.
812. M. Z. Akrawy et al., Phys. Lett. B **244** (1990) 135.
813. D. Buskulic et al., Phys. Lett. B **385** (1996) 445.
814. P. Abreu et al., Phys. Lett. B **393** (1997) 245.
815. M. Acciarri et al., Phys. Lett. B **401** (1997) 139.
816. K. Ackerstaff et al., Euro. Phys. J. C **1** (1998) 45.
817. I. Adachi et al., Phys. Lett. B **228** (1989) 553.
818. S. K. Kim et al., Phys. Lett. B **223** (1989) 476.
819. K. Ackerstaff et al., Phys. Lett. B **391** (1997) 221.
820. D. Buskulic et al., Z. Phys. C **59** (1993) 215.

821. H. Kroha, Phys. Rev. D **46** (1992) 58.
822. R. E. Shrock, Phys. Rev. D **24** (1981) 1275.
823. J. Maalampi and M. Roos, Phys. Rep. **186** (1990) 53.
824. A. Djouadi, D. Schaile, and C. Verzegnassi, Extended gauge models, in *Proceedings of the Workshop $e^+e^-$ Collisions at 500 GeV: The Physics Potential*, edited by P. Zerwas, 1992, DESY 92-123 A+B.
825. A. Djouadi, Z. Phys. C **63** (1994) 317.
826. A. Djouadi, J. Ng, and T. G. Rizzo, New particles and interactions, in *Electroweak Symmetry Breaking and Beyond the Standard Model*, edited by T. Barklow et al., World Scientific, Singapore, 1996.
827. J. C. Pati and A. Salam, Phys. Lett. B **58** (1975) 333.
828. R. N. Mohapatra and J. C. Pati, Phys. Rev. D **11** (1975) 566.
829. R. N. Mohapatra and J. C. Pati, Phys. Rev. D **11** (1975) 2558.
830. G. Senjanovic and R. N. Mohapatra, Phys. Rev. D **12** (1975) 1502.
831. J. Maalampi and K. Enqvist, Phys. Lett. B **97** (1980) 217.
832. M. Gronau, C. N. Leung, and J. L. Rosner, Phys. Rev. D **29** (1984) 2539.
833. F. J. Gilman, Comments Nucl. Part. Phys. **16** (1986) 231.
834. J. Bagger, S. Dimopoulos, E. Massó, and M. H. Reno, Nucl. Phys. B **258** (1985) 565.
835. J. L. Hewett and T. G. Rizzo, Phys. Rev. **183** (1989) 193.
836. C. T. Hill and E. A. Paschos, Phys. Lett. B **241** (1990) 96.
837. W. Buchmüller and C. Greub, Nucl. Phys. B **363** (1991) 345.
838. W. Buchmüller and C. Greub, Nucl. Phys. B **381** (1992) 109.
839. A. Datta and A. Pilaftsis, Phys. Lett. B **278** (1992) 162.
840. M. Acciarri et al., Phys. Lett. B **412** (1997) 189.

# Index

accelerators 11
acollinearity 17, 26, 27, 129, 131
– KORALB 41
– plot 4
acoplanarity 17, 26, 27, 58
Adler function 176–177
ADONE 1
$A_f$ 136
$A_\ell$ 100, 121, 130
ALEPH 13
– CVC 188
– $K_L^0$ identification 37
– muon identification 30
– neutrino mass 209
– one-prong puzzle 89
– $\pi^0$ reconstruction 20, 33, 34
– polarization 123–127
– quark–hadron duality 191
– spectral function 186
– $\tau \to \omega \pi \nu_\tau$ 170
– $\tau \to \pi \pi \nu_\tau$ 164
– transverse spin correlations 131
$a_1$ meson 152
– spin analysis 213
$a_1'$ meson 169
AMY 13
anapole moment 65
angular distribution
– Bhabhas 24
– $\tau$ pairs 105
anomalous magnetic moment 65, 67, 81
aplanarity 17, 131
ARGUS 13
– baryon number violation 275
– Kühn/Santamaria model 154
– neutrino helicity 214
– neutrino mass 205
– pseudorest frame 238
– second-class current 271
– $\tau$ mass 49

axial-vector current 140, 149
axial-vector form factor, pion 193

BABAR 13
background
– at hadron colliders 196
– electron pairs 23, 26, 103
– muon pairs 21, 26
– $q\bar{q}$ charged multiplicity 24
– $q\bar{q}$ multiplicity 25
– $q\bar{q}$ production 23, 26, 205, 271
– signal/background ratio 25
– two-photon events 25, 27
background sources 21–25
beam polarization 105, 240, 263, 265
beam spot 20, 53, 54, 57
BEBC
– neutrino magnetic moment 220
BELLE 13
BES 13
– $\tau$ mass 47
boost see Lorentz boost
branching ratios 72–92
– global analysis 87
– hadronic 82–85
– leptonic 75
– seven-prong 73
– summary 90–92
– topological 72
Breit–Wigner resonance 151, 164

Cabibbo angle 138
Cabibbo suppression 143
CDF
– charged Higgs 198
– leptoquarks 199
– multiplicity 197
– $t \to b\tau\nu_\tau$ 198
– trigger 196
CELLO 13
– branching ratios 87
– electron identification 29

312  Index

- one-prong puzzle  89
- topological branching ratios  73
charge radius  65, 71
- pion  193
charged Higgs boson  81, 198, 250
- $\mathcal{CP}$ violation  256
charged weak current  223–250
chiral anomaly  148–150
chiral perturbation theory  146–150
chiralities  224
CHORUS  8
CLEO  13
- $\mathcal{CP}$ violation  265
- $f_1$  85
- lifetime  56
- neutrinoless decays  273
- $\pi^0$ reconstruction  32
- pseudorest frame  239
- second-class current  270
- seven-prong  73
- $\tau$ mass  51
- $\tau \to 3\pi\nu_\tau$  169
color factor  173
compositness  71, 279
Compton polarimeter  108
cone angle $\psi$  15, 16, 18
conserved vector current  see CVC
constituents  64, 71
contact interaction  281
contour-improved perturbation theory  177, 190
correlators  174–175
Coulomb correction  46
couplings
- axial-vector  97
- left/right-handed  98
- photon  97
- vector  97
$\mathcal{CP}$ odd observables  259–260
$\mathcal{CPT}$  253
$\mathcal{CP}$ violation  253–268
- beyond SM  253–258
- indirect limit  259
- limits  263, 268
- Standard Model  255
- $\tau$ decay  264–268
- $\tau$ production  253–264
cross section  45, 46
cross section at threshold  6
Crystal Ball  13
current–current correlators  174–175
CVC  83, 141, 143–145, 161–163, 187
- branching ratios  162

Das–Mathur–Okubo sum rule  84, 192
DASP  2, 13
decay
- minimum energy  15
decay kinematics  12–16
decay length  12
- plot  14
DELCO  13
DELPHI  13
- angular distribution  105
- Čerenkov identification  35
- lifetime  61, 62
- $\tau$ cross section  102
- $\tau \to 3\pi\nu_\tau$  169
- $\tau$ selection  26
- transverse spin correlations  131
dipole form factor  258
discovery of $\tau$  1–3
DONUT  8
DØ
- W $\to \tau\nu_\tau$  201

effective couplings  64, 66, 279
effective Lagrangian  66, 258
electric dipole moment  65, 67, 81, 253–259
- classic  253
electromagnetic current  64
electron–muon problem  1
energy–energy correlations  128
event display
- discovery of $\tau$  3
- $\nu_\tau$ interaction  10
- $\pi^0$ reconstruction  22
- $\tau$ pair  20
- $\tau\tau\gamma$ event  67
- $\tau$ track  9
- vertex  21
excited leptons  279–281
- decay  280
- exclusion plot  281
- production  280
exclusive decays  87
exotic states  140, 143

$f_1$ meson  85
fixed-order perturbation theory  177, 190
flavor-changing neutral currents  234, 250, 256, 277–278
form factors  43, 64–71, 151, 152, 258

generalized weak current  225–226

Index 313

gluon condensate 181–183
Gottfried–Jackson angle $\theta^*$ 14–16, 110
Gounaris/Sakurai model 166
$G$ parity 142, 149, 269
grand unification 283

hadron colliders 195–201
hadron spectrum 173
hadron universality 237
hadronic current 140–142, 144–146, 235, 269
hadronic events  see background,$q\bar{q}$ production
hadronic tensor 146
hadronic $\tau$ decays 137–193
heavy leptons 283–284
helicity suppression 138–140, 143, 146
Higgs boson
– mass 96
HRS 13
– second-class current 270

impact parameter 58, 196
inclusive hadronic decays 173–193
initial-state radiation 102, 219
Isgur/Morningstar/Reader model 155, 169
isospin partitions 156
isospin relations 155–161
isospin suppression 139–140, 143

JADE 13

kaon identification 34–37, 265
kinematics 49, 51, 203
KORALB 38–44
KORALZ 38–44
Kühn/Santamaria model 152, 154, 166, 169, 215

L3 13
– acollinearity distribution 130
– acoplanarity 27
– lifetime 58
– magnetic moment 66
– Michel parameters 242–243
– multiplicity 25
– neutrino magnetic moment 219
left–right asymmetry  see $\tau$ production
left–right symmetry 250, 257, 273, 283
lepton number violation 272–278
lepton universality 79–82, 234

lepton-flavor-violating neutral current  see flavor-changing neutral current
leptonic branching ratio 75
– corrections 76
leptoquarks 199, 257
lifetime
– decay length method 53
– impact parameter difference 58
– impact parameter method 56
– impact parameter sum 59
– miss distance 59
– summary 62
– three-dimensional impact parameter sum 62
likelihood method 29–31, 77
Lorentz boost 12, 14, 117
Lorentz structure 223–250
– complete determination 232–234
– $\eta$ parameter 230, 232
– hadronic decays 235–237
– interference effects 230
– limits 233, 249
– neutral current 247
– pure vector current 248

MAC 13
– Michel parameter $\rho$ 238
magnetic form factor 71
magnetic moment 65–68
MARK I 13
– discovery of $\tau$ 1, 2
MARK II 13
– neutrino mass 204
MARK III 13
MARK J 13
– Bhabhas 24
mass effects 40, 182
maximum likelihood fit 48
Michel parameters 81, 227–234
– $\eta$ 230, 232, 238–239, 250
– geometrical interpretation 230
– results 246–248
– $\xi_{a_1}$ 237
– $\xi_\pi$ 235
– $\xi'$ 233
– $\xi_\rho$ 236
missing energy 26, 195, 199, 201, 207, 277
missing momentum  see missing energy
moments 64–71
Monte Carlo simulation 38–44

narrow width approximation 83

neural nets  31
neutral kaons
– identification  35–37
neutrino  8, 203–222
– electromagnetic moments  217–222
– fourth generation  80
– helicity  212–217, 235, 236
– magnetic moment  217
– mass  203–210
– – endpoint two-dimensional  209–210
– – endpoint, energy spectrum  204
– – endpoint, mass spectrum  205
– – kinematic reconstruction  207–208
– – limits  210
– massive  80, 272
– oscillations  210, 272
– weak dipole moment  222
neutrino beam  220
neutrino mixing  255
neutrino scattering  220–222
neutrinoless decays  272–275

$\omega$ meson  271
one-prong puzzle  89
OPAL  13
– $\mathcal{CP}$ violation  261
– electron identification  28
– excited leptons  281
– $K_S^0 \to \pi^+\pi^-$  36
– lifetime  55
– magnetic moment  66
– neutrino helicity  216
– neutrino mass  207
– power corrections  187
– quark–hadron duality  191
– running $\alpha_s$  192
– structure functions  167
– $\tau$ direction  19
– $\tau \to e\,\nu_e\nu_\tau$  77
operator product expansion  181, 191
optical theorem  174
optimal observables  118–119, 124, 126, 242, 260
ortho-lepton  7

para-lepton  7
parity violation  212–214
partially conserved axial-vector current  84, 141, 143
phase space suppression  137
photon reconstruction  see $\pi^0$ reconstruction

pion decay constant  193, 235
$\pi^0$ reconstruction  22, 31–33
PLUTO  2, 13
power corrections  182–183, 187
preselection  25
prong  72
propagator correction  225
pseudorest frame  238

QCD  173–193
quark condensate  182–183
quark–hadron duality  173, 188, 190

radiative corrections  40, 46, 183–184, 203, 245, 255, 278
renormalization group equation  177
renormalon chain perturbation theory  178, 190
$\rho$ meson  151
– spin analysis  114
$\rho'$ meson  164, 166
$R_\tau$
– contour integral  175
– definition  173
– electroweak corrections  183–184
– nonperturbative corrections  181–183, 187–188
– perturbative corrections  179, 182
– running $\alpha_s$  177
– theoretical prediction  179–184
– theoretical uncertainty  179–181
running of $\alpha_s$  190–191

scalar current  141
second-class currents  142, 143, 269–271
semihadronic decays  49, 51
sequential lepton  7
Shifman, Vainshtein, and Zakharov approach  181
SLAC–LBL collaboration  see MARK I
SLD  13
– $\mathcal{CP}$ violation  263
– left–right asymmetry  107
– lifetime  60
– Michel parameters  240
– vertex display  20
soft-pion theorem  84
spectral functions  144, 151, 162, 184–193, 209
spectral moments  187
spin analysis  108–119, 260

spin correlations   40, 69, 127–132, 240–243
– acollinearity   129
– energy–energy   128
– transverse   131–132
strong coupling constant $\alpha_s$   173–193
structure functions   167, 214
substructure   64, 279
sum rules   192–193
SUSY   256, 273, 277

t channel   102, 103
t quark   197–198
– mass   95
tagging   27, 205, 238
TASSO   13
$\tau$cf factory   286
$\tau$ decays
– classification   28–31
– dynamics   144
– hadronic   137–193
– inclusive   173–193
– strange decays   84, 172
– $\tau \to 3\,\pi\,\nu_\tau$   20, 27, 83, 115, 145, 148, 152, 156, 166–169, 207, 209, 237
– $\tau \to 4\,\pi\,\nu_\tau$   83, 158, 161, 169–171
– $\tau \to 5\,\pi\,\nu_\tau$   83, 158, 209
– $\tau \to 6\,\pi\,\nu_\tau$   83
– $\tau \to$ e $\nu_e \nu_\tau$   27, 75, 242
– – Michel spectrum   227
– – spectrum   6, 116–118, 238, 278
– $\tau \to \eta\,\pi\,\nu_\tau$   149
– $\tau \to K^+\,K^-\,\pi^\pm\,\nu_\tau$   149
– $\tau \to K\,\nu_\tau$   82, 235–236, 265
– $\tau \to K^\pm\,\pi^+\,\pi^-\,\nu_\tau$   149
– $\tau \to \ell\,\nu_\ell\nu_\tau$
– – Michel parameters   247
– – spectrum   241
– $\tau \to \mu\,\nu_\mu\nu_\tau$   27, 75
– – Michel spectrum   227
– – spectrum   5, 116–118, 123, 238
– $\tau \to \omega\,\pi\,\nu_\tau$   170
– $\tau \to \pi\,\nu_\tau$   19, 27, 82, 109–112, 123, 130, 145, 204, 235–236, 242, 244, 261
– – spectrum   241
– $\tau \to \pi\,\pi\,\nu_\tau$   83, 112–115, 123, 125–126, 145, 148, 151, 163–166, 212–217, 236–237, 244
$\tau$ direction   17–19, 51, 124, 207, 260
– plot   19
$\tau$ lifetime   53–63
$\tau$ mass   45–52
TAUOLA   40, 164

$\tau$ polarization   69, 96, 112, 134, 242
– angular dependence   126
– forward–backward asymmetry   96, 98, 120–131
– longitudinal   70, 98, 120–131, 240
– normal   70, 128
– transverse   70, 128
$\tau$ production   45, 46, 97–102
– angular distribution   71
– cross section   71, 96, 98, 102–105
– differential cross section   98
– forward–backward asymmetry   96, 98, 102–105, 132
– left–right asymmetry   96, 106, 134
– polarized forward–backward asymmetry   106
– $R_\ell$   132
$\tau$ selection   21–27
– at hadron colliders   196
$\tau\tau\gamma$ events   66
technicolor   199
tensor couplings   224
tensor observables   259
TEVATRON   195
thrust axis   17, 54, 57, 104, 207
TOPAZ   13
topology   72
TPC/$2\gamma$   13
– d$E$/d$x$   35
trigger   195
Tsai's formula   143–144
two-body decays   12, 204

vector current   140
vector dominance model   149, 151–152
vector spectral function   185
VENUS   13
vertex   20

weak current   64
weak dipole moment   65, 258–259
weak magnetic moment   69–70
weak mixing angle   see Weinberg angle
Weinberg angle   121, 122, 136
Weinberg sum rule   83, 193
Wess–Zumino anomaly   148–150
Wilson coefficient   181
W $\to \tau\,\nu_\tau$   201

Young tableau   156

$Z^0$ boson
– $\mathcal{CP}$ violating decay   258

- invisible width  222
- mass  96
- partial decay width  258
- $Z^0 \to \tau\,e$  277

# Springer Tracts in Modern Physics

140 **Exclusive Production of Neutral Vector Mesons at the Electron-Proton Collider HERA**
By J. A. Crittenden 1997. 34 figs. VIII, 108 pages

141 **Disordered Alloys**
Diffusive Scattering and Monte Carlo Simulations
By W. Schweika 1998. 48 figs. X, 126 pages

142 **Phonon Raman Scattering in Semiconductors, Quantum Wells and Superlattices**
Basic Results and Applications
By T. Ruf 1998. 143 figs. VIII, 252 pages

143 **Femtosecond Real-Time Spectroscopy of Small Molecules and Clusters**
By E. Schreiber 1998. 131 figs. XII, 212 pages

144 **New Aspects of Electromagnetic and Acoustic Wave Diffusion**
By POAN Research Group 1998. 31 figs. IX, 117 pages

145 **Handbook of Feynman Path Integrals**
By C. Grosche and F. Steiner 1998. X, 449 pages

146 **Low-Energy Ion Irradiation of Solid Surfaces**
By H. Gnaser 1999. 93 figs. VIII, 293 pages

147 **Dispersion, Complex Analysis and Optical Spectroscopy**
By K.-E. Peiponen, E.M. Vartiainen, and T. Asakura 1999. 46 figs. VIII, 130 pages

148 **X-Ray Scattering from Soft-Matter Thin Films**
Materials Science and Basic Research
By M. Tolan 1999. 98 figs. IX, 197 pages

149 **High-Resolution X-Ray Scattering from Thin Films and Multilayers**
By V. Holý, U. Pietsch, and T. Baumbach 1999. 148 figs. XI, 256 pages

150 **QCD at HERA**
The Hadronic Final State in Deep Inelastic Scattering
By M. Kuhlen 1999. 99 figs. X, 172 pages

151 **Atomic Simulation of Electrooptic and Magnetooptic Oxide Materials**
By H. Donnerberg 1999. 45 figs. VIII, 205 pages

152 **Thermocapillary Convection in Models of Crystal Growth**
By H. Kuhlmann 1999. 101 figs. XVIII, 224 pages

153 **Neutral Kaons**
By R. Belušević 1999. 67 figs. XII, 183 pages

154 **Applied RHEED**
Reflection High-Energy Electron Diffraction During Crystal Growth
By W. Braun 1999. 150 figs. IX, 222 pages

155 **High-Temperature-Superconductor Thin Films at Microwave Frequencies**
By M. Hein 1999. 134 figs. XIV, 395 pages

156 **Growth Processes and Surface Phase Equilibria in Molecular Beam Epitaxy**
By N.N. Ledentsov 1999. 17 figs. VIII, 84 pages

157 **Deposition of Diamond-Like Superhard Materials**
By W. Kulisch 1999. 60 figs. X, 191 pages

158 **Nonlinear Optics of Random Media**
Fractal Composites and Metal-Dielectric Films
By V.M. Shalaev 2000. 51 figs. XII, 158 pages

159 **Magnetic Dichroism in Core-Level Photoemission**
By K. Starke 2000. 64 figs. X, 136 pages

160 **Physics with Tau Leptons**
By A. Stahl 2000. 236 figs. VIII, 315 pages

161 **Semiclassical Theory of Mesoscopic Quantum Systems**
By K. Richter 2000. 50 figs. IX, 221 pages

Printing: Mercedes-Druck, Berlin
Binding: Stürtz AG, Würzburg